A BIOINFORMATICS GUIDE FOR MOLECULAR BIOLOGISTS

ALSO FROM COLD SPRING HARBOR LABORATORY PRESS

At the Bench: A Laboratory Navigator, Updated Edition

At the Helm: Leading Your Laboratory, Second Edition

Experimental Design for Biologists, Second Edition

Introduction to Protein–DNA Interactions: Structure, Thermodynamics, and Bioinformatics

Lab Math: A Handbook of Measurements, Calculations, and Other Quantitative Skills for Use at the Bench, Second Edition

Lab Ref: A Handbook of Recipes, Reagents, and Other Reference Tools for Use at the Bench, Volumes 1 and 2

Next-Generation DNA Sequencing Informatics

Statistics at the Bench: A Step-by-Step Handbook for Biologists

A BIOINFORMATICS GUIDE FOR MOLECULAR BIOLOGISTS

SARAH J. AERNI
MARINA SIROTA
Biomedical Informatics Program
Stanford University School of Medicine

CSH PRESS
COLD SPRING HARBOR LABORATORY PRESS
Cold Spring Harbor, New York • www.cshlpress.org

A BIOINFORMATICS GUIDE FOR MOLECULAR BIOLOGISTS

Printed in the United States of America

Publisher	John Inglis
Acquisition Editors	Ann Boyle, Kaaren Janssen
Director of Editorial Development	Jan Argentine
Developmental Editors	Kaaren Janssen, Gabrielle Leblanc, Lori Martin, Christin Munkittrick, Kimberly Prohaska
Project Managers	Maryliz Dickerson, Inez Sialiano
Production Editor	Rena Springer
Production Manager	Denise Weiss
Cover Designer	Michael Albano

Front cover: Art courtesy of Ryan Katkov.

Library of Congress Cataloging-in-Publication Data

A bioinformatics guide for molecular biologists / [edited by] Sarah Aerni and Marina Sirota, Biomedical Informatics Program, Stanford University School of Medicine.
pages cm
ISBN 978-1-936113-22-4 (hardcover : alk. paper)
1. Molecular biology--Data processing. 2. Molecular biology--Computer simulation. 3. Bioinformatics. I. Aerni, Sarah, 1983-II. Sirota, Marina.

QH506.B534 2013
572.80285- -dc23

2013017412

10 9 8 7 6 5 4 3 2 1

All World Wide Web addresses are accurate to the best of our knowledge at the time of printing.

For a complete catalog of all Cold Spring Harbor Laboratory Press publications, visit our website at www.cshlpress.org.

Contents

Section 4: Augmenting Your Data

Preface

Due to a staggering increase in the abundance and availability of molecular data during recent years, bioinformatics has enabled scientists to ask questions previously impossible to answer. The purpose of this book is to arm molecular biologists with the knowledge to enable their use of bioinformatics tools. The idea of the book was born when John Inglis and Kaaren Janssen visited Stanford in 2008 to discuss the future of the *Bioinformatics* textbook long used by students. From this meeting came the idea of a new project, to be led and executed by graduate students and postdocs, to create a bioinformatics "handbook," making computational and statistical techniques more accessible for molecular biologists. The sometimes oblique (to molecular biologists) language of bioinformatics would also be made more accessible by the introduction of significant terminology (within the text, these boldfaced terms are now defined in the context in which they appear). Such a book would drive the adoption of new techniques and technologies by providing a fresh perspective from the graduate students and create a natural source of new content with the new students entering the program, who would use this as an opportunity to "pass the torch" from one group of students to the next.

Without the important contributions and support of many individuals, this book would not have been possible. First, we thank all of the authors for their hard work. We are also grateful for the support of the Biomedical Informatics Program at Stanford. In particular, we thank the members of the executive committee of the Program, including Russ Altman, Atul Butte, Teri Klein, Larry Fagan, Mark Musen, Amar Das, David Paik, Dan Rubin, Mary Jeanne Oliva, and Darlene Vian, who were critical in enabling us to act as student representatives. Without their support and encouragement, this book would not have been possible. The authors and editors have chosen to donate all of the royalties from this publication to the Stanford Biomedical Informatics Program in support of future student endeavors. In addition, we thank all of the staff at Cold Spring Harbor Laboratory Press, who facilitated the creation of this book from its inception to publication, in particular, Kaaren Janssen, who initially presented the idea for the book to us while visiting Stanford University. Her ever-positive attitude and enthusiasm made this an amazing experience. We are grateful to many more at the Press, including John Inglis, Rena Springer,

Maryliz Dickerson, and Inez Sialiano, who were also instrumental in making this book possible. We also sincerely thank Greg Cooper and Sarah Elgin who provided thoughtful and critical reviews of many of chapters, as well as others who provided valuable comments on individual chapters, including Russ Altman, Steve Briggs, Betty Cheng, Mario Roederer, Gary Stormo, Jason Swedlow, and Michael Weiner. In addition, we are individually indebted to past and present mentors and teachers who, often unknowingly, inspired each of us to pursue biomedical informatics as a field of study. Lastly, and most importantly, we are eternally grateful to our families and friends, who provided us with limitless encouragement and support for our endeavors, including this one.

MARINA SIROTA AND SARAH J. AERNI
June 5, 2014

1

Introduction to Computational Approaches for Biology and Medicine

Sarah J. Aerni and Marina Sirota

Stanford University School of Medicine, Biomedical Informatics Training Program, Stanford, California 94305

The past decade has witnessed a staggering increase in the abundance and availability of molecular data produced, for example, from experiments in gene expression, proteomics, and DNA sequencing. With the expansion of data generation across these various areas of biomedicine, developing computational techniques and approaches has become instrumental not only in enhancing and advancing scientific discovery, but also in the analysis of these discoveries. However, many biomedical researchers are unaware of the ways in which informatics can improve their progress, finding it inaccessible because of what we term "informatics anxiety." We believe that this concern results from the improper communication of the intuition behind the approaches as well as the value of tools available.

The goal of this book is to enable biomedical researchers to use and develop computational tools by making these tools more understandable and accessible. To accomplish this goal, we approach the field from the viewpoint of a biomedical researcher: (1) arming the biologist with a basic understanding of the fundamental concepts in the field; (2) presenting tools from the standpoint of the data for which they are created; and (3) showing how the field of informatics is quickly adapting to the challenges and advances in biomedical technologies. After summarizing these points, we provide a series of brief descriptions for each chapter and a set of questions that the reader should be able to answer based on the material presented.

OBJECTIVES

Many wet-bench biomedical researchers experience *bioinformatics anxiety*. The various existing tools are often described in extensive mathematical detail to show the rigor of a model to fellow bioinformaticians. However, in the absence of explanations that are more accessible to scientists without extensive training in the field, the tools often go unused. The intended end users are therefore unable to understand how or why this tool should be applied to their own data.

Many informatics tools, although designed for use by the biomedical researcher, rarely have the impact hoped for by the bioinformaticians. This is largely due to the inability to communicate the methods and utility of the tools generally in a way that a biomedical researcher can properly understand. Typically, the tools described in the literature are applied a single time on a benchmark data set that the bioinformatics researchers used in development. However, the goal of these papers is largely to show a proof-of-principle to the biomedical research community in general, with the intention of adoption of the method in the community as a standard tool. Yet, many biomedical researchers appear to use only the tools they know, finding the process of understanding and incorporating new methods into their analysis a daunting task they prefer not to tackle.

The purpose of this book is to arm molecular biologists with the knowledge to enable their use of bioinformatics tools. These tools will undeniably enhance their work and lead to important discoveries. Over the years, bioinformatics has enabled scientists to pose and ask questions that would not be possible otherwise. Many of these have yielded success in various fields of biomedicine such as novel drug discoveries (Gleevec in the treatment of some leukemias), genetic interactions with various drug and disease phenotypes (Warfarin, an anticoagulant used to prevent clotting), and drug repositioning (applying known drugs in new treatment scenarios).

ORGANIZATION

In this volume, we provide biomedical researchers with the principles that form the basis of many bioinformatics methods and examples of how these techniques can be used to analyze diverse data sets. The book is organized in three sections: First, we tackle the fundamental concepts underlying most bioinformatics methods. Next, we present a series of data-specific techniques available to biomedical researchers. We conclude with integrative techniques that augment the biomedical researcher's primary analysis with additional complementary data.

Fundamental Concepts for Bioinformatics Methods

This section serves as a reference for how the basic principles are used in bioinformatics approaches. Here, we discuss the underlying technical knowledge in the fundamental fields of bioinformatics in the fields of computer science, statistics, and machine learning. These very technical fields are presented at a level of detail that provides the necessary intuition for understanding the methods without requiring expert knowledge. With this level of detail, the reader will be able to understand the inner workings of the tools that are currently available as well as new approaches beyond those described in the book.

- ***Chapter 2: Introduction to Computer Science.*** Chapter 2 covers the practicality of methods, algorithm development, and understanding what computational complexity means for the user. This chapter gives the framework for using and understanding existing published tools. What is a computer? How is information stored?
- ***Chapter 3: Probability and Statistics.*** This chapter presents the techniques from the field that reappear in many bioinformatics papers. How can the biomedical researcher assess whether the results obtained are significant? Which metric is appropriate for a given data set? These fundamentals will help biomedical researchers speak and understand the language of informatics. The chapter also discusses some basic and commonly used techniques to answer the following questions: What kind of correction needs to be applied to assess significance? How do I assess the quality of an informatics approach?
- ***Chapter 4: Machine Learning.*** Application of various machine learning techniques are presented in the context of biological data geared toward the noncomputationally trained life scientist. For various algorithms, the intuition is presented to help biologists understand when it is appropriate to use each one. What is the difference between a supervised and an unsupervised approach? Which algorithm should I use on my data?

Techniques for Analyzing Your Data

In the second section, we present a set of methods to help researchers understand what tools are available to perform an automated analysis on their data. The goal is to allow the reader to grow comfortable with these tools, encouraging their use and advancing research. We define the research in informatics at the cusp of biotechnological advances in various fields. We introduce the strengths of each technology and encourage its use by describing the opportunities that it offers. We also discuss computational approaches and how they can be applied to various types of data, ranging from the traditional sequence and expression analyses to image-based and proteomics data.

- ***Chapter 5: Image Analysis.*** Image based-analyses are a growing media for biomedical research ranging from basic biology produced in laboratories to patient images obtained in clinical settings. In this chapter, we present computational techniques used to accomplish and automate a plethora of tasks traditionally performed through manual curation. How can I automatically count the number of cells on my plate? Which segmentation algorithm is appropriate for my image?
- ***Chapter 6: Expression Data.*** Microarrays have become commonplace for measuring levels of gene expression in a biological sample of interest. In this chapter, we describe the technology behind microarrays and how it can be used effectively by

a researcher. This chapter will help answer the following questions: How do I normalize and preprocess my data? Which gene or group of genes is significantly up-regulated or down-regulated in my sample? When do I use other technologies like RNA-seq or qRT-PCR to validate my findings?

- ***Chapter 7: A Gentle Introduction to Genome-Wide Association Studies.*** With the advance of genotyping technology, it has become possible to study the effects of genetic variation on various phenotypes of interest. In this chapter, we introduce genome-wide association studies (GWASs) and show how the approach can be used to find novel gene–disease relationships. You will be able to answer the following: Which statistical techniques do I use to detect a genetic association? How can population stratification affect my ability to detect real signal? How do I distinguish between a genetic marker and a causal variant?
- ***Chapter 8: Next-Generation Sequencing Technologies.*** In this chapter, we introduce and describe the broad spectrum of sequencing platforms developed in recent years. We discuss the benefits and trade-offs of various experimental technologies and new computational challenges that have arisen with the advent of next-generation sequencing. How can we use sequencing technologies to perform different types of variant calling? How and when is sequencing used beyond primary sequence analysis (i.e., the transcriptome, ChIP-seq)? Which tools are available for analysis of my sequencing data?
- ***Chapter 9: Proteomics.*** Direct measurement of protein levels provides the researcher with a true snapshot of a biological system. Mass spectrometry and flow cytometry are two common ways of measuring levels of protein expression in a sample. In this chapter, we describe these technologies as well as computational methods used in analysis of such data. How can I measure the abundance of a protein of interest in my sample? What technique can I use to identify subpopulations of cells that alter protein expression in different experimental conditions?

Augmenting Your Data

The final section of this book provides the reader with ways of augmenting their primary data that lead to better scientific understanding. Integrating different data types can provide a deeper insight into biomedical questions. Although these "data mashups" have been presented in a variety of fields, the adoption by biomedical researchers can only be accomplished by providing an understanding of the basic principles used by these methods. The series of chapters that follows shows the reader how to enrich his or her data set with available data repositories. By providing ideas and tools for meta-analysis and data integration, we help the reader identify relevant publicly available resources to further support their findings and generate novel hypotheses.

- ***Chapter 10: Knowledge Base–Driven Pathway Analysis.*** We introduce the notion of biological networks and define a computational framework to model complex systems by applying Bayesian techniques. This type of modeling can enhance primary analysis by aggregating the observed signals and help identify the underlying biological processes. Which available databases can I use to study my pathway of interest? What computational techniques can I use to build a protein interaction network from my data?

- ***Chapter 11: Learning Biomolecular Pathways from Data.*** Our primary mission here is to educate the reader regarding the general principles and concepts underlying knowledge-base-driven pathway analyses. We consider the challenges presented by high-throughput profiling technologies, for example, used in protein microarrays and metabolomics studies; here we focus on gene expression analyses. How can knowledge-base-driven pathway analysis be used to extract biological meaning from a list of differentially expressed genes and proteins?
- ***Chapter 12: Meta-Analysis and Data Integration of Gene Expression Experiments.*** The research community maintains a rapidly increasing number of repositories containing data from previously published studies. Researchers are able to increase the power of their studies by leveraging these resources through combining several data sources together with their primary data. How can an existing experiment be used to validate my findings without performing any further biological replicates? How can I ask a novel biomedical question by aggregating signal from various data sources?
- ***Chapter 13: Natural Language Processing: Informatics Technique and Resources.*** Knowledge of the past 50 years of biomedical research is captured in an enormous body of literature. Computational approaches have been developed to mine literature sources effectively to enhance and validate findings. This chapter covers the underlying principles of natural language processing used to extract and store knowledge from free text. What controlled terminologies are relevant to my research? Which tools are available for performing natural language processing in the context of my research?

As we have seen, informatics can vastly assist advancement in research and development in biology and medicine. We hope you will find this guide to bioinformatics both accessible and informative regarding existing tools and resources. Furthermore, we believe that this book will build a foundation that allows you to identify and apply additional computational methods to enhance your research. It should exist as a resource allowing you to understand and evaluate cutting-edge bioinformatics approaches encountered in the literature.

2

Introduction to Computer Science

Eugene Davydov and Olga Russakovsky
Stanford University, Computer Science, Stanford, California 94305

Human beings have a large pool of knowledge about the world and are very good at identifying the steps to be taken to solve a problem. For example, a human looks at a product on the store shelf that costs \$5.84 and knows that in practice s/he will have to pay 8.25% sales tax on the item. S/he knows the steps that are needed to come up with the total price (multiply \$5.84 × 1.0825), but computing the right answer is a bit more challenging (try doing this in your head!). This is where computers come in. A simple handheld calculator cannot reason about the world or the product on the shelf, but it can compute 5.84 × 1.0825 in milliseconds. For a human to effectively use the computer, s/he must identify what set of instructions must be executed, how to effectively specify these instructions, and how to communicate the input and output with the computer.

During the past several decades, computers have transformed every scientific discipline by revolutionizing our ability to store, visualize, and manipulate data. This effect has been especially profound in biology, in which the sequencing of the human genome has raised new questions and opened entirely novel directions of research, based on insights gleaned from massive data sets that cannot be processed manually. Today, biologists need a basic understanding of computers and computer programs to determine the best approach for investigating new data.

One of the central ideas in engineering is abstraction—hiding complexity behind an interface that is easy to work with. Most people can drive a car or use a computer without knowing how it works under the hood. Yet, to use computers to process biological data we must pull back this veil of abstraction and understand a little about how computers work internally, in particular how they store and interpret data and program instructions.

WHAT IS A COMPUTER?

Computers consist of multiple hardware components, the most important of which are described in Table 1: central processing unit (CPU), memory, disk, input/output (I/O), and network. This hardware allows the computer to interact with the user and execute programs, or software. At the most basic level, each program is a list of instructions that tells the program to "do something," i.e., manipulate a piece of data in a certain way. A simple instruction might tell the CPU to multiply two numbers or tell the computer to read a value from some address in memory and send it to the CPU. There are also instructions that conditionally alter the flow of the program to respond differently to mouse clicks and keyboard input, for example.

Both data and program instructions are stored within the computer in the same way. A basic block of data is a byte, which consists of 8 bits of information—each is 0 or 1. Thus, each byte can represent an integer between 0 and 255 (zero is counted) if the bits are interpreted as a binary number. If we assign each letter a unique numerical code, e.g., "a" = 97, we can also interpret each byte as a character. This is the most common way of storing text—as a sequence of bytes, each corresponding to a single character (including punctuation, spaces, etc.). Alternatively, groups of bytes (typically 2, 4, or 8 bytes at a time) can represent integer or real-valued numbers, memory addresses, or program instructions. The versatility of data representation makes it important to keep track of what format a file (or any piece of data) is stored in—the same set of bytes can be interpreted differently depending on whether it is text, numbers, or program instructions.

When a program is ready to run, the computer loads its instructions (in the form of a file called the executable) into memory and then executes them. These instructions are designed to be simple for the computer to interpret but as a human, writing even a simple program one instruction at a time would be tedious and error producing. Thus, programs are written at a much higher level than individual instructions. A single line of code might call for the computer to print a string (piece of text) to the screen—although conceptually a simple operation, this will involve executing hundreds or thousands of instructions that tell the computer exactly how and where to

TABLE 1. Hardware components of a typical computer

Central processing unit (CPU). This "brain" of the computer executes program instructions and performs arithmetic computations, typically billions per second. It is common in modern computers to have multiple CPUs that work in parallel, allowing independent computations to occur at the same time.

Memory. Providing storage for programs and data, random access memory (RAM), for example, allows fast read and write access at any memory location.

I/O devices. Short for input and output devices, these include keyboard, mouse, and display.

Hard drive. Providing storage for data that is slower to access than memory, the hard drive typically has much higher capacity and persists when the computer is turned off.

Network device. This allows the computer to communicate with other computers via the internet.

display each character, one pixel at a time, depending on which windows are open, font type and size, color scheme, and so on.

Most modern computers feature processors with multiple cores or actual processing units, gigabytes of memory, and terabytes of disk space. Despite these advances, the most demanding applications could run for years on a single computer. Thus, multiple computers can be networked together within a distributed system (either via the internet or within a computer cluster), communicating and collaborating to solve a large task in pieces. Programs for distributed systems are often much more complex than their single-machine counterparts because of the considerable overhead involved in coordination and synchronization. Additionally, such systems must be robust to individual machine failures. Because of the complexity of writing programs for distributed systems, many companies offer cloud computing services that store data and execute programs that allow clients to avoid dealing with most of the system details.

WHAT IS A PROGRAM?

Here, we explore what it takes to go from an abstract idea of an algorithm to a specific implementation—a computer program. Programs are typically written in a human-readable language such as Python, R, MATLAB, C++, or Java and then either converted to machine-specific instructions (compiled) or executed via another program (interpreted). We highlight some key differences among these languages toward the end of this section, but first we discuss the most important elements of programming that are common to all of them.

The main "building block" of most programs is a function—a reusable block of code with a set of input parameters. When called (invoked), the function performs a computation and returns the result, potentially triggering other effects (e.g., drawing on the screen, playing a sound, writing to a file). Organizing programs into functions makes them easier to write, read, understand, and modify. It also means common tasks such as reading user input can be written once and reused by anyone who needs them. Reimplementing this functionality from scratch in every program would be tedious, error-prone, and make the task of writing even a simple program harder and longer than necessary. In fact, it is common to organize related functions for doing low-level tasks into collections called libraries that are then used by many programs.

Let us look at a simple example: writing a program in Python to print the words "Hello, World!" on the screen. Here is the entire program:

```
print("Hello, World!")
```

This one line hides an astonishing amount of complexity. This program simply calls a function in the Python standard library, "print," that will print the string representation of its input parameter to the screen. The actual implementation of this simple function is incredibly complex because the computer must determine which windows are active on the screen, where to put the text, how to render each character (determine the default font/size), etc.

Let us look at a slightly more complex example: reading a number from a file, taking its logarithm, and printing the result to the screen.

Here is a basic program:

```
import math
f = open("~/one_number.txt")
x = int(f.read())
y = math.log(x)
print y
```

In this example, *x*, *y*, and *f* are variables—named placeholders for storage locations that allow the programmer to refer to specific pieces of data without having to think where exactly in memory (or in a CPU register) they must be stored. Variables, like functions, are used in almost every major programming language; often, they are associated with a specific type, such as an integer or string, for safety and error checking.

Another critical concept is the notion of control flow or executing blocks of code for a variable number of times (0, 1, or more). For example, "if" statements will execute a block of code only if a specified condition is true when evaluated. By using "else if," it is possible to combine several conditions. Only the statements following the first condition that is found to be true will be executed. All other statements will be skipped. **For-loops** and **while-loops** offer slightly different ways of repeatedly running through the same set of instructions until a termination condition is satisfied (in the case of a for-loop, this condition usually involves a counter variable). At each iteration of the loop, values of variables may be different, so each iteration of a loop may proceed differently.

Let us look at an example.

```
sum = 0
for i in range(1, 100):
    sum = sum + i
    if (sum % 2 == 0):
        print "Sum is even."
    else:
        print "Sum is odd."
print sum
```

This block of code computes the sum $1 + 2 + \cdots + 100$. It adds up all of the numbers from 1 to 100 and checks whether the sum is even or odd. The variable *i* changes values from 1 to 100, and at each iteration the value of *sum* will increase by *i*.

Then, if the sum is divisible by 2, the program will print the line "Sum is even."; otherwise, it will print the line "Sum is odd." Can you predict how many times (and during which iterations) each line will be printed?

Millions of computer programs have been written during the last several decades, and most of them share a lot of basic functionality; user interactions, for example, usually take the form of mouse clicks and/or keyboard input, and many programs must be able to handle this input in a similar fashion. This is where the central notions of computer science—abstraction and reusability—come into play. There are many kinds of high-level programming languages, each with its own advantages and disadvantages. Common languages include Python, MATLAB, R, C++, and Java. We begin by discussing the similarities and then highlight some key differences among these programs.

All of these programming languages contain the basic notions described above: variables, functions, conditional statements, and loops (with some differences in syntax). However, each language has attributes that facilitate the writing of certain types of programs but not others. For example, C is a relatively low-level language that gives the programmer a lot of control over performance, at the expense of making it difficult to write large, modular programs. In contrast, C++ and Java are **object-oriented languages** that, unlike C, provide built-in support for classes—types of objects that encapsulate data and functionality. Perl and Python at their core are excellent scripting languages, making it easy to read and process text input (they have built-in support for regular expressions) and interact with the operating system, for example, to process a number of files in a directory. R (open source) and MATLAB are high-level languages with extensive built-in mathematical and statistical capabilities and many libraries, e.g., Bioconductor, for sequence and microarray analysis.

What Is an Algorithm?

There are many ways to describe and formulate a real-world problem in a computer-friendly way. Let us say that we want to perform a certain operation on a set of numbers. As we just discussed above, there are many programming languages that can be used to express these numbers, many options for inputting the numbers (do you have the computer [1] read them from a text file or a compressed file, such as a zip; [2] input the numbers one at a time; or [3] ask the computer to look online to find these numbers?), and many options for returning the output (such as [1] print out the number to screen; [2] save it to a file; or [3] pass it into a different program that will then use the number to do some other task). In all cases, one must first define the right set of steps to ask the computer to take and determine the practical feasibility of asking the computer to complete this procedure. We discuss the options above in later sections.

An algorithm is a set of instructions. When faced with a problem or question, the programmer must define a step-by-step procedure for the computer to solve this task. For example, suppose a programmer wants to find the longest region from a set of genomic regions. The algorithm might be to

1. Look at the first region and remember its length; call it X.
2. Look at the second region and compute its length. If this length is bigger than X, let X be this new length instead.
3. Continue looking at every region in the list, computing its length, and updating X if the region's length is larger than X.

This task is simple for a human when there are less than, say, 10 or 100 regions. However, if there are a thousand regions in the list, it would be useful to write a computer program to do it instead.

Although it is understood that computers are very powerful, it is also important to understand their limitations. Suppose you have a sequence of 1 million genomic regions. Is it still feasible to write a program to find the longest sequence among 1 million regions using this algorithm? What about 1 trillion regions? What if instead of simply finding the one longest region you want to sort these regions by length or find the best alignment among all of these regions? And what happens if instead of small regions you have full genomes?

Before beginning to write the problem, it is important to understand the algorithm that you want to implement and to ask yourself some analysis questions. For example,

1. How long would this program take to run?
2. How much space would this take?
3. Can the program be parallelized[1]?
4. How easy would it be to implement?

Running Time Analysis: How Long Would the Program Take?

Running time depends on the kind of computer used, the programming language used to specify the instructions, how long it takes for the computer to read in these regions, and many other factors. However, you can ask questions such as "if it takes 2 minutes to process 1000 regions, how long will it take to process 2000 regions?" This is called running time analysis of the algorithm. Once the programmer becomes familiar with different programming languages, s/he can roughly estimate how long it would take to process a small-sized input. Understanding how the algorithm scales with the size of the data will help the programmer to understand whether the program would be applicable to problems of the size s/he wants to solve.

On one end of the spectrum are constant-time algorithms that take the same amount of time to run regardless of input size. For example, if the genomic sequences

[1]Parallelized is when the computation is split up among more than one computer or thread.

are already sorted in order of increasing length, returning the shortest one would take constant time: The algorithm would be simply to return the first sequence in the list. Such algorithms are **constant-time algorithms**, denoted $O(1)$.

If the list is not sorted, finding the longest region can be performed using the algorithm listed above. A list of 2 million sequences will take about twice as long to process as a list of 1 million sequences, because each sequence will take about the same amount of time to process, and each sequence only needs to be considered once. Such an algorithm is called a **linear time algorithm**, denoted $O(n)$. In linear time algorithms, the running time of the algorithm scales linearly based on the size of the input, so if the size of the input doubles, the running time of the algorithm doubles as well.

Suppose instead that there are only two sequences but you need to find the optimal alignment between these two sequences. At minimum, each nucleotide within one sequence must be compared with every nucleotide in the other to determine whether there is a match. Thus, if there are N nucleotides in each sequence, the number of operations is N^2. As the input size doubles, the number of operations required quadruples. Thus, aligning two strings of length 1000 takes T seconds, aligning two strings of length 2000 takes $4T$ seconds, and aligning two strings of length 4000 takes $16T$ seconds. This type of algorithm is called a **quadratic algorithm** or $O(N^2)$. This doubling of the size of the input increases the runtime by a factor of 4.

Finally, on the far end of the spectrum are **exponential-time** algorithms. For example, suppose that instead of aligning two sequences, we need to align M sequences. Multiple genome alignment scales as $O(N^M)$, where N is the length of the genome and M is the number of genomes. Thus, aligning two genomes of length N takes T computer operations and aligning four genomes takes $T \times T$ operations.

What does this mean in practice? As a rule of thumb, algorithms between constant and linear time can generally scale to genome-sized data. Quadratic or cubic algorithms can generally be applied to sequences of length a few thousand (short genomic sequences) for quadratic algorithms or a few hundred (amino acid sequences) for cubic. Exponential-time algorithms are extremely time-consuming and generally require approximations to be applicable to genomic data sets.

Space Complexity Analysis: How Much Space Would It Take?

The space requirement for an algorithm depends on how data is stored, how much intermediate computation must be stored, and various other factors. However, as with running time analysis, approximations can be made, such as "if 10 Mb of storage are required when analyzing 1000 regions, how much space is needed to process 2000 regions?" Modern computers have large amounts of memory (100 Gb or more on the most powerful computers) for storing intermediate computation and disk space (terabytes), but when operating on the scale of genomes it is easy to overtax these resources. It is important to remember that reading and writing files from the hard disk is many orders of magnitude slower than operating on data stored in memory; thus, it is best to assume that the storage space to which the algorithm has access is limited by the size of memory.

A constant amount of storage, known as **constant-time storage** is required in the case of, for example, finding the longest length of the regions in a list. (Here, we consider **auxiliary space complexity** or the amount of space required by the algorithm beyond the initial input.) Whether there are 1 million or 1 billion sequences in the list, only one number needs to be remembered by the computer as the algorithm processes—the length of the longest sequence seen so far. Similarly, if the goal is to find the length of the three (or even 1000) longest sequences, the storage requirement is still constant or $O(1)$. Only three (or 1000) numbers must be remembered regardless of input size.

In contrast, **linear time storage** is required, for example, to sort the regions in order of length. Because the output is the sorted list of regions, the space requirement would be $O(N)$. One thousand regions require T bytes of storage; 2000 regions require $2T$ bytes.

There are algorithms with quadratic, cubic, exponential, etc., space complexity. Similarly to runtime, if an algorithm requires cubic space complexity it is unlikely to scale to genome-sized data. Thus, in the implementation, it is important to consider the amount of space that would be required to run the algorithm.

Parallelizability: How Easy Is It to Parallelize on a Computing Cluster?

With the growth of cloud computing resources and computer clusters, one important factor in designing algorithms is to consider whether the algorithm is parallelizable, i.e., able to be broken down into independent computation chunks. If so, these computation jobs can be run in parallel on multiple machines and their results later aggregated. Some overhead is involved in distributing the input data to multiple machines (or multiple threads within the same machine) and later combining the results, but this will generally be insignificant compared with the gains from distributing the core computation.

Consider again the task of finding the longest region in a set of n regions. We can break up this task into smaller subtasks, i.e., split the set into 10 sets of $n/10$ regions in each and independently find the longest region within each set. Then, we combine the outputs of the 10 subtasks to get a list of 10 candidate lengths and return the longest of these. We know that finding the longest region takes linear time, so doing this on a list 10 times smaller than the original would take 10 times less time.

With the growth of data, it is becoming increasingly important to keep in mind the concept of parallelizability when designing algorithms. Generally, when designing an algorithm to run on genomic-level data, consider how the task can be split across machines both for running time and space requirements. Of course, when writing simple algorithms to parse text files, for example, this is unlikely to be easy. Parsing is read over process.

Ease of Implementation: How Much of a Programmer's Time Will It Take?

One important (and often not considered) factor to consider when writing a computer program is how much time it will take the programmer to write. As one becomes more proficient at programming, it takes less and less time to implement familiar algorithms. However, there is always a trade-off to consider between implementation time and algorithmic runtime.

Some algorithms are inherently more complex than others, meaning that they take longer to implement, and this makes it more likely that the programmer will introduce errors into the program. Programmers should consider whether a simpler algorithm is preferable, even if it takes slightly longer to run. For example, the decision of whether to parallelize an algorithm will ultimately depend on how much of a time investment is involved for the programmer versus how important would the computation time savings be.

Some programming languages are better than others for certain tasks. For example, Python is a better choice for parsing files than C. However, if a programmer is very familiar with C, it might be faster to use C to write a parser instead of investing time into learning a new language just to write one program. On the other hand, if the programmer is going to be writing many similar programs to solve similar tasks, it might be worth the time investment. Furthermore, choosing to use one programming language over another requires the programmer to implement some algorithms from scratch instead of using preexisting libraries.

USING THESE TOOLS TO SOLVE BIOLOGICAL QUESTIONS

We can use the considerations described above to help us solve various biomedical questions. For instance, let us say a researcher has been browsing the human genome on the UCSC genome browser and has found a region of the human genome that is conserved across multiple species. Because conservation often implies function, let us say s/he is interested in finding all such regions in the genome. The researcher must first determine whether existing tools are available to perform such an analysis. How efficient are the tools? Are the results precomputed or will there be a need to rerun the analysis? If it is the latter, how long does that take? Is it feasible to run the computation on one machine or would it be necessary to parallelize the computation? What are the inputs and outputs of the tool? What assumptions does it make? Does it answer the biological question of interest or would it be necessary to develop a program? Is it something that the researcher can tackle on his/her own or would it be necessary to reach out to a bioinformatician or a software engineer?

3

Probability and Statistics

Alexander A. Morgan and Linda Miller

Stanford University School of Medicine, Biomedical Informatics Training Program, Stanford, California 94305

Statistics is essential in all areas of biomedicine. However, it is especially important when dealing with large amounts of data. When a single experiment is being performed to test a single hypothesis, the techniques for analysis are fairly straightforward and commonly taught. However, in biomedical informatics, very large data sets, perhaps with lots of missing data, with very many variables that interact with one another in both known and unknown ways are becoming the rule, not the exception. These kinds of data require much more sophisticated analysis techniques. At the same time, the language of probability and statistics is full of many terms such as likelihood, random, independent, and significant that are commonly used in everyday language but have a very specific, carefully defined meaning in probability and statistics. It is important to understand and use these terms correctly. In this chapter, we give a basic overview of some of the key ideas that are important in bioinformatics. We hope that this piques enough interest that you can continue your own investigations into methods of the quantitative analysis of data.

We start by defining notions of probability and expectation. After introducing descriptive statistics, which are used to characterize a distribution of data, we then dive into some examples of how to perform statistical analysis on continuous and categorical data. We touch on concepts such as sampling, permutation testing, and multiple hypothesis correction; talk about how to deal with missing data; and end with a few examples of creative visualization.

PROBABILITY AND DATA

Most individuals have a well-developed sense of what probability and likelihood mean, and that is sufficient for many daily tasks. Unfortunately, our intuitions are often wrong, even when we think that we have an airtight logical argument. Many of these errors are associated with problems involving conditional probability and/or questions spanning many orders of magnitude. Both are very important areas in biomedical informatics, so although you may feel as if you have a good, intuitive understanding of what probability is, it can help to be somewhat more formal in our thought and analysis.

One well-known puzzle that causes considerable confusion is the *Monty Hall* problem, named after the host of the television game show *Let's Make a Deal*. In this puzzle, the contestant is faced with three doors. Behind one of the doors is an expensive sports car and behind each of the other two doors is a goat. The contestant is allowed to select one of the doors, and the host then opens one of the other doors, revealing a goat. The contestant then has the option of choosing another door and taking home whatever is behind that other door or taking home whatever happens to be behind the door they have originally chosen. Does the contestant gain any advantage in switching to a new door over keeping their original choice?

You may have already heard this problem in this form or another, and you may have been convinced that the contestant gets no advantage by switching, because there may seem to be an equal chance of the car being behind each door. If this is what you believed, then you are in the company of a large number of mathematicians and scientists when they initially heard the problem. However, the contestant should always switch because this boosts the chance of winning the car to 2/3. If this is not clear to you, then you should make a diagram listing all the possibilities and/or do simulations with a live partner. Unfortunately, if you are interested in trying to use this result to win a car, the actual game show had a slightly different version in which the host was given more control, as explained in an excellent interview with Monty Hall himself, where he discusses the puzzle with *The New York Times* reporter John Tierney (July 21, 1991).

To help us be more exact in our understandings of probability and avoid mistakes in problems such as deciding which door to choose, we will define some basic notation. Although we can formulate the concept of probability in different ways, the way that is probably the most intuitive and makes the most sense when analyzing experimental data is to define the relative frequency of an occurrence or set of occurrences if the same experiment were repeated many times. With a set of distinct events, A_i out of a set of s possible events, we can define some basic notation for the probability of A_i. By event or occurrence, we mean anything of interest that is measurable or identifiable, such as an actual event in time or an attribute under study. For example, the set A_i could consist of all the different shades of the *Biston betularia* moth that we consider black out of all the possible colors and variations in color s. Probability (P) is defined using the following equation.

$$P(A_i): P(A_i) \geq 0, \ \sum_{A_i \in S} P(A_i) = 1, \ P(S) = 1. \tag{1}$$

This simply says that the probability of any of the events must be nonzero (so negative probabilities are not allowed) and the probability of all the possible events is 1. Often, all possible events may be considered equally likely. As an example, consider all distinct hands drawn from a deck of cards. In this case, once you have enumerated all the possible outcomes, the probability of any subset of outcomes is just the ratio between this number and all possibilities. For many of the common ways in which you can select equally likely events, there are some commonly used formulas, often called **counting rules**. You may recall the simple formulas for permutations and combinations. If you draw r cards from a deck of n total cards, then you can calculate the total number of different possible sequences of cards using the total number of permutations of n choose r via this formula:

$$\frac{n!}{(n-r)!}. \tag{2}$$

As a simple example, what is the number of 2-number permutations of the set {1, 2, 3}? Here, we get $3!/(3-2)! = (3 \times 2 \times 1)/1! = 6$. One must remember that Equation 2 pertains only to ordered combinations (the number of distinct subsets of size r that can be selected from n distinct objects where $r \leq n$).

However, if you do not care about the ordering and consider any hands containing the same card as equivalent, then you can calculate that there are n choose r different combinations using the following:

$$\binom{n}{r} = \frac{n!}{(n-r)!r!}. \tag{3}$$

If we consider the above example, we have

$$\begin{bmatrix} 3 \\ 2 \end{bmatrix} = \frac{3!}{(3-2)!(2)!} = \frac{3 \times 2 \times 1}{(1)(2 \times 1)} = 3.$$

We are often particularly interested in outcomes that influence one another. The **joint probability** is defined as the probability of two events both occurring and is given by

$$P(A, B) = P(A \cap B). \tag{4}$$

The joint probability that an animal simultaneously has a particular genotype and a particular phenotype can also be expressed this way. If two events are statistically **independent**, then the product of their probabilities is equal to their joint probability, or

$$P(A, B) = P(A)\, P(B). \tag{5}$$

To illustrate joint probability, if we roll two dice, what is the probability of rolling two 5s? The probability of rolling one 5 is 1/6, so the probability of rolling two 5s is (1/6)(1/6) = 1/36 = 0.0278.

However, when events are not independent, we are often interested in the **conditional probability**. In our example, the relationship between the animal's phenotype and genotype may be the most interesting when they are not independent. The probability of A given B is expressed as $P(A \mid B)$. For example, the probability of a phenotype π given a genotype g is $P(\pi \mid g)$.

The conditional probabilities are related to the joint probability in a very simple way because we know that the probability of A must be equal to the sum of the probabilities of all of the different probabilities when different Bs occur:

$$P(A) = \sum_{All\ B} P(A, B), \quad P(B) = \sum_{All\ A} P(A, B),$$
$$P(B \mid A) = \frac{P(A, B)}{P(A)}, \quad P(A \mid B) = \frac{P(A, B)}{P(B)}. \tag{6}$$

In the case of continuous outcomes that we are not able to easily partition into discrete events, we can generalize the summations to integrals, and our general relationships will continue to apply. We can relate these conditional probabilities to another via some simple algebraic manipulation:

$$P(A, B) = P(A \mid B)P(B) = P(B \mid A)P(A) \longrightarrow$$
$$P(B \mid A) = \frac{P(A \mid B)P(B)}{P(A)}. \tag{7}$$

This last relationship is called **Bayes' Rule** and relates the odds of events A_1 to A_2 before and after conditioning to another event B. Although it may seem like some very basic algebra, it is a very important relationship among probabilities that underlies much of statistics, machine learning, and decision theory.

As an example, suppose you want to be able to diagnose a disease using a genetic test. You can use Bayes' Rule to determine your diagnostic criteria. If you know the base allele frequency in the general population $P(Genotype)$, the disease prevalence $P(Disease)$, and the probability of the genotype in those with the disease $P(Genotype \mid Disease)$, you can estimate the probability of the disease in a new person:

$$P(Disease \mid Genotype) = \frac{P(Genotype \mid Disease)P(Disease)}{P(Genotype)} \longrightarrow$$
$$\frac{P(Genotype \mid Disease)P(Disease)}{\sum_{Disease\ Status} P(Genotype \mid Disease)} \longrightarrow$$
$$\frac{P(Genotype \mid Disease)P(Disease)}{P(Genotype \mid Disease)P(Disease) + P(Genotype \mid Healthy)P(Healthy)}. \tag{8}$$

To put this into practice, let us say that the probability of having a specific genotype when the disease is present ($P(Genotype \mid Disease)$) is 0.35 (i.e., 35% of people with the disease have that phenotype) and the probability of having the disease ($P(Disease)$) is 0.05. From these values, we can calculate the probability of having the phenotype and being healthy (i.e., not having the disease; $P((Genotype \mid Healthy))$ as $1 - 0.35 = 0.65$ and the probability of being healthy ($P(Healthy)$) as $1 - 0.05 = 0.95$. Plugging these values into Equation 7, we get

$$P(Disease \mid Genotype) = \frac{(0.35)(0.05)}{[(0.35)(0.05)] + [(0.65)(0.95)]} = 0.039.$$

In practice, we must be careful that there are no biases in the populations that we examine for disease or measure genotype, so that we do not preferentially enrich one group over another. Additionally, if we are examining many genotypes together, we need to be careful regarding multiple hypothesis correction (considering a set of statistical inferences simultaneously, drawing conclusions from data that are subject to random variation, or inferring a subset of parameters selected based on the observed values). We also must consider whether genotypes are independent of one another or if there is some epistasis (see Chapter 7).

Expectation

We can extend our basic notion of probability to the idea of **expectation**, E.

$$E[g(x)] = \sum_{x \in A} g(x)P(x). \tag{9}$$

Intuitively, we can think about the number that we would expect to come up on a die. If each number, 1 through 6, is equally likely, we would expect an average number of 3.5. We can use this idea that defines the expectation of a numeric outcome value as value multiplied by the probability of that value, summed over all possibilities (i.e., a weighted average of all possible values). For example, for the die, there are six possible events $x \in A$. For each event, we observe a number of pips on the die, $g(x)$. The expected value is then

$$E[g(x)] = 1 \cdot \frac{1}{6} + 2 \cdot \frac{1}{6} + 3 \cdot \frac{1}{6} + 4 \cdot \frac{1}{6} + 5 \cdot \frac{1}{6} + 6 \cdot \frac{1}{6} = 3.5.$$

Descriptive Statistics and Bias

It is often useful to summarize a set of values by a single number or a small set of numbers as a set of characteristic numbers that represents the values and their distribution. For example, sample **mean** and **median** are very common, but there are additional measurements. These **summary statistics** are all called **statistics of location** (or measures of central tendency), meaning that they give some indication of where the values appear on a number line. Additionally, numbers that summarize the distribution of data and how it is spread are called **statistics of dispersion** (or measures of spread) and include **range** (difference between the **minimum** and the **maximum**), **interquartile range**, and **variance**. Even fancier statistics of this type, including **skew** and kurtosis (not defined here), attempt to give a general overview of the properties of the data, but all can be called summary statistics or **descriptive statistics**. Such statistics provide simple summaries regarding a sample and the observations that have been made. It is important to emphasize that descriptive statistics does not enable one to make conclusions beyond the data analyzed or reach conclusions regarding hypotheses made. The importance of descriptive statistics lies in its ability to present data in a more meaningful way. For example, a fluorescence-based assay in a 384-well format will return fluorescence values for each individual well, most likely for numerous plates. Looking at hundreds, possibly thousands, of numeric values does not allow one to get a clear picture of the meaning of the data. That is where descriptive statistics allows us to understand our data. For example, the **sample mean** (not to be confused with the **population mean**, discussed below) from a group of observations is an estimate of the population mean:

$$\hat{x} = \frac{\sum_{i=1}^{n} x_i}{n}. \tag{10}$$

Additional descriptive statistics include **median** and **percentile**. The median of a set of values is the middle value, when they are sorted high to low. If there is an even number of values, the median is the mean of the middle two. It is a robust statistic, meaning that it is much less sensitive to outliers than is the mean. This can also be referred to as a resistant statistic. The percentile is the fraction of points that lies below the given value. The median is the 50th percentile. To calculate percentiles, first order the values. The 50th percentile is the value at position $(n + 1)/2$. In general, the pth percentile (of n ordered values) is

$$\frac{(n+1)}{100/p}. \tag{11}$$

The **variance** of a data set is a very important concept that describes how close the data points are to one another and how far the values lie from the mean. If we have data values x coming from a population that has a mean μ, then we can define the variance σ^2 using our conception of expectation:

$$\sigma^2 = E[(x - \mu)^2] = E[x^2] - \mu^2. \tag{12}$$

Now, if we go back to our example population (fluorescence values) and collect some data from that (i.e., a subset of the fluorescence values from the plate wells as representative of the whole population), we have a sample. We can use that sample to estimate the mean and variance of the population. The arithmetic mean is then a very good estimate of the population mean, and $\hat{x}$ is a very good estimate of μ. If we want to estimate the variation in population, we can estimate the population variance σ^2 by calculating a sample variance s^2. However, to calculate a variance for the population, we need to know the population mean μ, and we only have an estimate of this quantity. Using the population mean that we have calculated from the data, $\hat{x}$, we will have a biased estimate of the variance.

Bias is a term that is overloaded in statistics and has slightly different meanings in different contexts. In this context, it means the difference between an estimator's expected value and the true value of the parameter, for example, $Bias = E[s] - \sigma = E[s - \sigma]$. Biased estimators often have desirable properties such as reducing the mean squared error. This is an important basic idea: Bias is not always bad. Sometimes, you want a biased estimate, but it is important to know when your estimates are biased, in which direction, and (ideally) by how much. A way to think about the bias that enters into our attempts to estimate the population variance using our sample is to notice the term $(x_i - \mu)$. Our estimate of μ is $\hat{x}$ chosen to minimize this difference by taking the arithmetic mean. Therefore, the x_i are in a sense closer to $\hat{x}$ than to μ, so the difference between the sample values and the estimate must be lower; thus, it is likely that

$$\sum_{i=1}^{n} (x_i - \hat{x})^2 \leq \sum_{i=1}^{n} (x_i - \mu)^2.$$

That means that our naive calculation for population variance is biased toward a smaller value. We can correct this by dividing by a slightly smaller number. This leads to the commonly known formula for the population variance:

$$s^2 = \frac{\sum_{i=1}^{n} (x_i - \hat{x})^2}{n - 1}. \tag{13}$$

Many people are often confused by the division by $(n - 1)$ instead of n. This is just a slight correction for this bias. More sophisticated approaches use a more complicated formula for computing an even less biased estimate, so if you are using statistical software you may notice a slight discrepancy in the sample variance that you calculate compared with the one that the system returns. However, the difference is only noticeable when n is very small, and, in this case, the estimation of population statistics (mean and variance) from the sample is not very reliable, which should actually match your intuition of the poor ability to estimate a value such as the population mean from a small sample. In practice, people often interchange σ and s because in most problems involving an analysis of real data, you are usually trying to estimate σ using samples.

The **standard deviation** is a term for the square root of the variance and indeed why the variance is expressed as a square in the first place: $s = \sqrt{s^2}$. A useful property of the standard deviation is that, unlike variance, it is expressed in the same units as the data. The **standard error** of the mean is the standard deviation divided by square root of the sample size:

$$s_{\overline{X}} = \frac{s}{\sqrt{n}}. \tag{14}$$

Another useful measure of variability is the **interquartile range (IQR)**. This is the difference between the third (75%) and first (25%) quartiles; thus, it is also called the midspread or middle 50. Unlike the variance, it is a robust statistic, so it is not sensitive to extreme values. A single extreme value that is many orders of magnitude larger than others may inflate variance beyond the dispersion defined by the middle range of values. For this reason, in practice, many analyses remove values outside of the interquartile range, also known as **outliers**.

There is also a measure of symmetry in the distribution of values. In a unimodal, symmetric distribution (e.g., normal), the mean and median are identical. However, in a situation in which the distribution is skewed rather than symmetrical, these three measures are not equal. The **skewness** is the extent of asymmetry in a distribution (see Fig. 1). Highly skewed data may often be dealt with using various data transforms, such as taking a logarithm, or by using nonparametric statistics that perform their tests on the ranks of the data values. Such transformations of the data for analysis purposes are often very helpful because many statistical techniques perform better with more symmetrical data.

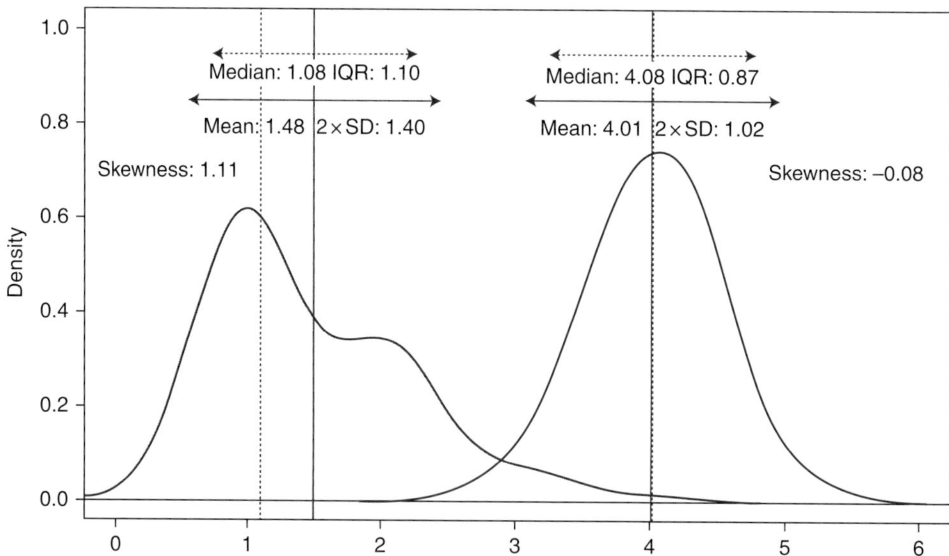

FIGURE 1. Comparison of summary statistics for two different distributions. The distribution line in red shows how data that are highly skewed will have a larger difference between the mean and median of the data and therefore a larger IQR and standard deviation. The distribution indicated in blue is more symmetric, and the mean and median are very close together.

STATISTICAL ANALYSIS OF DATA

When analyzing data, we are often interested in knowing how **significant** something is, often in a quest for the Holy Grail, the p value < 0.5. What does this mean? When we are looking at a comparison in measurements, such as the amount of mRNA (messenger RNA) for IL-4 produced by T cells exposed to two different drugs, how do we know which differences are likely to be important and repeatable in further studies? When we are looking at the association between the level of low-density lipoprotein (LDL) in the blood and milkshake consumption, how do we know that any association we find is believable? When looking at a difference between the rates of Crohn's disease in individuals with genotypes of AA, AT, or TT at a locus of interest, how do we know that any differences we find are meaningful? How do we know these findings are likely to be reproducible?

The methods of statistical hypothesis testing were developed to try to put such questions into a framework that is quantitative and consistent among researchers. Here, we take what is called a **frequentist** approach and describe experimental testing in terms of reproducibility. This approach focuses on the probability of the data given the hypothesis. The data are treated as random and the hypothesis as fixed. This is termed the frequentist approach because it focuses on the frequency with which one expects to observe the data based on the hypothesis. We can never know with certainty that a particular relationship will always hold—all we can do is increase our confidence if we repeatedly and reproducibly observe that relationship. The statistical framework we describe below involves six steps that we will discuss briefly and then go into greater detail with examples.

1. Simple question
2. Null hypothesis
3. Assumptions
4. Summarizing data to test the statistic
5. Testing the hypothesis
6. Interpretation

1. Simple Question

The formulation of a question is the first item, and it may seem rather silly to make note of this explicitly. However, the point is that we are trying to be careful and quantitative, and this requires being explicit regarding what question is being asked. When presented with very large data sets with thousands of different variables, it becomes important to ask very specific questions. For example, if we are doing an analysis of memory B cells and expose populations of the cells to different signaling molecules and then do RNA sequencing (RNA-seq), our general question might be,

How are the different drugs affecting the metabolism of the B cells? However, that question is not quantitative or specific. We need to turn it into a much narrower, more specific question such as, Are any genes increased in expression in the cells treated with CD154 compared with those not treated?

2. Null Hypothesis

Once we have a specific question, we can create a **null hypothesis**, which simply means that there is no relationship between the two measured phenomena. The null hypothesis is a default position that can potentially be proven false, or at least we can collect enough evidence from our data to reject it as being unlikely (meaning that there is a relationship between the phenomena). The null hypothesis for our B-cell example is that the expression level of every gene is the same when treated with CD154 and when untreated.

3. Assumptions

When performing a statistical test, it is important to know what **assumptions** we are making. Some of these assumptions have to do with the design of the experiment or data collection. For instance, in our example, we may assume that the two populations of B cells are representative samples, and the only difference is the exposure to CD154. We might assume that there is a particular bias in our measurements of expression level based on GC content of the transcript. Other aspects may have to do with the nature of the data and the difference that we observe. We might assume that the variance in expression of any single gene between subsampled groups is equal, or we might assume that it is unequal. We might be able to test some of these assumptions in our data before we examine our null hypothesis, or we might have so few samples that it is difficult to test accurately. Whatever assumptions we make, we need to be very clear and explicit regarding what they are, ensure that they are appropriate for the kind of statistical test that we are using, and determine how any violations of these assumptions will affect our result. Failure to do so will generate inaccurate conclusions.

4. Summarizing Data to Test the Statistic

When doing a statistical test, we create a **test statistic** that we derive from the data. It is a numerical summary of a data set that reduces the data to one value that can be used to perform a hypothesis test. A test statistic is generally selected and defined in such a way as to quantify behaviors that would distinguish the null from the alternative hypothesis. The field of statistics derives its name from working with these values. Values that we discussed before such as sample mean, sample variance, median, or interquartile range might be used as test statistics, or we can use more complicated values such as the F statistic, the difference in medians, or, in theory, any number that we choose to calculate from our sample data. In practice, we will likely only ever use a handful of very useful statistics as test statistics, such as the F statistic. What is important is that the test statistic is explicitly named in our null hypothesis.

5. Testing the Hypothesis

Our statistical **hypothesis testing** comes when we compare the derived test statistic with the values that we would expect to obtain if the null hypothesis were true. We can use our assumptions to construct a set of likely values of the test statistic and assign a probability that a similar experiment would return that value or an even more extreme value if the null hypothesis were true. That is our *p* value, and if that value is very low, such as below 0.05, we might reject the null hypothesis and decide that our result is significant and likely to be replicated in future experiments.

6. Interpretation

An important point to remember is that the *p* value is an estimate of probability that a test statistic would be observed in repeated experiments. It therefore relates more to reproducibility than it does to magnitude of difference. For example, ionizing radiation exposure increases as a function of altitude because there is less atmosphere to absorb or scatter incoming radiation from the sun and from space. However, the difference in radiation exposure is negligible and typically has little part in influencing decisions regarding where people live or work. In a similar way, with large enough sample sizes, we may be able to detect small, subtle differences among attributes of groups and get a significant *p* value. However, effect size (the actual amount by which the groups differ) is often equally important, whether it is in the magnitude in the difference in mean expression of IL-4 levels among different exposures or the difference in disease risk attenuated by the amount of blueberries in the diet. It is important to keep in mind the magnitude of the actual difference and not just whether the difference has a low *p* value. At the same time, although statistical tests aim to be about reproducibility, they are more about measuring **precision** than they are about **accuracy**. Our statistical tests use the variability in the samples to approximate variability in the population, for example, by using sample variance. That gives a measure of how precise the measurements are (i.e., how much they cluster together within the experiment). That does not necessarily correspond to accuracy and reproducibility across experiments. The classic example is a dartboard. If all the thrown darts cluster in a single square inch on the board, then the throwing was very precise. If that cluster is nowhere near the bullseye that was being aimed for, then the throwing was not very accurate. Hopefully, these ideas will begin to be more clear as we look at examples of statistical tests and as you gain experience in their application.

Statistical Tests on Continuous Data

One of the most basic statistical tests is a comparison of the means between two groups. Suppose we are comparing the lengths of adult red and blue fish; then we might collect a distribution of lengths for each group (Fig. 2). This is an example of continuous data, which is defined as any data that have infinite values with connected data points (often, a measurement). We are not interested in relative length of the longest or shortest of each fish; we are interested in the group average. Does it differ? How do we ask this question a little more formally?

We can formulate a null hypothesis that states: The mean length of each kind of fish is the equal. However, we can also make a slight variation of that null hypothesis and instead state that the t statistic is zero, where the t statistic is defined for the two samples of size n_1 and n_2 with means $\hat{x}_1$ and $\hat{x}_2$ with sample variances s_1 and s_2. The t statistic is used to determine if two sets of data are significantly different from one another:

$$t = \frac{\hat{x}_1 - \hat{x}_2}{s_{x_1 x_2}}, \tag{15a}$$

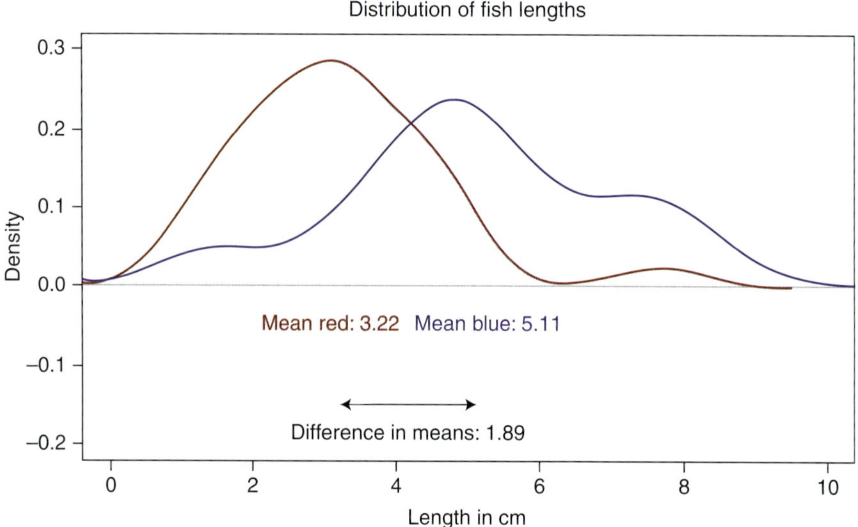

FIGURE 2. Distribution of fish lengths. The range in fish lengths is similar between red and blue fish; however, the distributions differ widely, with the mean length of red fish being 1.89 cm less than the mean length for blue fish.

where

$$s_{x_1x_2} = \sqrt{\frac{s_1^2}{n_1} + \frac{s_2^2}{n_2}}. \tag{15b}$$

This test statistic is Welch's version and can be used when the two samples are of unequal size and have unequal variance. The t value can be compared with the ***t* distribution** to get a p value. Values that are largely positive or largely negative indicate that the relative difference in means may be larger than expected by chance.

The t distribution has an interesting history. While William S. Gosset was employed for the Guinness Company doing research into ways to improve the process of growing barley and making beer in the early 1900s, he began looking into ways to analyze differences in small samples and, along with Karl Pearson and Ronald Fisher, developed what is now called the t-test. He was unable to publish under his own name because the brewery did not want any trade secrets released, so he published under the alias "A Student." Today, the t distribution and the t-test are thus often referred to as the "Student's t distribution" and "Student's t-test."

Graphically, the t distribution looks very similar to the normal distribution, but it is different in a subtle, but important way: It has slightly fatter tails or, put another way, increased probability for more extreme values. Simplifying a bit, the t distribution gives us the expected difference between two estimates of the average taken by two different samplings. It is a family of continuous probability distributions that arises when estimating the mean of a normally distributed population where the sample size is small and the population standard deviation is unknown. Because our estimate of the average of one sample is likely off by a bit and our estimate of the other sample is off by a bit in the opposite direction, we can get a larger difference between the two. However, what is important is that, if we sample enough and our sample sizes become large enough, the average will converge to the normal distribution as expected by the central limit theorem. This theorem describes the characteristics of the "population of the means" that has been created from the means of an infinite number of random population samples, all of them drawn from a "parent population." We discuss the t distribution in greater depth when we discuss the t-test.

To put this more mathematically, the deviates $\overline{Y} - \mu$ of sample means from the true means of a normal distribution are also normally distributed. These deviates divided by the true standard deviation $(\overline{Y} - \mu)/\sigma$ are still normally distributed, with $\mu = 0$ and $\sigma = 0$ (standard normal). The distribution of deviates of i samples, each with mean Y_i and standard error $s_{\overline{Y}_i}$, $(\overline{Y}_i - \mu)/s_{\overline{Y}_i}$ is not normally distributed. As noted, it is wider because the denominator is the sample standard error instead of the population standard error. It will sometimes be smaller and sometimes larger

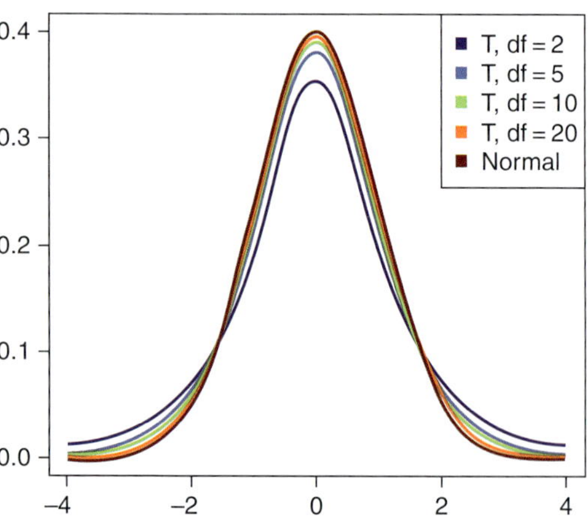

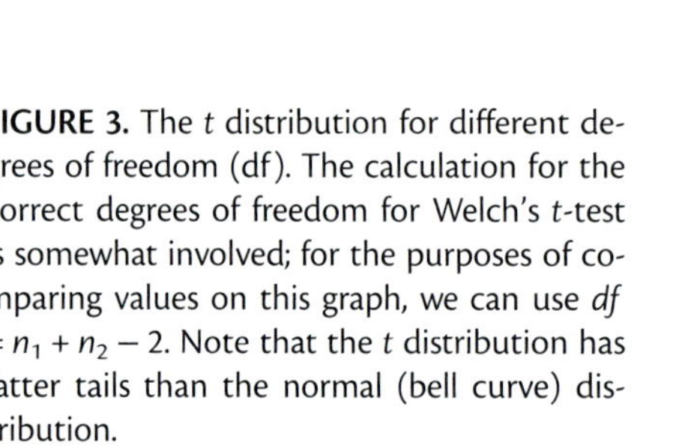
FIGURE 3. The *t* distribution for different degrees of freedom (df). The calculation for the correct degrees of freedom for Welch's *t*-test is somewhat involved; for the purposes of comparing values on this graph, we can use $df = n_1 + n_2 - 2$. Note that the *t* distribution has fatter tails than the normal (bell curve) distribution.

than expected, so the variance is greater. The *t* distribution's shape is dependent on the degrees of freedom ($n - 1$), where n is the sample size. As the degrees of freedom increase, the *t* distribution approaches the normal distribution and is equal to it when $n = \infty$ (and close to it when $n > 25$), as illustrated in Figure 3.

Note that the *t* statistic takes the form of a measure of variability. This measure can reflect the variability *between or among* different groups, or it can reflect the function of the variability *within* the sample groups. This idea can be generalized, and we can look at the difference in the "between group" variability compared with the "within group" variability. If we are looking at multiple groups, we can use a generalization of the *t*-test, called ANOVA (analysis of variance), which compares the variability between groups with the variability within groups to a generalization of the *t* distribution called the **F distribution**. This is a statistical distribution that arises in the testing of whether two observed samples have the same variance. The F distribution is the ratio of the variation due to an experimental treatment (i.e., the effect) to the variation due to experimental error. The null hypothesis would state that this ratio equals 1, meaning that the effect of the given treatment is the same as the experimental error. Thus, the null hypothesis is rejected if the F ratio is significantly large enough that the chances of it being 1 are less than some preassigned criteria (i.e., 0.05).

If the data are highly skewed or there are major outliers, we can try to use a more robust form of statistics based on the ranks of the data instead of the actual continuous values, typically called **nonparametric statistics**. Such statistics do not assume that a given data set or population has any characteristic structure or parameters. These statistics are required when data have a ranking but no obvious numerical interpretation. The rank-based version of the *t*-test is called the **Wilcoxon rank-sum test** or **Mann–Whitney U test** and is conceptually very similar. A rank best test instead of the F test for ANOVA is called the **Kruskal–Wallis test**. A good book on applied statistics will give examples of all of these approaches.

Statistical Tests on Categorical Data

Categorical data refers to data in which there are a finite number of categories into which the data fall. We can think of a pond with a fixed proportion of red and blue fish in it. As we draw fish from the pond, it can change the likelihood of the color of the next fish drawn. For example, if the pond has 100 blue fish and one red fish, if we have caught four blue fish and one red fish already, the likelihood that the next fish caught is red is zero. The sampling of fish is not independent from one another. This is considered a **hypergeometric experiment**, where a sample size n is randomly selected without replacement from a population of N items.

To use the example used in almost all descriptions of the hypergeometric, imagine an urn containing N balls: m are white and $(N - m)$ are black. If we draw out n balls, we can use the **hypergeometric distribution** to model the probability that i of them will be white and thus $(n - i)$ of them will be black. It is important to note here that the hypergeometric distribution is performed without replacement.

$$P(X = i) = \frac{\binom{m}{i}\binom{N-m}{n-i}}{\binom{N}{n}}, \tag{16}$$

where m is the number of white balls in the urn; i is the number of white balls selected; n is the sample size, the total balls selected; and N is the population size, the total number of balls in the urn. This hypergeometric distribution is used for statistical tests when we are interested in detecting if there is overall enrichment of one category over another and underlies **Fisher's exact test**. For example, if we believe that the pond contains 10,000 blue fish and five red fish, we catch six fish, and half of them are red, intuitively that seems to suggest that our original assumption regarding the distribution of fish is wrong. The hypergeometric distribution tells us just how unlikely (or likely) that is. We can use the hypergeometric distribution to provide all the enumerated probabilities of getting three or more red fish when drawing six and use that to decide whether or not to reject our null hypothesis regarding the distribution of fish in the pond. Alternatively, if we know that the distribution of fish is 10,000 blue and five red, then we can use the test to try to reject the null hypothesis that our fishing method does not preferentially catch one color fish over another. Figure 4 shows another application of the hypergeometric distribution.

For an example of **Fisher's exact test**, consider that we have data from a genome-wide association study (described in further detail in Chapter 7) of cardiac arrhythmia. We have two categories of people: individuals with arrhythmia and healthy controls, and we determine the genotypes of many single-nucleotide polymorphisms (SNPs) for each individual. We might summarize the data for one SNP in a **contingency table** such as that shown by Table 1, in which each cell contains

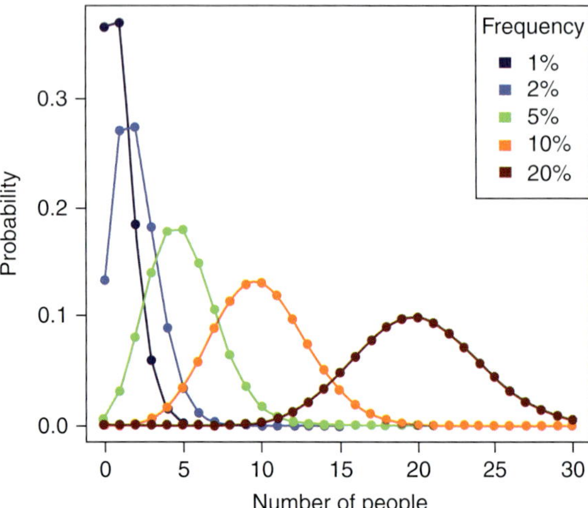

FIGURE 4. Hypergeometric distribution. Consider the case of testing 100 individuals out of a population of 100,000,000 for various genotypes of different frequencies in the population. The graph shows the probability (*y*-axis) of finding a number of people (*x*-axis) having the genotype with a given frequency in the population (indicated by the different colored lines).

the count of individuals with that particular genotype and phenotype. In other words, a contingency table displays the frequency distribution of all variables. If the proportion of individuals in each column is variable among rows (or vice versa), there is a contingency among them (i.e., they are dependent). If there is no contingency among variables, they are independent.

What if we want to test whether the two groups, disease and control, differ in allele frequency at this locus? This is the basic idea of a genome-wide association study, which aims to find the SNPs with the most extreme differences in genotype distribution between disease and control groups (see Table 2). One commonly used test statistic is χ^2 (pronounced "Kai squared"), described in many biostatistics books, which was useful before computers were able to rapidly calculate all of the factorials used in the hypergeometric distribution. The two tests return almost the same result, but Fisher's test is better for small sample sizes and is overall preferable to χ^2.

TABLE 1. Example 2 × 2 contingency table

	Disease	Control	Row totals
Genotype AA or AT	O_{11}	O_{12}	R_1
Genotype TT	O_{21}	O_{22}	R_2
Column totals	C_1	C_2	N

TABLE 2. Expected contingency table

	Disease	Control	Row totals
Genotype AA or AT	$(C_1/N) \times R_1$	$(C_2/N) \times R_1$	R_1
Genotype TT	$(C_1/N) \times R_2$	$(C_2/N) \times R_2$	R_2
Column totals	C_1	C_2	N

TABLE 3. Contingency table for one SNP from example data

	Disease	Control	Row totals
Genotype AA or AT	600	800	1400
Genotype TT	200	200	400
Column totals	800	1000	1800

Let's walk through an example of performing a test of disease association. Using the same scenario as above, suppose we have 800 patients with cardiac arrhythmia and 1000 healthy controls, with genotype distributions for an SNP as shown in Table 3. The question we would like to answer with the data in Table 3 is whether there is a significant relationship between genotype and disease occurrence. We first compute the expected frequency for each cell under the assumption of the null hypothesis (i.e., that there is no relationship). Let us first calculate the expected frequency for the AA or AT genotype/disease combination. Cell C1 (refer back to Table 2) shows that 800 people with AA or AT developed cancer, so that proportion is 800/1800 = 0.44. Because 1400 people had the AA or AT genotype, we would expect that (0.44)(1400) = 622.22 cases of cancer with this genotype. Now, looking at the TT genotype and disease occurrence, we would expect that (0.44)(400) = 177.78 cases of cancer with this genotype. For the control (healthy) individuals with AA or AT, the expected frequency is (0.56)(1400) = 777.78 healthy individuals with the AA or AT genotype. This gives us (0.56)(400) = 222.22 healthy people with the TT phenotype. We may then use these data to perform a significance test, such as the χ^2 or Fisher's exact test. These values are all displayed in Table 4.

To calculate the probability of a table *at least* that extreme, compute the probabilities for all tables with the same row/column totals, but that have more extreme values, and sum these probabilities.

TABLE 4. Expected contingency table for one SNP

	Disease	Control
Genotype AA or AT	622.22	777.78
Genotype TT	177.78	222.22

BOX 1. Mixing Categorical and Continuous Data

There are a variety of approaches for working with mixes of categorical and continuous data. For example, you may be examining how gender and country of origin (forms of categorical data) are associated with lipid level, body mass index, and expression level of the *PPARG* gene (all continuous variables).

(Continued.)

More complex regression models may be found in an introductory text on applied statistics, and you are referred to such a reference on the topic of logistic regression and design matrices. In the Machine Learning chapter (Chapter 4), many methods (i.e., decision trees) that allow for an interpretable mixture of continuous and categorical variables are discussed. However, it is a common mistake with large data sets to use categorical data (i.e., using a country of origin encoded as a number) as continuous data, so it pays to be careful. It is important to check with the statistical software that you are using to make sure that the statistical tests and the data that you are examining are in the right format. It is also important not to let slight differences in vocabulary among domains cause you confusion. In statistics, the observed quantities are typically called dependent variables or observables, whereas in machine learning these may be called features. Bioinformatics and medical informatics tend to use a mix of these two terminologies, whereas chemoinformatics uses the term chemical descriptors. They all mean essentially the same. However, it is important that you know which type of data the software that you are using expects and whether the data need to be transformed in any way.

Multiple Hypothesis Testing

The Problem with Testing Multiple Hypotheses

When making multiple comparisons or testing multiple hypotheses from the same data (e.g., we have 500,000 SNPs and we want to see which of them are associated with a particular disease), there is a chance of false discovery, because some effects may be significant purely due to chance when you are making many comparisons. For example, if we have performed 200 statistical tests, the null hypothesis is true for each test and we have a p value of <0.05, we would expect 10 of these results to be significant just owing to chance ($0.05 \times 200 = 10$). Thus, we would have 10 "significant" results that are really false positives. This can lead to serious problems in future experiments if strong conclusions are based on these false data, especially if an outside laboratory performs the same experiment(s) and cannot reproduce the initial data.

A similar problem on a smaller scale is testing effects in multiple groups in a study. For example, say we are testing a drug's efficacy in premenopausal women, postmenopausal women without hormone replacement therapy, and postmenopausal women with hormone replacement therapy. We cannot simply test the drug separately in each group and claim that it works in one group without correcting for testing three times. The most appropriate analysis to do here is ANOVA or, at least, correct for multiple testing (discussed below).

A related issue is called data snooping. This occurs when you take an initial look at the data and discover relationships that you did not explicitly specify as hypotheses to test in advance. For example, you are examining the relationship between smoking and heart disease and, in the process, notice a correlation between BMI and heart disease. You *cannot* report this association on the original data—you can test it in a new data set. One way to avoid this is to set aside half the data in advance so that you can test hypotheses like this subsequently.

Error

When doing multiple testing, the type 1 error increases. We use the **family-wise error rate** (FWER) to describe the experiment-wide significance level α. The FWER is the probability of making one or more false discoveries (type 1 errors) among all the hypotheses.

$$\alpha = 1 - \left(1 - \alpha_{\text{per comparison}}\right)^{\text{number of comparisons}}. \quad (17)$$

The **false discovery rate** (FDR) is the rate at which features that are called significant are truly null. For example, an FDR of 5% means that among all features called significant, 5% of these are truly null, on average. Thus, as the number of comparisons increases, α is closer to 1, decreasing the overall significance of the experiment. The **false positive rate** is the rate at which truly null features are called significant. For example, a false positive rate of 5% means that, on average, 5% of the truly null features will be called significant. Simply put, the false positive rate is a situation in which the null hypothesis was incorrectly rejected.

The FDR, defined as the expected proportion of false positives among all significant tests, allows you to identify a set of "candidate positives," of which a high proportion are likely to be true. The false positives within the candidate set can then be identified in a follow-up study. It is less stringent than FWER, with more power, but at the cost of increasing the likelihood of type 1 error.

Correction Methods

Multiple-testing correction means recalculating *p* values from a statistical test that was repeated multiple times. The *p* value for each individual comparison must be much more stringent, in order to keep the FWER low.

Bonferroni

The Bonferroni method is used to control the FWER and is the most stringent correction. It simply involves dividing each *p* value by the number of comparisons. As an example, if we have performed 100 statistical tests and have a *p* value of 0.05, the Bonferroni method gives us a new *p* value of 0.0005 (0.05/100). Thus, only tests with $p < 0.0005$ would be considered significant. This controls type 1 error but increases the probability of type 2 error (calling truly significant features nonsignificant, i.e., erroneously accepting the null hypothesis).

Tukey

The Tukey method is used when the family is all pairwise comparisons of means. For example, you have several treatment groups (drugs A, B, C, and control), and you want to know which drug has the largest difference from the control. It is a single-step correction for multiple testing and finds which means are significantly different. The test statistic is similar to a t-test except that it corrects for the FWER:

$$q_s = \frac{Y_A - Y_B}{SE}, \tag{18}$$

where Y_A is the larger mean, Y_B is the smaller mean, and SE is the standard error of the data. If q_s is larger than $q_{critical}$ (from a studentized range distribution), the two means are significantly different (i.e., it identifies any difference between means that is greater than the standard error). Always start by comparing the largest mean with the smallest mean. If these are not different, there is no need to test the other combinations.

Assumptions made when using this method include observations are independent, the means are from normal distributions, the variance of the observations is equal (homoscedasticity), and observations are equal across the groups associated with each mean in the test.

Benjamini–Hochberg False Discovery Rate

This is the least stringent of correction methods. We rank the p values from smallest to largest and let $p(k)$ be the kth smallest (out of n) p values. Then the FDR δ_k for hypothesis k is bounded by $\delta_k \geq np(k)/k$. If we want to control the FDR for the entire experiment at δ, then we can call all hypotheses that satisfy $p(k) \leq \delta(k/n)$ significant. Unlike other multiple comparison methods such as Bonferroni, this procedure controls the expected proportion of falsely rejected hypotheses.

q Values

The ***q* value** is the FDR analog of the p value and can be described as a function of the p value. The q value of an individual feature is the minimum FDR at which this feature can be called significant.

$$\hat{q}[p(i)] = \min_{\tau \geq p(i)} F\hat{D}R(\tau). \tag{19}$$

The ***q* value** for a particular feature is the expected proportion of false positives that we incur if we call this feature significant. In other words, the q value of a feature is the expected proportion of false positives among all features that are more extreme than the current feature. If we calculate q values for each feature and threshold them at some level α, we get a set of significant features, with proportion α, that are expected to be false positives.

Whenever multiple hypothesis tests are performed, one of the correction methods described above should be used, and corrected p values should be reported.

BOX 2. Odds Ratios

Oftentimes, just knowing the p value for association is not enough. What may be more important is the size of the effect. **Odds ratios** (ORs) measure the effect size and describe the strength of association.

Odds is defined as

$$odds = \frac{p}{1-p}. \tag{20}$$

For example, if 9/10 people in a room are male, the odds of being male is 0.9/0.1 = 9, or 9 to 1. The odds of being female are 0.1/0.9 = 0.11, or 1 to 9.

An odds ratio is a sample-based estimate of this ratio or the ratio of the odds of an event occurring in one group to the odds of it happening in another. Let us continue with our previous example and compute the odds ratio of possessing the A allele at the SNP of interest and its association with disease status (see Table 5).

The odds ratio compares the odds of having the allele in the disease with the odds of having the same allele in healthy controls:

$$\begin{aligned} OR &= \frac{\text{odds of allele in disease}}{\text{odds of allele in controls}} = \frac{\dfrac{\text{allele A and diseased}}{\text{no allele A and diseased}}}{\dfrac{\text{allele A and control}}{\text{no allele A and controls}}} \\ &= \frac{a/c}{b/d} = \frac{ad}{bc} \\ &= \frac{600 \times 200}{800 \times 200} = 0.75. \end{aligned} \tag{21}$$

An odds ratio of 1 means "no effect" because the odds of exposure in disease is the same as in controls. For our example, because the odds ratio is <1, exposure to the A allele

TABLE 5. Odds ratio table

	Disease	Control
Genotype AA or AT	a	B
Genotype TT	c	D

(Continued.)

protects individuals from disease (diseased individuals are *less* likely to have allele A). Conversely, the T allele puts individuals at risk for disease. The odds ratio with respect to the T allele is 1/0.75 = 1.33. In other words, this means that individuals possessing the T allele are approximately 1.3 times more likely to have the disease than individuals without the T allele. Note that the odds ratio is a measure of effect size, whereas the *p* value is a measure of significance and the reproducibility of the effect.

Resampling Methods and Permutation Testing

In many cases, we want a way to validate a model built on a data set that does not involve acquiring vast quantities of additional data, which may be time-consuming and costly. In these situations, we can use randomly generated subsets of the original data to create many representative sets that we can use to evaluate the model. These techniques are called **resampling methods** and include bootstrapping, jackknifing, and permutation testing.

Bootstrapping

Bootstrapping is a method for estimating properties of a distribution or estimator (such as its mean, variance, etc.), and it allows the researcher to assign measures of accuracy to these estimates. Bootstraping is performed by sampling with replacement from the original data set many times to generate a large number of sets that we *might* have gotten as the original data set. In this way, we generate many random sets that, taken together, represent the empirical distribution of the observed data. We recalculate the statistic (or model) of interest in all of the new sets, and in this way we get an estimate of the distribution of the statistic. Sampling with replacement is more accurate than sampling without replacement. It is important when doing bootstrapping to take as many bootstrap samples as is feasible given the computing power, in order to reduce the effects of random sampling errors from the procedure. Examples of instances in which bootstrapping is useful include when (1) the theoretical distribution of a statistic of interest is complicated or unknown; (2) the sample size is insufficient for straightforward statistical inference; or (3) power calculations must be performed and a small pilot sample is available (Ader and Mellenbergh 2008).

The Jackknife

The Jackknife technique is similar to bootstrapping in that it estimates the variance and reduces bias of a statistic based on random samples from the original population. However, in this case, we sample without replacement from the original sample. Each jackknife sample is generated by leaving out one or more observations at random and computing the statistic of interest from the remaining data points. From this, one can calculate an estimate for the bias and variance. The jackknife is computationally less intensive than the bootstrap because it only estimates the variance of a statistic rather than estimating the entire distribution of the statistic. The main practical difference between the bootstrap and jackknife methods is that the bootstrap gives different results when repeated on the same data, whereas the jackknife gives exactly the same result each time. Thus, the jackknife method is the preferred method when the estimates need to be confirmed several times before publication.

Permutation Testing

Before discussing the details of permutation testing, let us review the important features that we need to consider in order to perform these calculations. First, it is important to know that permutation testing does not refer to a single test, but rather a class of tests. Permutation refers to any of a specified class of rearrangements or modifications of the data. We must also have a null hypothesis that would state that all of the permutations are equally likely. The sampling distribution of the test statistic under the null hypothesis is computed by forming all of the permutations, calculating the test statistic for each, and considering these values all equally likely.

In situations in which we are trying to show the effect of a treatment or a difference between two groups, we can use permutation testing to estimate the empirical distribution of differences based on random groupings of the data. Then we can see how extreme the original difference is when compared with the distribution. For example, say we have a cohort of individuals with diabetes and a cohort of controls, and we observe a difference in their mean cholesterol levels of 20 mg/dL. Is this an extreme difference? To test this, we can randomly shuffle the disease status labels of who is diabetic and who is a control, thus achieving a random partitioning of the data into two groups of the same size as the original groups. We then compare the two groups that we have relabeled as "diabetics" and "controls," even though each group will have individuals from each category. We do this many times, each time calculating the difference in mean cholesterol levels between the two groups, obtaining a distribution. We can see how extreme our original observation is compared with this distribution and obtain a p value. If the p value is small, we can safely say that the difference observed between the two groups is more extreme than would be expected by chance. It is not possible to give a detailed example of permutation testing because they are computationally intensive.

CORRELATION AND COVARIANCE

When analyzing data, we are often interested in whether two variables are associated. This usually means some sort of measure or test for correlation. Correlation is a concept that describes statistical dependence between two or more random variables.

For example, there is a positive correlation between height and shoe size: Taller people generally have larger feet. On the other hand, supply and demand are negatively correlated: When there is less supply of a product, there is greater demand for it. To understand the mathematical definition of correlation, we first need to define **covariance**, which measures how two variables change together.

Covariance

$$Cov(X, Y) = E[(X - \mu_X)(Y - \mu_Y)], \tag{22}$$

$$Cov(X, Y) = E(X\,Y) - E(X)\,E(Y). \tag{23}$$

For example, if the greater values of one variable correlate with the greater values of the other variable (and the same for the smaller variables), meaning that they show similar behavior, the covariance is positive. Conversely, if each variable shows opposite behavior, the covariance is negative. Thus, the sign of the covariance shows the tendency in the linear relationship between the variables.

The most commonly used measure of correlation is the **Pearson correlation**, which measures linear dependence between two quantities. It is defined as

$$Corr(X, Y) = \frac{covariance\,(X, Y)}{stdev\,(X)\,stdev\,(Y)}, \tag{24}$$

where the denominator normalizes the covariance to be between −1 and 1. A perfect increasing linear relationship has a Pearson correlation of 1, and a perfect decreasing linear relationship has a Pearson correlation of −1. An important distinction must be noted here because (1) the covariance of two random variables is a population parameter and can be seen as a property of the joint probability distribution and (2) the sample covariance is an estimate of the parameter.

For a population, the Pearson correlation is

$$\rho_{xy} = \frac{cov(X,Y)}{\sigma_X \sigma_Y} = \frac{E((X - \mu_X)(Y - \mu_Y))}{\sigma_x \sigma_y}, \tag{25a}$$

where

$$\sigma_x = \sqrt{\frac{1}{N}\sum_{i=1}^{N}(x_i - \bar{x})^2}. \tag{25b}$$

For a sample, the Pearson correlation is

$$r_{xy} = \frac{\sum_{i=1}^{n}(x_i - \bar{x})(y_i - \bar{y})}{(n-1)s_x s_y}, \tag{26a}$$

where

$$s_x = \sqrt{\frac{1}{n-1}\sum_{i=1}^{n}(x_i - \bar{x})^2}. \tag{26b}$$

There are other ways to identify an association between two variables. Another form of correlation, the **Spearman rank correlation**, measures how much two variables tend to increase together without requiring a linear relationship and is thus more robust to outliers. Note that this also means that Pearson is more sensitive. Spearman detects monotonically increasing functions (such functions between ordered sets preserve the given order) by computing the correlation between the rank of data points in two sets. For example, in an S-shaped curve, Spearman will give correlation of ≈ 1, whereas Pearson will give a much lower value. Spearman's rank correlation is an example of a **nonparametric statistic**. Recall that nonparametric statistics are those that do not assume that the data have any characteristic structure or parameters. (They are used for populations that take on a ranked order but have no clear numerical interpretation.) By looking at the ranks of the values, it is more robust to outliers but also makes a less strong assumption of how the data are related. The difference between the Pearson's and Spearman's methods can be seen by looking again at the figure showing Anscombe's quartet (Fig. 5).

Spearman correlation is calculated as follows:

1. Convert the raw values X_i, Y_i to ranks x_i, y_i.
2. Calculate the differences $d_i = x_i - y_i$ between the ranks of each observation on the two variables.
3. If there are no tied ranks, then the Spearman correlation is

$$\rho = 1 - \frac{6\sum d_i^2}{n(n^2 - 1)}. \tag{27}$$

4. Otherwise, if there are tied ranks, use the Pearson correlation between ranks for the calculation:

$$\rho = \frac{\sum_i (x_i - \bar{x})(y_i - \bar{y})}{\sqrt{\sum_i (x_i - \bar{x})^2 (y_i - \bar{y})^2}}. \tag{28}$$

Missing Data

In many problems involving the analysis of biomedical data, some of the data are missing. In some types of the analysis, the data may be considered **censored**, which

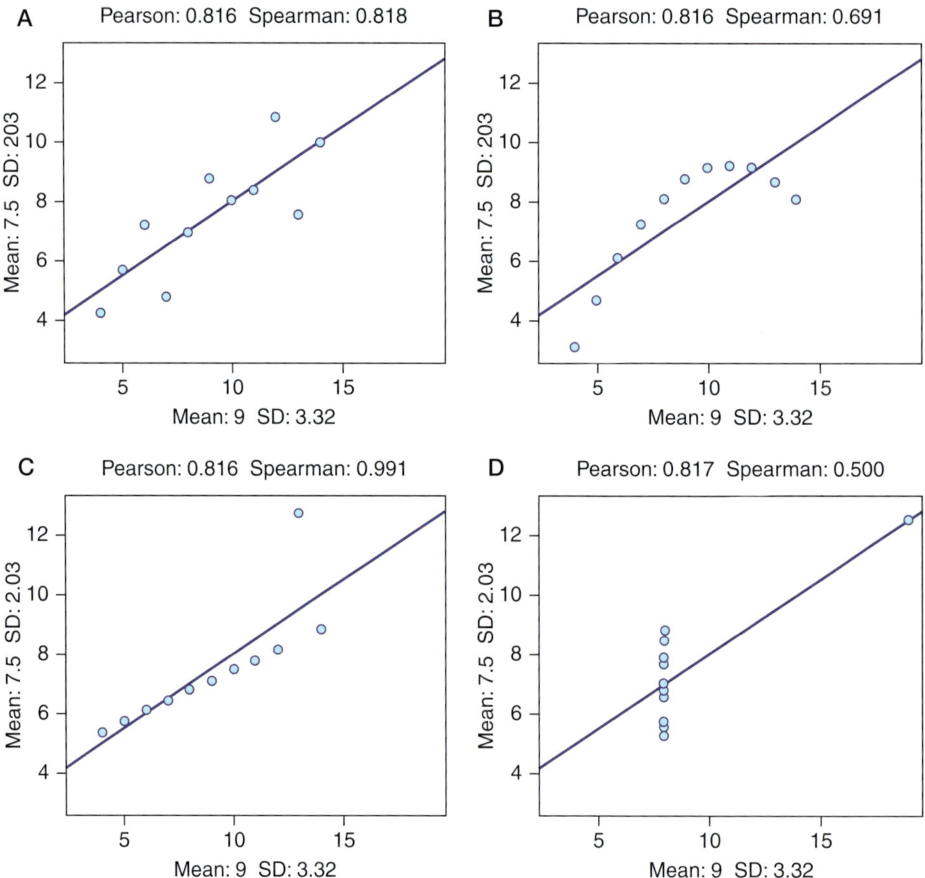

FIGURE 5. Anscombe's quartet. The figure shows a classic illustration of four data sets (A–D), all having the same summary statistics (mean and variance) as well as the same value of Pearson's correlation coefficient. However, visual inspection or calculation of Spearman's correlation coefficient reveals very different patterns of distributions.

implies that the data are missing in some systematic way. This could include instances such as when individuals in a clinical trial are lost for follow-up or measurements are outside the range of the instrument. Censoring implies that the data are missing in some systematic way, such as patients who may be lost to follow-up in clinical trials or a measuring device that only returns a value within a given range. There are a variety of statistical methods that have been developed to address censored data; probably most widely used are **Cox proportional hazard models**, which should be discussed in any good book on epidemiology or biostatistics. In general, proportional hazard models are a class of survival models, which relate the time that passes before some event occurs to one or more covariates (variables that are possibly predictive of the outcome under study).

Alternately, data may be missing at random. When there is a very large amount of data and only a few entries are missing, a very simple approach is to remove those samples, which is what is commonly done by default in most statistical packages. However, with very-high-dimensional data sets with a lot of missing data, this may mean removing too many samples. One approach is to view the missing data as a type of categorical data, with "missing" as a category, and then use an approach that allows the mixing of categorical and continuous data when necessary. Another approach is to attempt to **impute** the missing data in some way. This means using the existing data to create a model and use the nonmissing data to fill in the missing values.

VISUALIZATION

Although summary statistics can be very useful for characterizing a set of values, a graphical representation can often be extremely useful. Simple plots that group values together such as histograms and their smoothed version, density estimates, can all be powerful tools in analyzing data. A nice example showing some of the inherent limitations of common summary statistics are four sets of values known as the **Anscombe's quartet** (Fig. 5). Each of the four data sets shown in the figure has the same population mean and variance (in both the horizontal and vertical dimension); however, they have very different patterns of distribution that are obvious by eye. The human brain is very good at finding patterns in data, and using visualizations can take advantage of this ability.

In Figure 5A, this plot appears to be a simple linear relationship corresponding to two variables, and it follows the assumption of normality. The plot in Figure 5B is not normally distributed, although there is an obvious relationship between the two variables. It is not a linear relationship, and the Pearson correlation coefficient is irrelevant here. Figure 5C shows a graph where the distribution is linear but has a different regression line as compared with that in Figure 5A. This regression line is offset by one outlier that has enough influence to change the regression line and lower the correlation coefficient. The graph in Figure 5D shows an example in which one outlier is sufficient to give a high correlation coefficient, although there is no linear relationship among the variables. Anscombe's quartet is an excellent illustration of the importance of looking at a set of data graphically before beginning analysis according to a particular type of relationship and, in addition, the importance of the effect of outliers on statistical properties.

Many different ways have been devised to help in the visualization of high-dimensional data, providing a range of ways to project high-dimensional data onto lower dimensions. One common method is simply to plot the pairwise relationships between variables in what is known as a **draftsman's plot** (Fig. 6). Here, we show the relationship among three different kinds of measurement on three different species of iris. One can investigate other examples with evocative names such as box and whisker plots or mosaic plots. The ability to visualize and present data in an

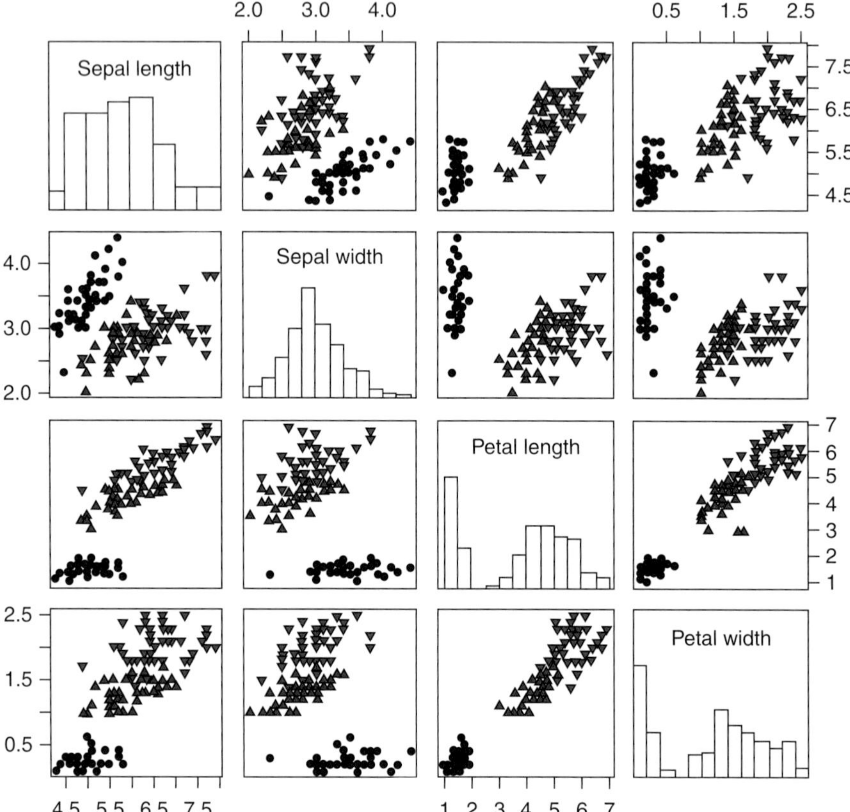

FIGURE 6. Draftsman's plot showing pairwise relationships between sepal length, sepal width, petal length, and petal width in three species of iris. Plots on the diagonal display the distribution of values for a single variable. Off-diagonal plots display the relationships between pairs of variables. The row and column location of each plot determines which pairwise relationship is displayed. For instance, the plot in the top row and third column shows the relationship between sepal length and petal length. (Dark blue circles) Species 1; (red triangles) species 2; (purple triangles) species 3.

accessible way is instrumental in the ability to explain your experimental results and conclusions.

CONCLUDING THOUGHTS

In this chapter, we introduce many basic statistical concepts that will aid the biomedical researcher in data analysis. We provide real life examples to accompany the techniques described both inside and outside of the biomedical domain. First, we define notions of probability and expectation; we talk about descriptive statistics used to characterize distributions of data and discuss how to perform statistical analysis on continuous and categorical data. We touch upon concepts such as sampling and permutation testing and multiple hypothesis correction. We talk about how to deal with missing data and the importance of correlation and covariance when performing association testing. These are all important considerations, and we urge you to continue to expand your statistical knowledge beyond the scope of this chapter.

REFERENCE

Ader HJ, Mellenbergh GJ, eds. 2008. *Proceedings of the 2007 KNAW Colloquium. Advising on Research Methods: A Consultant's Companion.* Johannes van Kessel, Huizen, The Netherlands.

4

Machine Learning

Marc A. Schaub and Chuong B. Do

Stanford University, Computer Science, Stanford, California 94305

During the last decade, advances in high-throughput molecular techniques have led to the generation of biological data sets of unprecedented size. Examples of high-throughput tools regularly used by biologists include gene expression analysis, genotyping assays, and sequencing. In these approaches, thousands or even millions of measurements are performed for each studied sample (such as a cell or an individual, depending on the assay), and studies of several thousands of samples are no longer uncommon. The analysis of such large data sets presents new challenges but also offers opportunities for new, data-driven approaches. When analyzing smaller-scale experiments, researchers often look for interesting patterns in their data that can explain the observed results and, in doing so, generate new hypotheses regarding the underlying biological process. The ability to comprehend the experimental data is thus a key component of the scientific process. Finding patterns in data sets that include billions of measurements would, however, be intractable without the help of computational approaches. **Machine learning**, in particular, is the study of the theory and application of mathematical and statistical algorithms in order to computationally identify complex patterns in large data sets. Machine learning methods are commonly applied to a wide variety of areas in science and engineering, ranging from the computer vision methods used to automatically read handwritten labels on packages (Hull 1994) to automated controllers that learn how to drive cars (Thrun et al. 2006) or fly model helicopters (Coates et al. 2008).

Biomedical research, however, represents a particularly challenging area of application given the speed at which high-throughput assays are currently growing. The explosion of information technologies during the last few decades has been largely fueled by the exponential rate at which computer hardware has evolved. In particular, Moore's law, stated by Intel founder Gordon Moore in 1965, predicted that the number of transistors per chip would double approximately every two years. This prognosis has held true until recently, meaning that the performance of computers has been improving exponentially, and costs have kept decreasing. This impressive progression is, however, currently outperformed by the even more rapid rate of improvement of biochemical assays such as sequencing technologies. As the size of the data sets generated by biological experiments keeps growing, successful analyses of these data will undoubtedly rely more and more on computational approaches. Given that biological data sets are growing faster than computer hardware performance, current computational approaches in general—and machine learning methods in particular—must be improved in order to be able to analyze these data efficiently. This makes biomedical applications of machine learning a particularly exciting field of research.

Machine learning methods are likely to gain even more importance in the analysis of biological data in the future and will become part of the toolkit that biologists regularly use in their studies. Biomedical applications of machine learning methods are usually an interdisciplinary endeavor involving teams of researchers from diverse backgrounds. The development of the machine learning algorithms themselves is usually the domain of statisticians and computer scientists. Although certain applications will benefit greatly from algorithms specifically tailored to the problem, most often, existing algorithms can be used to answer the question of interest. Such interdisciplinary applications are often handled by computer scientists with some training in biology or by biologists with some expertise in machine learning. Finally, experimental biologists will be involved in projects that include machine learning components, and their domain knowledge will be extremely useful for the researchers working directly on the machine learning application.

This chapter presents an overview of machine learning aimed at researchers in the biomedical field and is organized into two broad sections. The first half of the chapter presents a high-level overview for researchers who collaborate with computer scientists, to provide an understanding of the key concepts and terminology of machine learning from an application perspective. In the second half of the chapter, we discuss how to apply machine learning to a problem. In particular, we focus on how biological data can be used, discuss some of the most commonly used algorithms, and point to some available tools and resources for data analysis. Whereas the second part is directed more toward the researcher who is interested in applying machine learning approaches to a specific problem, the goal of the first part is to provide an introduction that will be broadly useful to biomedical researchers in general.

INTRODUCTION TO MACHINE LEARNING

The specific goals of the first part of this chapter are to (1) provide a high-level overview of what can be done with machine learning, (2) introduce some commonly used machine learning terminology, (3) discuss how to evaluate the performance of a machine learning approach, and (4) discuss how to properly design a computational experiment using machine learning. Here we assume that the described algorithms are black boxes that perform the described task. This part will thus be useful to a broad readership—both to researchers who are interested in a high-level understanding of machine learning and to those interested in applying a machine learning approach themselves, because it constitutes the foundation on which the second part of the chapter is built. This overview section will therefore be of interest to experimental biologists who collaborate with computer scientists and would like to understand what the computer scientists are doing (or could be doing) in terms of machine learning. It will also be useful for experimental biologists reading a mainly biological article whose methods make use of machine learning. In particular, we cover some common pitfalls in experimental design and metrics used for performance evaluation, thus providing the tools required to critically interpret the results obtained using machine learning methods.

Overview of Machine Learning Terminology

In this section, we introduce some key machine learning terminology by walking through an example of the application of various machine learning methods to biological data. Suppose that we are interested in studying gene expression in cancer. We have obtained cancer cell samples from 100 patients with a given type of cancer and have used a microarray chip to measure the expression of every known gene transcript in each sample. We call the collection of gene transcript measurements for a given sample the *gene expression profile* of that sample. The 100 samples represent a small fraction of the larger set of patients across the world suffering from the studied type of cancer. Our goal is to learn more regarding the biology of the cancer by studying this data set using machine learning methods.

Let us assume that the cancer we are studying has two subtypes: one that has a fast progression and high mortality and one that has a slow progression. Furthermore, suppose that the two subtypes are hard to distinguish at the time of diagnosis. An interesting question to ask in this context is whether gene expression profiles can predict outcome. We can formulate this problem as a **classification** task in which the algorithm (called a classifier) assigns a class label (either fast or slow progression) to each sample based on a set of **features** of the sample (the gene expression profile). The number of features is often very large and they can be seen as vectors in a high-dimensional space (called the feature space), where each feature represents one dimension. For example, if our microarray assays for the expression levels of 10,000 distinct genes, then the features for the sample corresponding to a single patient might be represented as a 10,000-dimensional vector, where each component of the vector corresponds to the expression level for a single gene.

The **input** of the classifier would be the values of each feature (i.e., the gene expression profile) for a given sample, and the **output** of the classifier would be the predicted class label (i.e., slow or fast progression). If we have such an algorithm, then we can measure the gene expression profile for a new patient and immediately predict the cancer subtype. In some situations, the relationship between the features and the class label is already known, in which case, devising such a classifier would be fairly trivial. This is unlikely to be the case in our example, however, and we will therefore use machine learning methods to derive a classifier from the data to which we have access.

Now, suppose that in addition to measuring the gene expression profiles for our 100 patients, we performed a long-term follow-up on these patients and acquired outcome information telling us which subtype of cancer each patient eventually developed. This means that for each sample, we would know both the features and the class. During the **training** phase, a machine learning algorithm attempts to identify the differences between the features of samples with fast and slow progression. Once the training has been completed, the machine learning algorithm produces a trained classifier that can then be applied to new samples and predict an outcome based on gene expression.

A different question that could be asked using this same data set is whether the survival time of the patient from whom the sample was taken can be predicted based on the expression level of his or her gene expression profile. If we are only interested in a coarse approximation, we could use the same approach as in the first example and classify the samples into a few discrete classes, such as survival of <1 yr, 1–5 yr, or >5 yr. These classes can be fairly arbitrary, however, and we might want to obtain a continuous value instead: a real number that more specifically quantifies the estimated survival time of a patient. This problem is called a **regression** task, in which the algorithm does not separate samples into classes, but instead predicts a continuous value based on the features. Although this is an important distinction in terms of how the internals of the training algorithm work, the setup of the training phase is almost identical to the first example. For each sample, the survival time is the value the algorithm should predict, and the expression levels of all genes are the features the algorithm will use.

In both examples above, we had access to measurements of the features of each sample as well as the value we would want the algorithm to predict for that sample. Those problems are examples of the broader class of **supervised learning** tasks. There are, however, many situations in which we have samples for which we have measurements of the features but not of the desired output value. For example, we might have access to gene expression measurements of an additional 100 cancer samples for which no long-term follow-up data are available. Using these samples may be helpful when training the classifier because they provide additional information regarding the underlying distribution of gene expression in general. Problems in which the training set contains both samples for which the desired output value is known and samples for which the desired output is unknown are called **semisupervised learning** tasks.

Machine learning methods can also be applied in situations in which there are no samples with a known output variable. These classes of tasks are called **unsupervised learning**, which are not able to predict the same relationships as supervised approaches. Let us consider the cancer gene expression example and assume that we have the same gene expression data but do not know the outcome (fast or slow progression) for the samples. We cannot learn a classifier that predicts the outcome because there is no information linking gene expression to outcome from which to learn. However, we can use unsupervised learning approaches to group samples that appear similar based on the gene expression information alone. In a follow-up analysis, one could then get additional information in order to assess whether samples that are grouped together have similar outcomes or not.

A commonly used example of unsupervised learning is **clustering**. Gene expression analysis is an area in which clustering algorithms have been widely used. For example, we might be interested in finding out whether the type of cancer we are studying can be separated into subtypes based on the gene expression information alone. Unlike the supervised learning setting, here we do not know to which subtype the sample belongs. Most likely, we do not even know into how many subtypes the samples should be split. The clustering algorithm will group samples that have a similar gene expression profile into clusters, and each cluster can then be further analyzed in order to identify the biological properties those samples have in common. Similarly, clustering algorithms can be used to group genes that have similar expression levels across all samples.

Figure 1 illustrates these different types of learning approaches based on gene expression data (the features of the samples) derived from a synthetic data set.

Although the examples above have been simplified for the purpose of describing machine learning theory, they are actually based on real research results. Pioneering work by Todd Golub and colleagues in the early days of gene expression analysis showed that machine learning algorithms can be used to identify subtypes of leukemia using gene expression data (Golub et al. 1999) and to classify cancer samples (Ramaswamy et al. 2001).

We now consider an example that provides a more intuitive view of the difference between supervised and unsupervised learning. In this example, we have a large number of microscopy images showing different cells. The cell type is determined in a time-consuming process during which an experienced biologist looks at each cell and labels it. We would therefore like to use a machine learning approach in order to automate the process. A first question to be asked is what features should be used. Picking a set of features was straightforward in the gene expression examples above (we simply used the expression levels of all transcripts); however, the question is more complex for this image analysis problem. A simple solution would be to use the intensity of each pixel as a feature. However, there might be more informative

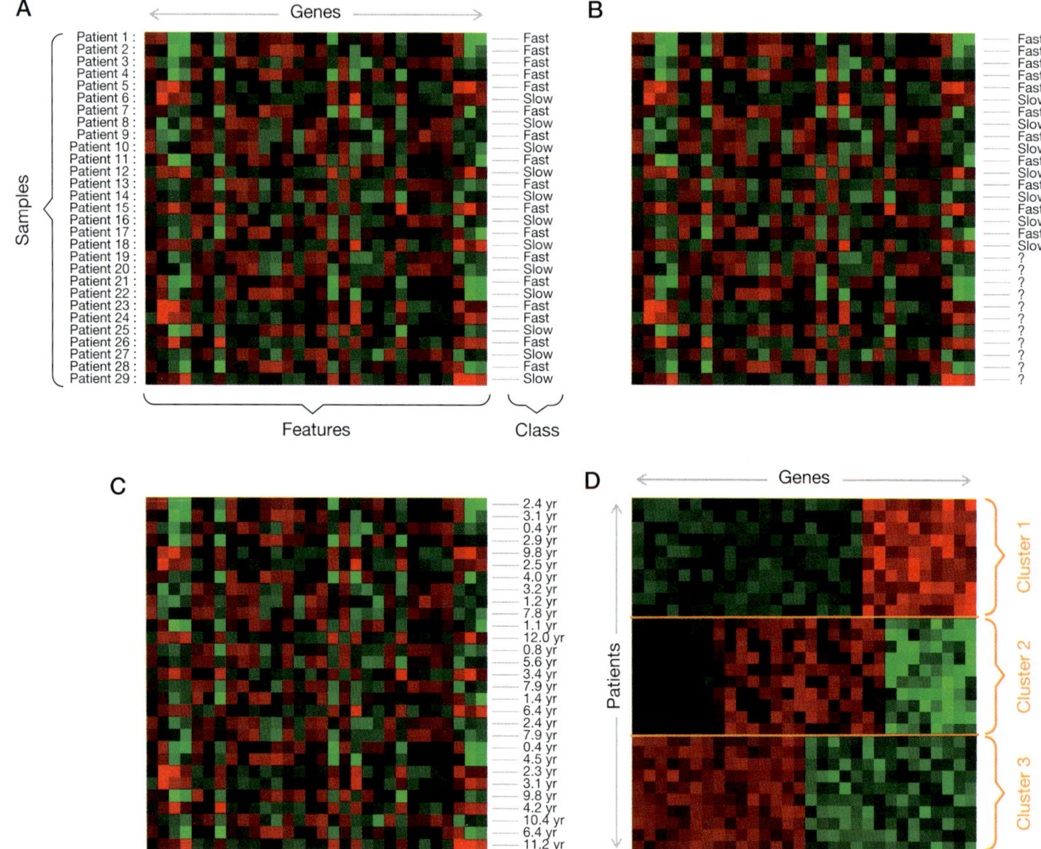

FIGURE 1. Examples of different types of learning in the context of a gene expression analysis (synthetic data set). Rows of the matrix represent patients from which a sample was taken, and columns represent genes. Each cell represents the level of gene expression of a specific gene in a given sample. The color brightness is proportional to expression intensity: red indicates overexpression and green indicates underexpression. (*A*) Supervised learning. A classification task with two classes: fast or slow progression. (*B*) Semisupervised learning. Question marks represent unknown labels. (*C*) Regression task. Rather than predicting discrete classes, the task of the classifier is to predict survival time, a continuous variable. (*D*) Clustering (unsupervised learning). Patients have been reordered on the vertical axis in order to visually highlight the three distinct clusters.

features that could be computed for each cell, such as the area of the cell, the average intensity of the cell, or its texture. **Feature selection** is often a nontrivial task that can have a major influence on the performance of the algorithm. This is an area that can benefit from suggestions from an experimental biologist who knows the problem well. Once we have selected a set of features, we can formulate this task as a supervised learning problem in which the training algorithm will use training samples that have been labeled by the human expert. During the learning phase, these labeled samples are used to understand the differences between the cell types based on the features. Metaphorically, the training process can be seen as the classifier learning from the labeling work performed by the expert, and this approach is therefore called supervised learning. However, it is important to note that the learning algorithm does not know why the expert classified a cell a certain way, but only knows the features and the label of the cell. If we consider the slightly different context in which we have access to the same data set but have no expert, we can still try to separate cells into different classes using an unsupervised learning method. Such an algorithm would be agnostic to the notion of cell types and will look for the best way to separate the data set, given the features of the images. We would then have to validate whether the separations made by the algorithm make sense from a biological perspective.

Performance Evaluation

The previous discussion covers a broad range of machine learning applications, but it is based on the blind assumption that the learning algorithm worked. In reality, learning is a difficult problem, and there are a myriad of reasons why applying a certain learning algorithm to a given problem might not work. To come back to the first example above, we hypothesized that there is a difference at the level of gene expression between the cancer subtypes with fast and slow progression. If there is no such difference, then a classifier will not be able to learn meaningful differences between examples of the two subtypes and will perform poorly when it comes to classifying a new sample. Even if there is a difference, the specific classification algorithm we choose to apply might not be able to detect it and would also do poorly on a new sample. Finally, even if a classifier can detect this difference and correctly predict the subtype for most samples, in practice, it is unlikely that a classifier will always be correct. It is therefore essential to properly evaluate the performance of any machine learning algorithm applied to a biological data set.

We now introduce several commonly used performance metrics and, in doing so, will expand on the example of the classifier that distinguishes between cancer subtypes with fast and slow progression. First, let us make the goal of the approach a little more specific: We want to use the classifier in order to identify patients with a fast-progressing cancer so as to treat them with more aggressive drugs. Let us assume that those drugs are the only potential cure for the fast-progressing subtype and have significant side effects, but have no effect on the slow subtype. We do have a classifier and are interested in determining how well it can make this distinction. Based on the gene expression data of a patient, the classifier will determine whether or not the patient has the fast-progressing form of the disease. Note that this classifier will always return a prediction (fast- or slow-progressing form) and cannot abstain from making a decision. At this point, there are four possible scenarios. First, let us consider the case in which the classifier predicts that the patient does have the fast progression subtype. If this is the case, then it is called a **true positive** and if it is not, it is called a **false positive**. Let us now consider the case in which the classifier predicts that the patient does not have the fast progression subtype. If the classifier is correct, then this is called a **true negative** and if the classifier is incorrect (the patient does have the fast progression subtype), this is called a **false negative**.

The simplest metric that can be used to evaluate the performance of the classifier is **accuracy**, which represents the fraction of samples for which the classifier will make the correct decision. Formally, accuracy is defined as the ratio of true positives plus true negatives over the total number of predictions made by the classifier. There is, however, a major limitation to using accuracy as the sole performance metric of a classifier: The definition of accuracy makes no distinction between false positives and false negatives, even though in practice the implications of the two are often very different. In our example, the consequences for a patient who is a false positive would be the side effects of the unnecessary treatment, whereas the consequences for a patient who is a false negative would be the absence of treatment even though it is needed. It is therefore important to be able to make this distinction when evaluating the performance of a classifier for this task. This can be done by considering two separate metrics: sensitivity and specificity. In our example, **sensitivity** (also called **recall**) measures the fraction of patients who do have the fast subtype that are correctly identified by the classifier. Formally, sensitivity is defined as the ratio of true positives over false negatives plus true positives. If, as in this example, false negatives have a high cost, then one will want an algorithm that has very high sensitivity. **Specificity** is the exact opposite to sensitivity, because it measures the fraction of patients who do *not* have the fast subtype that are correctly identified by the classifier. Formally, specificity is defined as the ratio of true negatives over true negatives plus false positives. In our example, the more significant the side effects, the more important it will be to have a high specificity as well. In summary, sensitivity and specificity allow you to independently assess how well the classifier detects true positives and true negatives, respectively. One question that neither metric directly addresses is how reliable a positive or negative prediction is for an individual patient. To assess the reliability of positive predictions, one could compute the **precision**, the fraction of true fast progressors among those predicted to be in the fast progression class. Formally, precision is defined as the ratio of true positives over true positives plus false positives. Precision is also called **positive predictive value** (PPV). To assess the reliability of negative predictions, one could use the **negative predictive value** (NPV), defined as the ratio of true negatives over true negatives plus false negatives.

Figure 2 provides an overview of these metrics. An important caveat to keep in mind when using these metrics to determine the performance of a classifier is that some can be affected by the proportion of individuals in each of the two classes within the set of examples on which the metrics are computed. For example, the precision of a classifier may appear artificially high in situations in which the evaluation data contain far more positive than negative examples; in contrast, the precision will be low when the evaluation data have very few positive examples. In the same way, accuracy and negative predictive value can also be affected by the proportion of positive and negative examples in the evaluation data. Because our goal is to estimate

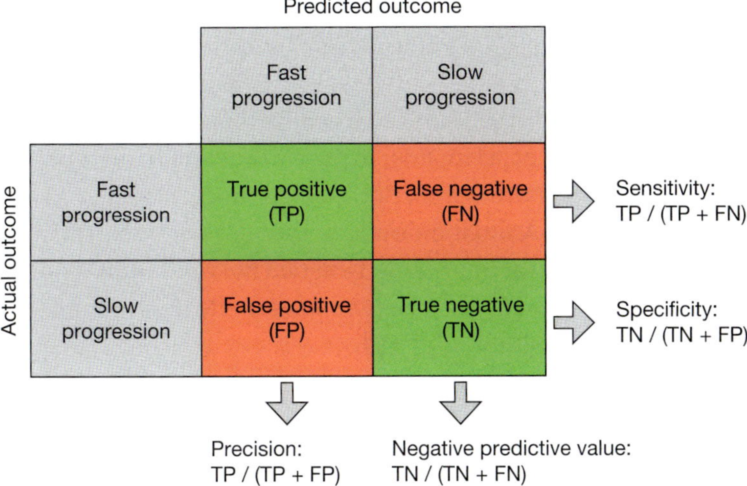

FIGURE 2. Overview of the different metrics used in evaluating the performance of a classifier. The goal of the classifier is to identify patients with the fast-progressing form of the disease. The two green boxes represent situations in which the prediction made by the classifier is correct, and the two red boxes represent situations in which the classifier is wrong.

how well the classifier would do on a new patient, it is important to compute these performance metrics on a set of samples that have the same distribution as the general patient population or to adjust accuracy estimates appropriately to account for differences in the proportion of positive examples between the evaluation set and the intended population for the test.

In an ideal world, a classifier would have perfect accuracy (and thus both perfect sensitivity and perfect specificity); however, in reality, the performance of a classifier is often a trade-off between sensitivity and specificity. Let us consider the extreme situations first. If we do not care about specificity at all (i.e., whether a patient has the fast- or slow-progressing subtype), then a classifier with perfect sensitivity can be trivially obtained by predicting that every patient is in the fast progression class. Similarly, if we do not care about sensitivity, then a classifier that never predicts that a patient is in the fast progression class would have perfect specificity. Although those extreme examples are not particularly useful classifiers, the idea behind them can be generalized: If specificity is most important, then the algorithm can be tuned so as to predict a patient to be in the fast progression class only if the algorithm has a very high confidence that the prediction is correct. This confidence threshold can be lowered to improve sensitivity at the expense of specificity.

Most machine learning algorithms do have more complex parameters that can be adjusted so as to favor sensitivity or specificity. How to choose an appropriate trade-off for a particular application will depend on the consequences of a false negative, respective to a false positive. Clearly, this design decision could benefit greatly from the input of an experimental biologist. Let us now consider the example of a machine learning algorithm that predicts new binding sites for a certain protein that are then validated experimentally in the wet laboratory. If, on one hand, the specificity of the approach is low, then the wet laboratory biologist will waste a lot of time and energy on unsuccessful experiments. If, on the other hand, the specificity is high but the sensitivity low, then the number of predictions to validate will be low. If the experimentalist has a given validation budget, then it may be worthwhile to adjust the sensitivity of the algorithm in order to generate the right number of predictions. The positive predictive value can then be used to estimate how many of these predictions are likely to be successfully validated.

This duality between sensitivity and specificity makes it more difficult to assess the overall performance of a classification approach because parameter tuning will change the trade-off between the two metrics. However, assessing the overall performance of a classifier is often useful, in particular, when comparing different machine learning algorithms or in order to show that a new method outperforms the state of the art for a given problem. The most common approach is the use of **receiver operating characteristic curves** (ROC curves) (see Fig. 3). A ROC plot represents the sensitivity on the vertical axis (ranging from 0 to 1), and the false positive rate (the fraction of false positives among all negatives, which is equal to 1 – the specificity) on the horizontal axis (also ranging from 0 to 1). Each point on the curve represents the sensitivity and 1 – specificity of the method for a given parameter choice. A perfect classifier would be plotted at the upper-left corner of the plot (i.e., sensitivity and specificity both equal to 1). The trivial classifier that obtains perfect specificity by never predicting a patient to be in the fast progression class would be plotted in the lower-left corner of the plot, and the trivial classifier that obtains perfect sensitivity by always predicting a patient to be in the fast progression class

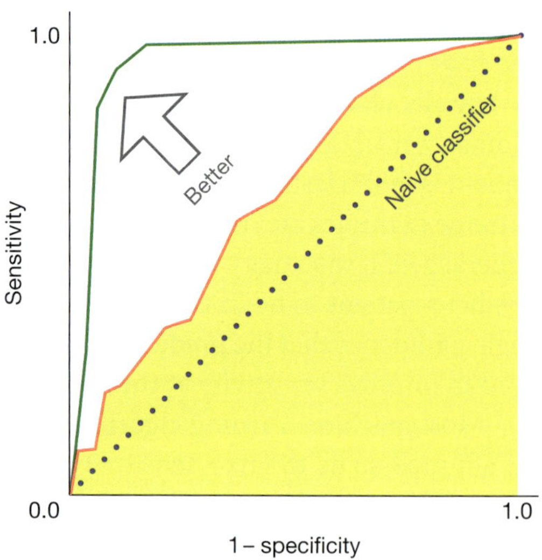

FIGURE 3. Example of a receiver operating characteristic curve (ROC curve). Sensitivity is plotted on the vertical axis and (1 – specificity) on the horizontal axis. The yellow area represents the area under the curve (AUC) for the red curve. The green curve has a better AUC than the red one, which indicates that the classifier evaluated to generate the green curve is performing better than the one used for the red curve. The dashed blue line represents a naive classifier that is unable to distinguish between the two classes.

would be plotted in the upper-right corner of the plot. When evaluating a classifier, parameter choices are modified in order to generate more (*sensitivity, specificity*) pairs that are added to the plot. A curve that represents the performance of the classifier in terms of both sensitivity and specificity is obtained by connecting these points. The performance of a classifier can be summarized by computing the **area under the curve** (AUC). The better the classifier, the larger is the AUC, and a perfect classifier would have an AUC of 1. On the other hand, a diagonal line from the lower-left corner of the plot to the upper-right corner would represent a baseline classifier that is unable to distinguish meaningfully between the two classes (represented by the "naive classifier" line in Fig. 3). Such a baseline classifier could, for example, work by generating a random number and then classifying a patient as being in the fast progression class if the random number is above a prespecified threshold. Varying the threshold varies sensitivity and specificity and will yield different points on the diagonal. Because the classifier does not even consider the features, it is obvious that its output is meaningless. It is important to note that this classifier has an AUC of 0.5, which means that when evaluating the performance of a classifier, one should consider how much above 0.5 the AUC is.

Although the AUC is formally defined as the area under the ROC curve, it can also be shown that the AUC has a simple, alternative interpretation as the probability that for a randomly chosen pair of positive and negative examples, the classifier would be able to correctly assign a higher chance of being positive to the positive example. For a perfect classifier, this probability would be 1.0, which is the maximum achievable, as stated above. In contrast, a baseline classifier that "flips a coin" in order to decide which among a pair of positive and negative examples is more likely to be positive would clearly have a 50% chance of being right and, hence, an AUC of 0.5.

This section focused on a binary classification task, which is one of the most commonly used applications of machine learning. Most of these error metrics can be extended to classification problems with more than two classes. The performance of a regression method is most often computed based on the difference between the predicted value and the actual value for each sample. A commonly used metric to quantify the error over an entire data set is the **root mean square error**, which is obtained by averaging the squares of the differences between predicted and correct output values for individual samples, and then taking the square root of this average. Alternatively, correlation metrics between the predictions and the correct values can also be used. Various methods exist to evaluate the performance of unsupervised learning approaches like clustering, but their description would go beyond the scope of this chapter.

Experimental Design

In this section, we discuss how to design an experiment that allows proper measurement of the performance of a machine learning approach. Designing a proper computational experiment is very likely the most critical component of any machine learning application. It should be given the same high degree of attention that the design of a wet laboratory experiment receives, both when conducting a study and when reading or reviewing an article in which machine learning is used. Subtle mistakes in experimental design can easily lead to performance estimates that are very misleading and could make a machine learning method look particularly promising even though the algorithm is, in reality, performing poorly. Even though these general issues are well known among the machine learning community, such mistakes unfortunately still appear fairly regularly in articles that apply machine learning techniques and are often significant enough to cast doubts on the whole methodology or the overall findings.

In the previous section, we discussed several metrics that can be used to evaluate the performance of a machine learning algorithm and, in particular, of a binary classifier. We did not, however, discuss which samples should be used to do this evaluation. To do so, a good first step is to look at this question from the perspective of how the algorithm will ultimately be applied. In the context of the classifier that identifies patients who have a fast progression cancer subtype, the ultimate application of the methodology would be to classify a newly diagnosed patient. This means that the sample from the patient was not even collected at the time the classifier was developed, and thus it would be absolutely impossible that any information regarding the gene expression of this patient could have been used in any manner when training the classifier. From the classifier's perspective, this is a brand-new sample that it has never seen before. The notion of using the classifier on a sample in general and not just on the subset of samples that were in the training set is called **generalization**. If the classifier is able to learn the true differences between the two cancer subtypes, then it will have good performance on the new, unseen samples and is said to generalize well. Therefore, we do need to evaluate the performance on samples that were never used before and, more specifically, samples that in no way contributed to any step of the classifier design or training. One approach would be to train the classifier on all the patients in our initial data set and then find an entirely new cohort of patients in order to measure the performance of the classifier. An easier solution is to emulate this process as part of the experimental design. At the very beginning, a subset of samples for which we know both gene expression level and cancer subtype are selected for the specific purpose of measuring classifier performance. These samples form the **test set**. The remaining samples then form the **training set** that is used to train the classifier. Once the classifier has been trained, then every sample in the test set is given as input to the classifier. Thus, the classifier is in the very same situation as it would be for a new patient: It has never seen this sample before and does not know its correct classification. The performance of the classifier on the test set is called the **testing error** and represents how the classifier would perform on unseen data. The classifier could also be applied to all the samples in the training set in order to compute a **training error**. The general approach of separating the available data into a training set and a distinct test set is called **cross-**

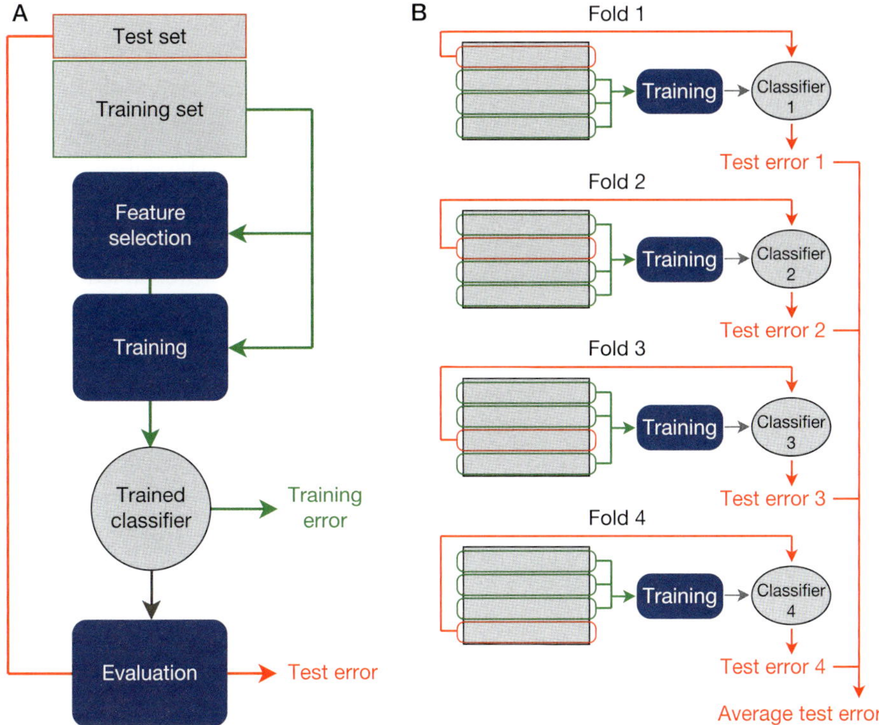

FIGURE 4. Flow overview of the experimental design. (*A*) Correct split between test and training set. The green path indicates how the training set is used to train the classification and compute a training error, and the red path indicates how the test set is held out until the classifier has been fully trained and is used only to compute the test error. (*B*) Example of *k*-fold cross-validation, with $k = 4$, showing how a different subset of the data is held out during training at each iteration. (Green) Data used for training; (red) data held out for testing. The average test error is then computed by averaging the test error of all folds.

validation. Figure 4A shows a flowchart depicting a correct split between the test and training sets.

Understanding why the test error is different from the training error is key to understanding why such a clean split between test and training set is absolutely necessary. The training error indicates how well the classifier performs on examples it has already seen. A classifier that has a high training error is not capturing the complex features of the data set well enough. During the learning phase, most machine learning algorithms do actually use the training error in one way or another in order to estimate how well they are doing. If a classifier has a high training error after training has been completed, then a different classifier type or a different feature set should be considered. Although a classifier that has a high training error will likely also have a high testing error, the converse is not true. It is actually very easy to design a classifier that is able to classify the training set almost perfectly. Given an input sample, this classifier would simply cycle through all the samples in the training set until it finds one whose features exactly match the feature of the input sample and then return the prediction that corresponds to this matching sample. This classifier will always make the correct prediction unless two samples in the training set have the same features but different outputs. Although this classifier does perform extremely well on the training set, it is unlikely to perform well on the test set or on any other sample that was not in the training set. The issue with this classifier is that it did not learn how to distinguish the cancer subtype with fast progression from the cancer subtype with slow progression in general, but, rather, it learned how to distinguish between the instances of the two classes in the training set.

Although the example we present here is obviously a caricature, all learning algorithms potentially suffer from this problem: By learning the characteristics of the training set extremely well, they become too specific and unable to do well in general, a problem called **overfitting**. The second part of this chapter discusses how to handle this problem from a learning perspective. From an experimental design perspective, overfitting is the reason why a classifier can do very well on the training set but perform poorly on the test set. It is thus absolutely essential to evaluate the performance of a classifier on data that were never used during the training process. This is probably the single most important take-away point in this chapter.

Now that we have established the importance of proper cross-validation, we discuss how this is commonly performed in practice. If we use the approach outlined above and keep a fraction of the available data as a single test set while using the rest of the data as training data, then we do end up with a fairly small test set. In our cancer classifier example, we start with 100 samples. We could use 20 samples in the test set and 80 in the training set. However, this would mean that the test error would be computed on only 20 samples, which is quite small to be considered a random sample of the general landscape of gene expression in this cancer. We could increase the test set at the cost of the training set, but this would make the learning step more difficult and likely result in worse performance. Ideally, we would like to use all the available samples as the test set. This can be done if, rather than training and evaluating a single classifier, we repeat the cross-validation process multiple times. This approach is called ***k*-fold cross-validation** (Fig. 4B). The available samples are first split into k groups. One group forms the test set, and the remaining $k - 1$ form the training set. A classifier instance is then trained on this training set and its performance evaluated on the test set. This procedure is performed a total of k times, with a different test and training set each time. At the end of the process, each sample has been used in a test set exactly once, yet performance is never evaluated using samples that were also used for training. The performance of the k classifier instances is then averaged to obtain an estimate of the overall performance of the approach. Generally, the choice of the value k is a trade-off between the size of the training set (which will be larger with a greater k) and the computation time required to train the k classifiers. The case in which k is equal to the number of samples, meaning that at each iteration a classifier is learned using all but one sample and then evaluated on that single sample, is called **leave-one-out cross-validation**.

A proper cross-validation setup is necessary for showing that a machine learning method works well and will hopefully avoid blatant mistakes such as measuring performance on samples that were used for training. There are, however, much more subtle ways in which testing data can be indirectly used when training a machine learning algorithm. One example is feature selection, which is sometimes incorrectly considered as a preprocessing step: All the information available is used in order to choose which features will be used by the classifier, and then k-fold cross-validation is performed to assess the performance of the classifier. This means that the test samples do indirectly contribute to the training, because they contribute to feature choice. It is intuitive to see why this is a problem if we go back to looking at this from the perspective of a sample from a new patient being given to the cancer subtype classifier. Because the gene expression from this sample was unavailable at the time the feature selection was performed, it is impossible that this information could have been used for feature selection. Because a sample in the test set cannot be used for anything more than a sample from a new patient, we cannot use it when doing feature selection. A correct experimental design would thus perform feature selection in each fold of the cross-validation, using only the samples in the respective training set (refer to Fig. 4 for the flowcharts of both cross-validation and k-fold cross-validation). Other challenging situations include pipelines in which multiple learning algorithms are used at various stages. The best approach in this case is to look at it again from the perspective of how a new sample would be used and to avoid using test set samples in a different way from how a new sample would be used.

Although it is clear that feature selection on the whole data set uses information to which it should not have access, one might wonder why this is such a big issue, especially given that the test data are never used when training the classifier. Imagine a thought experiment in which a feature is, by chance, associated with the output label (i.e., a high expression of the gene represented by this feature is correlated with fast progression of the cancer) but, in reality, has no association. Because there is an overall association in the combined data set, it is more likely than not that the association in the training samples is in the same direction as in the test samples (i.e., high expression is correlated with fast progression in both). Therefore, feature selection performed on the entire data set will tend to choose features that have associations that go in the same direction between the test and training sets. It therefore favors features that will behave similarly on the test set as on the training set, which leads to an incorrectly low error when evaluating the classifier on the test set. An example in which this problem can easily occur is the application of a machine learning method to the strongest associations identified in a genome wide association study (GWAS; see Chapter 7 for more information regarding GWASs). Limiting the feature set to the strongest associations is a form of feature selection performed on the entire data set, and one therefore needs to be particularly careful when evaluating the performance of a classifier using these features with a cross-validation scheme over the same data set (Hinds et al. 2012).

USING MACHINE LEARNING ALGORITHMS

This section first discusses how to go from data to features given as input to the learning program and then gives the higher-level intuition behind several commonly used machine learning algorithms. It builds on the general concepts and the experimental design aspects discussed in the previous section. The goal of this section is to allow the reader to use off-the-shelf machine learning packages to analyze their data, several of which are discussed at the end of the section. This section is also useful when collaborating with computational scientists, because it explains why certain requirements exist in terms of data format and provides an overview of different algorithms the computational scientist might suggest using.

Data

The most essential single factor determining the success of a biomedical application of machine learning is obtaining a good data set to analyze. Generally speaking, increasing the amount of data used to train a machine learning algorithm will lead to more significant performance improvements than using more sophisticated algorithms. However, in the context of biology, acquiring additional data is often expensive, time-consuming, or even impossible given the available project resources. An important aspect to keep in mind when working with data sets containing a limited number of samples is that the more complex the classifier is, the more samples will be needed to train it in a meaningful way. Ideally, one would want to have more samples than features, but this is often not the case. A simple rule of thumb is that if the number of features is the same or greater than the number of samples, then the learned classifier is likely to be too complex and will not generalize well (although there are some exceptions to this rule). Intuitively, one can think about this in terms of the number of possible samples, which is exponential in the number of features, and compare it with the number of samples in the training set. If the number of features is much higher than the number of samples, then the fraction of possible samples in the training set will be extremely small, and the classifier will be learning from a subset that is unlikely to be representative of the general feature space. The classifier will learn patterns that are specific to the training set, will not be likely to generalize, and will thus overfit. In practice, this rule of thumb is often too conservative, because classifiers might not necessarily use all of the features that are available, or may use other techniques for reducing the "effective dimensionality" of the feature space. Most machine learning algorithms allow for a way to control the complexity of the learned classifier, but the specific details of how this can be performed are beyond the scope of this chapter. Furthermore, some types of classifiers are more prone to overfitting than others when trained on a small set of samples, and this aspect should definitely be kept in mind when choosing which classifier to use.

To show the notion of overfitting, we present a very simple example of regression in Figure 5. This example has a single feature x represented on the horizontal axis, and we are trying to predict a continuous variable y on the vertical axis. Training samples are represented by blue circles, and one can easily guess the general relationship between x and y just from looking at the figure. For this example, we chose to model y as a polynomial function of x, which is plotted in red. We control the degree of the polynomial, which can be thought of as controlling the complexity of this model. The first plot (top left) in Figure 5 shows the situation in which the constraint is too stringent and the model does not capture the relationship between x and y. In this case, both the training error and the test error will be high. The second plot (middle) shows a more complex model (third-degree polynomial) that captures this relationship well and thus has a low training and test error. The model does not, however, exactly predict the correct value for each training example. The third plot (top right) shows an even more complex model (higher-degree polynomial) that exactly predicts each training example, which leads to a much smaller training error. However, one

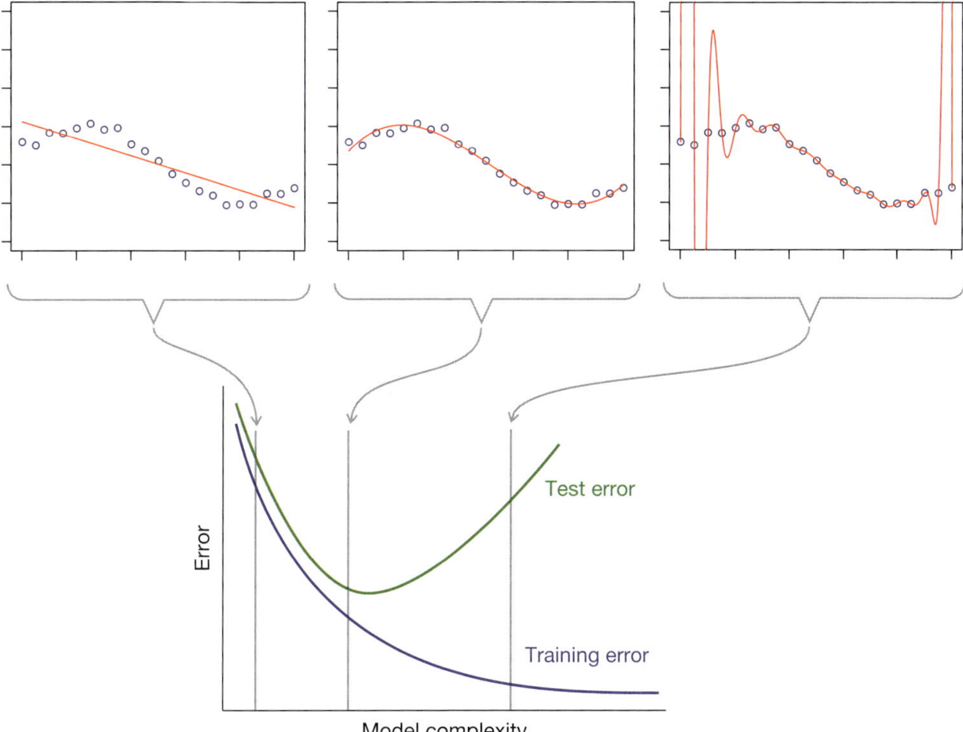

FIGURE 5. Example of regression. Data points (blue circles) as well as the curve learned from the data (red line) are plotted for models of increasing complexity. The first example (*top left*) shows a model that is too simple to represent the data. The *middle* classifier is able to capture general trends in data but does not overfit to noise. In the third example (*right* plot), the learned curve precisely fits each training example but does not capture the general trend. The figure illustrates how the training error keeps decreasing even though the test error increases when the model is too complex, and thus shows overfitting.

can easily see that this model also captures the noise in the training set and would not generalize well to other samples, hence causing a high test error. This example shows that although the training error decreases as the model complexity increases, the test error will start increasing again once the classifier starts overfitting. It is thus important to use cross-validation in order to try out various parameters of a classification algorithm to determine the right model complexity for a given data set.

The previous paragraphs explained why the number of samples available to train and evaluate an algorithm matter, but their quality is important as well. There are two main aspects to consider in terms of data quality. The first aspect is the importance of having independent samples in the data set. For example, let us assume that we use multiple samples from the same patient in the cancer subtype example discussed above. If one of the patient's samples appears in the training set and another appears in the test set, then it is easy to see how the classifier is predicting outcome on a sample that is much more similar to a sample it was trained on than a sample from a random patient would be. If all of the samples are in the training set, then the classifier will give too much weight to the features that are characteristic of this patient but not the cancer subtype in general, and will thus overfit. In practice, one might inadvertently end up with nonindependent samples, especially if the process of generating samples is very complex and involves multiple data sources. Alternatively, it may be possible to account for the fact that there are samples that are not independent, both during the learning phase and the performance evaluation process. The second important aspect is the use of reasonably clean data. Biological data are inherently noisy, and one concern is that the classifier might thus overfit and make decisions based on differences that are just noise. As discussed above, this is a concern, especially when the number of training samples is low. Machine learning algorithms are, however, designed to handle some amount of noise and might perform well even on unprocessed data. In fact, in some scenarios, it may even be better to have a large noisy data set than a small clean data set, simply because the increased sample size outweighs the penalty from having messier data.

Once we do have a set of biological data, the next question is how to represent this information as features that will be used by the machine learning algorithm. Features can have different types. A **Boolean feature** can take two values, true or false, and be used to represent the presence or absence of a property. A **discrete feature** can take multiple predetermined values. It is important to distinguish discrete features that represent categories but do not imply an order between them (e.g., the value of a DNA nucleotide can be A, T, C, or G) from **discrete ordered features** (e.g., a gene expression represented with discrete levels "low," "medium," or "high"). The latter types of features are also called **ordinal features**. Sometimes, an off-the-shelf package will assume the latter case and impose an arbitrary mapping between categories and numbers (to come back to nucleotides, A = 0, C = 1, T = 2, G = 3) and then treat the feature as a number, which is incorrect. If this is the case, one might have to represent one nucleotide by four Boolean features, respectively, representing whether the nucleotide is an A, a T, a C, or a G (or three Boolean features, by omitting one of the four nucleotide categories arbitrarily). In some situations, the same information can be represented by a discrete or a discrete ordered feature and one might want to determine which one works better in practice. An example of this is genotype data, where each single-nucleotide polymorphism (SNP) can take three values: homozygote with the major allele, heterozygote, and homozygote with the minor allele. This SNP could be represented as a categorical variable without making any implications on the relationship between the three categories (and the algorithm would have to find interactions that are useful). The SNP could also be expressed as a discrete feature counting the number of copies of the major allele (0, 1, or 2) that the individual has at that location.

A third type of feature is a **continuous feature**, which is represented by a real number. Examples of continuous features are the expression levels of transcripts used in the cancer subtype classifier. Often, such simple low-level features can be derived directly from the data. This is the case for expression levels or for genotype data. In other applications, it makes sense to generate higher-level features based on the data and incorporating prior knowledge that could be difficult for the algorithm to learn. An example here is to include the features of images such as the size or the texture of a cell. One particular situation in which using more elaborate features makes sense is sequence analysis. Although we could code each nucleotide as a discrete feature, it might also make sense to represent coding sequences using amino acids. Similarly, when building a machine learning algorithm that models gene regulation based on sequence information, one could first look for the presence or absence of binding sites for known transcription factors and represent those as features in addition to low-level features representing the sequence. Finally, one particular point to keep in mind is the handling of missing information in features, such as a SNP that could not be genotyped in a particular individual. Often, this would be coded as a special category in the input file, and when using a standard machine learning package, it is important to ensure that the package treats the "unknown" values properly rather than considering them to be just another category. Machine learning algorithms might be able to **impute** missing values based on other variables.

In the previous paragraph, we discussed a variety of methods that can be used to generate features from raw data. In practice, it is important to keep the number of features reasonably small and to focus on the most informative ones, a process called **feature selection**, which we briefly introduced in the "Machine Learning Terminology" section of the chapter. There are two main reasons for doing this. First, if the number of features is larger than the number of samples, then the learning algorithm will tend to overfit. Second, the running time of most learning algorithms heavily depends on the number of features. In such a case, adding many unnecessary features might cause the runtime of the training algorithm to become practically intractable. Most feature selection methods are based on information theory, and the specifics of how they work is beyond the scope of this chapter. The high-level intuition is that they are computing how much information a feature provides about the output variable. A feature that is highly correlated with the output variable will most likely be useful for the learning algorithm and should thus be kept, whereas a feature that always has the same value across the training set will not be useful. Furthermore, if there are perfectly correlated features (such as SNPs that are in perfect linkage disequilibrium), then keeping one of them is sufficient. A limitation of an approach considering features individually is that there might be specific combinations of multiple features that together correlate with the output even though the individual variables do not and are thus at risk of being filtered out at this stage. This can be avoided by considering subsets of features instead, and various search strategies can be applied to this task.

Intuitively, one can see that there is a trade-off between leaving a large number of features available to the training algorithm, which will then make the decision on how to use them, versus pruning features before the training step. This is another situation in which deciding what to do will depend on the problem, the learning algorithm, the number of features, and the number of samples in the training set. In practice, one can explore how much feature selection is necessary using cross-validation. Unless the number of features is too large for the training step to complete, then starting by using all features when training the algorithm is a good approach, because feature selection might not be needed for the specific problem at hand. As discussed earlier in this chapter, feature selection should always be performed on the training set, and one should not use samples that are kept out for performance evaluation.

The last aspect we must consider in terms of data is how to define the output variables. This task was fairly straightforward in the supervised tasks we discussed so far in this chapter, because it could be easily derived from the question asked and there was sufficient information available in the data to assign an output variable to each sample. One important point to keep in mind is that mistakes in the output variables (such as the human expert being wrong when labeling a cell) can lead to worse performance of the algorithm on the actual task, but this might not be noticed when evaluating the performance on the available data. A situation in which determining labels is more difficult is when we have examples of the presence of a phenomenon but no clear definition of its absence. For example, we might be interested in detecting splice sites based on sequence data and have a large compendium of known splice sites available. Those would all be in the same class, and thus we need to generate a good set of negative examples in order to be able to formulate this as a classification problem. One solution would be to look for sequences that share many properties of the observed sequences (e.g., length, overall nucleotide distribution, overall position on the chromosome) so as to limit the difference with the positive examples to the fact that there is no known splice site at those locations.

A final, yet critical point to consider with respect to output variables is that whenever an output variable is defined (even when the information is provided by a human expert), one should make sure that there is no hidden correlation between the output variable and the features that is caused by a **confounding factor**, something other than the question of interest. A typical example of such a hidden variable would be the situation in which we only measured gene expression on cancer patients with one subtype and used the data obtained by a collaborator who had a different preparation protocol for the second subtype. The most extreme case would be if the two experiments were performed on different gene expression assays. In this case, if there is any overall difference between the two platforms (which is commonly the case in practice; e.g., the average signal intensity might be higher on one platform than on the other), then the learning algorithm will find the affected features and heavily use them because they are correlated with the output variable. This results in a classifier that has good performance on this data set during cross-validation, even though it captures only artifacts that are not related to the question of interest.

Supervised Learning Algorithms

In this section, we present the general intuition behind several commonly used supervised learning algorithms and subdivide them into general categories. We cover the advantages and limitations of those algorithms and discuss the high-level assumptions on which they are based. Two separate sections are dedicated to a similar overview of probabilistic models and unsupervised learning. The goal is to present briefly how the algorithm works at a high level and to indicate in which situations these algorithms could be useful. You will be directed to appropriate references and websites that provide tools and a deeper discussion. However, describing the math behind each of these algorithms in detail or how they are implemented would go far beyond the scope of this chapter. It is, however, always advisable to familiarize yourself with the algorithm you plan on using because it may place specific requirements on the input data or make assumptions that are incompatible with the question you are trying to answer. The list of algorithms we present is by no means exhaustive, and there are many more approaches that are commonly used in practice. Another aspect that we will not discuss is the amount of computer resources needed to train these algorithms, both in terms of computation time and memory. In practice, there are significant differences between different methods, and, depending on the size of the data sets, training a method that would perform better for the task at hand might be intractable on the available hardware resources.

One of the simplest supervised learning algorithms is called ***k*-nearest neighbors** (often abbreviated k-NN). As its name indicates, this algorithm will predict the output of a new sample by finding the k closest samples among the training samples. Various distance metrics can be used to evaluate how close the features of two samples are. To compute the difference between two samples, one would first compute the differences between the values of each feature and then aggregate those differences. The **Manhattan distance** is the sum of the absolute values of the differences between each feature in the two samples. The name comes from the fact that to go from one point to another in a grid-like city where it is impossible to cut diagonally across city blocks, one has to first move in one direction (e.g., walk on 12th Avenue from 5th to 8th Street) and then in the perpendicular direction (walk on 8th Street from 12th to 15th Avenue). Although there are other path options, all will have similar lengths. The **Euclidean distance** is the square root of the sum of the squared feature-wise differences (and, to come back to the city example, would correspond to the shortest path if one could go in a straight line from one point to another through buildings). The *k*-nearest neighbors algorithm can be used as follows: (1) as a classifier, by determining which class is the most prevalent among the *k* neighbors; or (2) in a regression task, by computing the average output value for the *k* neighbors, usually weighed based on the proximity to the input sample.

The *k*-nearest neighbors approach does have several limitations. First, it works best if there are many training samples that are spread out throughout the feature space. If the number of training samples is small, then a new sample might be far away from any sample in the training set, and thus identifying which samples are the closest neighbors would not be particularly meaningful. This might happen if we have a large number of features, and therefore a very high-dimensional feature space, but only a small number of samples. A second limitation is that *k*-nearest neighbors does not prioritize features. Imagine that we have one feature that is a very good predictor of the output and many other features that are not related to the output at all. If we use a distance metric such as Euclidian distance, then all features will contribute equally to determining which samples are close to each other. The contribution of the useful feature will be outweighed by the contributions of the features that are not related to the output. In the next sections, we introduce algorithms that perform well when the amount of training data is limited and that prioritize features.

One algorithm that usually performs well with limited training data is **naive Bayes**. This algorithm is based on Bayes's theorem regarding conditional probability. In terms of conditional probabilities, the classification of a new sample can be seen as computing the conditional probability of being in each class given all the features *P*(*class* | *features*), also called posterior probability, and then predicting the sample to be in the class with the highest posterior probability. Using Bayes's theorem, we can rewrite this in terms of the prior probability of the class *P*(*class*), the probability of all the features given the class *P*(*features* | *class*), and the probability of the features *P*(*features*): *P*(*class* | *features*) = *P*(*class*) * *P*(*features* | *class*) / *P*(*features*). Because *P*(*features*) does not depend on the class, it does not need to be computed to determine which class has the highest conditional probability. In practice, however, computing the joint probability of all the features given the class will be impossible unless the exact sample was in the training set. The key assumption that naive Bayes makes is that all features are independent given the class. This means that the probability of all features given the class *P*(*features* | *class*) is assumed to be equal to the product of the conditional probability of every individual feature given the class: $P(features \mid class) = P(feature_1 \mid class) * P(feature_2 \mid class) * P(feature_n \mid class)$.

Because each conditional probability term is easily estimated from the training data, the naive Bayes model can be effective in situations in which the number of samples is fairly low. Naive Bayes does, however, make a strong assumption regarding the data, namely, that features are independent if the class is known. Although this is seldom the case in practice, naive Bayes will still perform reasonably well as long as the classification is not heavily dependent on feature interactions. For example, if we were to apply naive Bayes to the cancer subtype classifier, then it could perform reasonably well if the expression of individual genes is correlated with the outcome, even though it is well known that the expression of some genes is dependent on the expression of other genes. A case that naive Bayes could not capture, however, would be the situation in which a combination of several genes together is correlated with outcome, even though the individual genes are not.

This brings up the second factor to consider when choosing an approach, which is to ensure that the approach is able to capture the underlying structure of the feature space well. Choosing an approach often means making assumptions regarding the structure of the feature space, and the performance of the approach will depend on how good those assumptions are.

It turns out that naive Bayes can be shown to belong to the family of linear classification algorithms. In this section, we present the high-level intuition behind these algorithms. To do so, we consider a simple case in which we have only two features, x_1 and x_2 (see Fig. 6). Each sample can therefore be plotted on a two-dimensional (2D) figure. Figure 6A represents a case in which samples from two classes (represented by an O and an X, respectively) can be separated by a line. The equation for this line is $x_2 = 2x_1$. Let us now consider the following equation: $y = 2x_1 - x_2$. For any point on the line, y will be 0. For any point below the line, y will be positive,

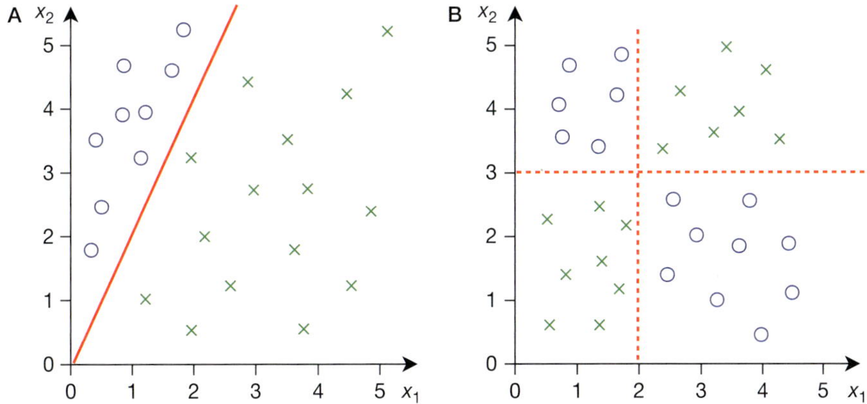

FIGURE 6. Classification in a 2D feature space. Continuous feature x_1 is represented on the horizontal axis, and continuous feature x_2 is represented on the vertical axis. The blue X and green O represent samples from two different classes. (*A*) In this example, a line with equation $x_2 = 2x_1$ (in red) separates the two classes. (*B*) Example of a situation that cannot be separated by a single line. The dashed lines represent a pair of lines that a partitioning classifier, such as a decision tree, could use to separate the two classes.

and for any point above the line, y will be negative. If we get a new sample to classify, we can therefore compute y based on the features x_1 and x_2 and assign a class to the sample based on the sign of y. In this example, y is called a linear combination of the features: the weighted sum of the features $y = w_1x_1 + w_2x_2$ (in this case, the weights are $w_1 = 2, w_2 = -1$). The training phase for a classifier using this approach consists of learning the weights based on the samples in the training set. Conceptually, if we have two features, this can be seen as finding the line that separates the classes. This concept scales to larger numbers of features and thus higher dimensions (where the separating line becomes known as a separating hyperplane or linear decision boundary). For example, if we have three features, we would want to find the plane that separates the three-dimensional (3D) space into two halves, one containing the samples of one class and one containing the samples of the other class. One important question is whether it is always possible to separate the positive and negative samples in the training set. Coming back to our 2D example, it is easy to construct an example for which this is not possible (Fig. 6B). In this example, it is necessary to consider interactions between features to determine the class of the samples. Such interactions cannot be represented in an equation of the form $y = w_1x_1 + w_2x_2$, and there is no line separating the classes. In practice, most real problems are not separable. Linear algorithms, however, still perform well unless the true difference between classes heavily relies on interactions between features. A commonly used approach in linear algorithms is the so-called **kernel trick**. Conceptually, this can be seen as a function mapping the feature space (in our example, the 2D plane) to a different high-dimensional space in which the samples are expected to be separable. Widely used machine learning algorithms such as **logistic regression** and **support vector machines** (SVMs) are based on linear combinations of features.

If we know that there are significant interactions between features, we might still be able to use a linear algorithm and explicitly model known interactions as additional features. Furthermore, linear algorithms such as SVMs do, in practice, tend to perform well even when there are interactions between features. However, we might be interested not only in training a classifier that achieves good performance, but also in learning more regarding the underlying relationship between features. In the case of the cancer progression prediction example used above, we might want to know which genes are good predictors of the outcome. If we train a linear classifier, we can look at the weights and discover which individual features play an important role in the trained classifier (usually, the absolute value of their weight will be higher than for less important features). Linear classifiers, however, will not identify interactions between features automatically, and there are cases in which we might be interested in finding such interactions. In such cases, new features representing the interactions must be added directly to the set of features used by the linear classification algorithm.

A different category of algorithms that can capture interactions between features without the need for explicitly adding those interactions to the feature is called

partitioning algorithms. These algorithms work by separating samples into different groups. A very commonly used partitioning algorithm is a **decision tree** classifier. The training of a decision tree starts by finding the feature that is most informative regarding the output and deriving a rule that splits the training samples into two subsets based on this feature. The process is then recursively applied to each subset. Thus, this algorithm can be seen as a tree, in which each node contains a rule that splits the incoming samples into two branches. The tree can be built until there is only one sample left in a branch, at which point a leaf node is created. A new sample is classified by traversing the tree from the root (which leads to the first rule) to a leaf node and then determining the class of the sample based on the class of the training samples that ended in this leaf node. Figure 7 represents a decision tree and illustrates how a new sample is classified. It is easy to see that training the tree until there is only one training sample left in each leaf node would make this algorithm very sensitive to individual training samples and thus prone to overfitting. It is therefore necessary to control the complexity of the tree. A simple way to do this would be to limit the depth of the tree (i.e., the number of rules that are applied to each sample until a leaf node is reached), but more elaborate methods exist as well. Although we discuss decision trees in the context of a classification problem, the ideas behind decision trees can be applied to regression problems as well, and the algorithm is then called a **regression tree** (Breiman et al. 1984).

The major drawback of a decision tree is that the algorithm is looking for the best single feature at each step and might miss situations in which a pair of features is

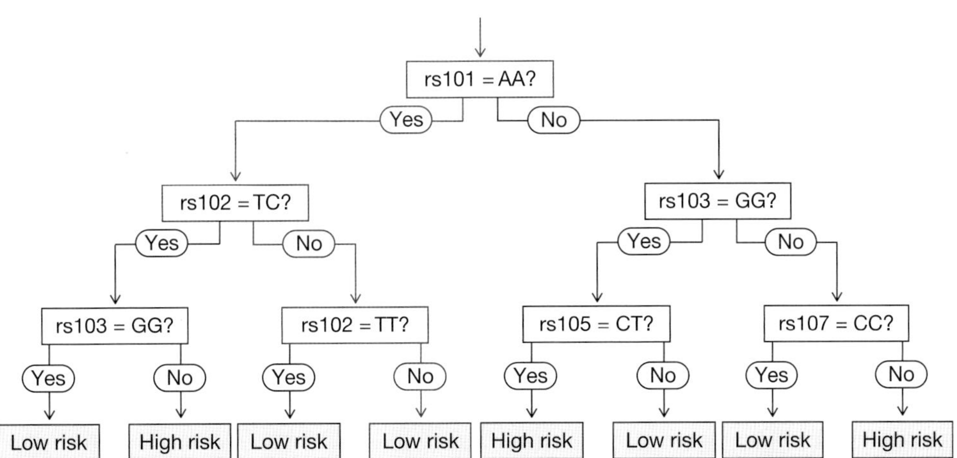

FIGURE 7. Example of a decision tree. This decision tree classifier uses the genotype of an individual at various single-nucleotide polymorphisms (rs101–rs105) to predict whether the individual is at low or high risk for a certain disease. An individual is classified by following the path from the root (at the *top* of the figure) to a leaf node. At each node, the genotype of the individual is used to decide which branch to follow. Each leaf node indicates which prediction the classifier makes for individuals whose path ends at that leaf node. The path in blue shows how an individual would be classified given that his genotype is AA at rs101 and CC at rs102.

jointly correlated with the class but none of the features is individually correlated. Algorithms based on taking the best decision at every step, even though it might not be the best decision overall, are called greedy algorithms. Although this is a fairly significant limitation of decision trees, they still perform well on many tasks and are among the most commonly used machine learning approaches. One reason for this is that the trained classifier provides a very intuitive representation of how the classification decision is made: The learned tree, with one simple rule per node, is easy to interpret and similar to the kind of decision process people commonly follow. This illustrates another aspect that is important when picking a machine learning algorithm: If the goal is not just performance but also the interpretation of how the classifier works, then one might want to pick a classifier like decision trees over a classifier that could perform slightly better but be harder to interpret. An example in which interpretation is important is the regression example presented at the beginning of this chapter, where the expression of a gene of interest is predicted based on the expression of other genes. In many cases, the interpretation is the output most desired.

The methods above consist of training a single classifier, which we expect to have a good performance. Another approach is to train multiple classifiers that together would have a better performance than a single classifier. A first approach is called **bagging** (Breiman 1996), which is short for "bootstrap aggregation." The idea behind bagging is to generate multiple different training sets by sampling with repetition from the actual training set. One classifier is then trained on each of those sampled sets, and a consensus prediction is made by combining the predictions of each classifier (either by averaging or voting). The use of multiple sampled training sets makes this approach less susceptible to the influence of individual samples and less prone to overfitting. Another approach that integrates multiple classifiers is called **boosting** (Schapire 1990; Schapire and Freund 2012). Boosting starts by training a simple classifier that might not perform particularly well on the training set. Samples that are misclassified by this first classifier are then reweighted, so as to be more important when training a new instance of the classifier. These training and sample reweighting steps are then repeated multiple times, generating one classifier each time. A new sample is then classified based on the combined output of all these intermediate classifiers.

Probabilistic Models

In the approaches described above, we made very few assumptions on the structure of the feature space and the relationships between features. There are, however, practical situations in which much more is known regarding those relationships. In such a situation, one might want to reason about the feature space as a high-dimensional probability distribution. This field of machine learning is called **probabilistic modeling**. In this section, we introduce the high-level intuition behind two types of probabilistic models commonly used in biological applications: Bayesian networks and hidden Markov models.

We start with an example in order to show how existing knowledge regarding the relationship between features can be used to build a particular type of graphical model called a **Bayesian network** (BN) (Pearl 1985). Let us consider a small biological pathway that consists of four genes: g_1, g_2, g_3, and g_4. The genes g_1 and g_2 encode two transcription factors: g_1 activates the transcription of g_3 and g_4, and g_2 inhibits the transcription of g_4. Figure 8A presents an overview of this pathway. We can use the information we have regarding the pathway to reason about the relationship between the expression levels of these genes. We can see that from a biological point of view, the level of g_3 only depends on the level of g_1. If g_1 is highly expressed, this will lead to the transcription of g_3, which will also have a high expression level. The level of expression of g_1 and g_3 will therefore be correlated. Similarly, the level of expression of g_4 depends on the expression level of g_1 and g_2. From a biological point of view, the

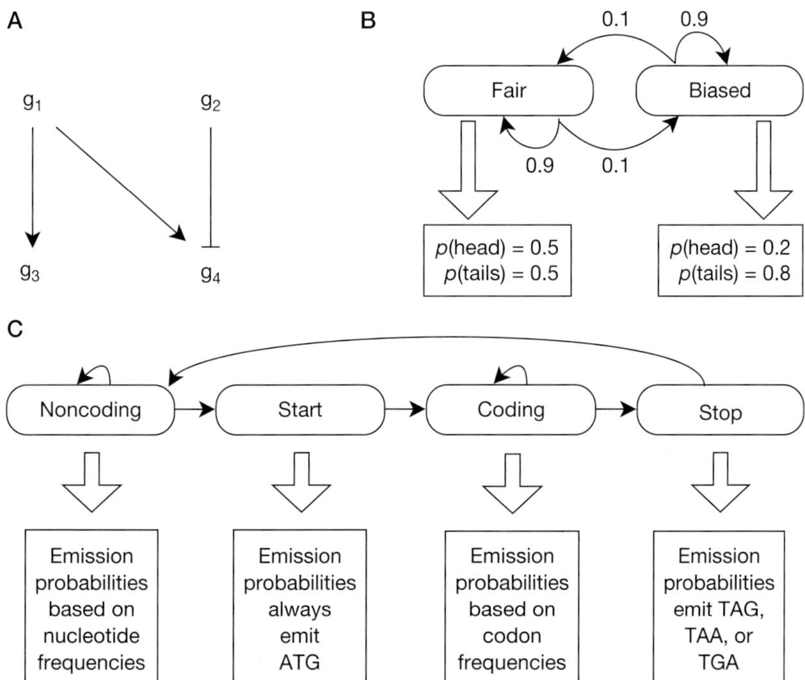

FIGURE 8. Probabilistic models. (*A*) Simple biological network used in the introduction of Bayesian networks. The model includes four genes: g_1, g_2, g_3, and g_4. Arrows represent activation, T-shaped edges represent inhibition. (*B*) The hidden Markov model (HMM) representing the coin toss example. Two states correspond to using the fair or the biased coins. At each step, the model can either transition to the other state or stay in the same state. Transition probabilities are represented on the edges. Emission probabilities are indicated for each state. (*C*) A simplified HMM representing coding regions. Each rounded rectangle represents a state in the model, with corresponding emission probabilities. Arrows represent possible transitions from state to state. This simple model only consists of four states that represent noncoding regions, the start codon, an uninterrupted coding region, and stop codons.

expression levels of g_3 and g_4 do not influence each other. However, they will be correlated: If there is no inhibition from g_2, then the expression levels of g_3 and g_4 will be similar—high if g_1 is high, low if g_1 is low. Using this knowledge, we could predict the expression level of g_4 knowing only the expression level of g_2 and g_3, because the expression level of g_1 can be inferred from the expression level of g_3.

More formally, we can represent the feature space as a joint probability distribution: $P(g_1, g_2, g_3, g_4)$. The expression level of each gene is now called a random variable. This means that any combination of gene expression values will be associated with a probability. Unlikely combinations, such as cases in which g_1 is high and g_3 is low, will have much lower probabilities than combinations that are consistent with our knowledge of the pathway. Furthermore, we can incorporate the information we know regarding the structure of the pathway into the way we represent the joint probability distribution. Because g_3 only depends on g_1, the level of g_3 can be represented as a conditional probability distribution: $P(g_3 \mid g_1)$. The expression level of g_4 can similarly be represented as a conditional probability distribution: $P(g_4 \mid g_1, g_2)$. We can therefore express the joint probability distribution as $P(g_1, g_2, g_3, g_4) = P(g_1)P(g_2)P(g_3 \mid g_1)P(g_4 \mid g_1, g_2)$. The structure of the joint probability distribution can be represented by a network, hence the name "Bayesian network."

The benefit of this approach is that we do not need to represent the joint probability distribution directly but can split it into smaller conditional probability distributions that are much easier to estimate based on the available data. If, as in this example, we already know the structure underlying the joint probability distribution, then we can learn all the conditional probability distributions from the data. If we only have a set of samples but no prior knowledge regarding the relationship between the features, then we can also try to learn the structure of the joint probability distribution. This is, however, a much harder task than learning only the conditional probabilities. If we were to do so on gene expression data, the resulting Bayesian network could provide meaningful information regarding the relationship between the expression levels of various genes. A trained Bayesian network can then be used to answer many types of questions, like the one we mentioned earlier: the prediction of the level of one random variable that has not been observed given some other random variables that have been observed.

A type of probabilistic model commonly used in biological sequence analysis is a **hidden Markov model** (HMM). An HMM is a probabilistic approach that models a process generating a sequence of events. In this section, we start with a simple example process, build an HMM for this process, and then show how the same approach can be applied to biological problems. Let us assume that I have two different coins I can flip. One is fair, meaning that the probability of obtaining heads (H) is 50% and the probability of obtaining tails (T) is also 50%. The second coin is biased: The probability of obtaining heads is 20%, and the probability of obtaining tails is 80%. I start by flipping one coin, and the result is either heads or tails. Then, either I use the same coin or I switch to the other coin, and this flipping process is repeated many times. Now let us assume that I switch the coins 10% of the time. The only information you obtain is the sequence of heads or tails (e.g., HTTTT). In particular, you do not know which coin is being used at any given point in time.

We can represent the coin being used as the state of the process: At each point in time, the process is either in the state corresponding to flipping the fair coin or flipping the biased coin. Furthermore, after the coin flip, the process will either stay in the same state (90% of the time) or switch states (10% of the time). Figure 8B provides a graphical representation of this process, with nodes corresponding to states and arrows corresponding to **transition probabilities** from one state to another. At each time step, the process is in a specific state, which is hidden from the observer (i.e., you do not know which coin is being used), and generates an observation. The probability of observing a head or a tail depends on the state in which the process is. These probabilities are called **emission probabilities**. The model just described is an HMM.

Although this model can be used to generate sequences according to the described process, this is of limited interest. A much more interesting question that can be answered using this model is the following: Given an observed sequence (e.g., HTHTTTTTH), which coin was most likely being used for each flip? Coming back to our model, we want to find the sequence of states that is most likely to generate the observations. Intuitively, one can see that the first part of the sequence (HTHT) is more likely to be observed when using the fair coin and the second part of the sequence (TTTTH) when using the biased coin. Thus, a fairly likely possibility is that I used the fair coin for the first four flips, then I switched to the biased one. A less likely possibility is that I used the biased coin the whole time. Efficient algorithms can be applied in order to find the sequence of hidden states most likely to generate a given observed sequence or to compute the probability of observing a given sequence for some HMM. Finally, although in this example the emission and transmission probabilities were known upfront, it is possible to learn them given a training set of sequences.

Hidden Markov models are often used for biological sequence analysis. The sequence of nucleotides is the sequence of observations, and the hidden states correspond to features one wants to identify. An example of such an application is gene finding. Let us start with a simple example in which there is no splicing. A sequence of DNA will therefore consist of a juxtaposition of coding and noncoding sequences. Those will be the two states of the HMM. Emission probabilities for a coding region can be derived from the frequency of codons in known genes (and in this case, at each time step, one codon of three nucleotides will be observed). Similarly, emission probabilities for the noncoding state can be obtained from the nucleotide frequencies in real noncoding regions. We also want to model the length of a coding or noncoding region, and we can use the transition probabilities to do this: The higher the probability of the process staying in the same state, the longer the region will be. We could use additional knowledge in order to improve our model. For example, if we know that each coding region must start with a start codon, then we can add a specific state to represent this. The state corresponding to the start codon will only emit the sequence corresponding to this codon, and the process will always transition to the coding sequence state, because we want exactly one start codon. Stop codons can be modeled similarly (this model is shown in Fig. 8C). This model obviously represents a very simplified view of the structure of a gene; in order to properly identify eukaryotic genes, we would also need to model exons and introns separately, as well as promoter regions, untranslated regions, and so on. Once such an HMM has been trained, it can be used to find the most likely sequence of hidden states for a new sequence of DNA and thereby indicate the positions in the sequence that correspond to different parts of genes and the positions in the sequence that are intergenic.

Unsupervised Learning Algorithms

In this section, we provide the high-level intuition behind several unsupervised learning algorithms. These algorithms are fairly commonly used in biological data analysis, and people might be familiar with a task like clustering even if they are not familiar with machine learning at all. Some algorithms used for general purpose clustering have even been specifically developed for other bioinformatics applications, such as phylogenetic tree reconstruction. Figure 9A provides an intuitive view of a clustering problem in 2D space (i.e., two features).

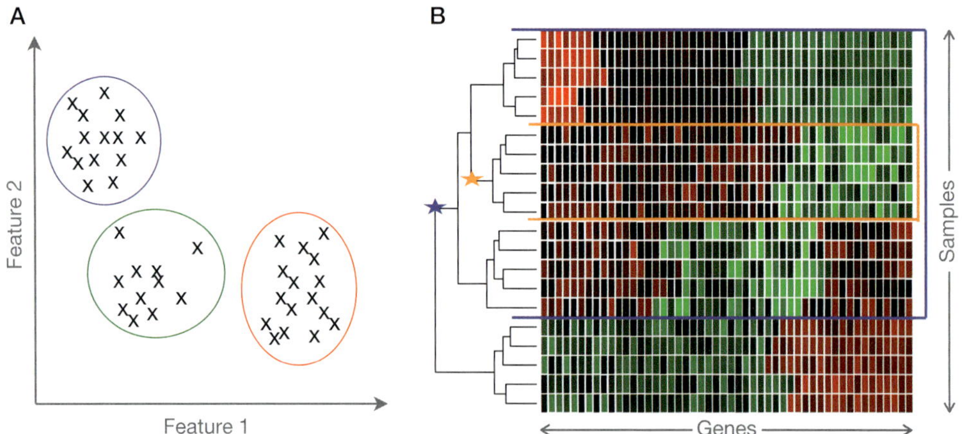

FIGURE 9. Example of clustering. (*A*) General example of clustering. Because there are only two features, the samples can be represented in a 2D view. The three clusters are visually apparent and circled in blue, green, and red. Clustering is generally used in a higher-dimensional feature space. (*B*) Hierarchical clustering of a mock gene expression data set. Rows are the matrix samples, and columns represent genes. Each cell represents the gene expression of a specific gene in a given sample. The color brightness is proportional to expression intensity; red indicates overexpression, and green indicates underexpression. This example illustrates the hierarchy of clusters that can be built. The blue bracket represents the samples grouped together by grouping all the leaf nodes reachable from the blue starred node and the orange bracket the subcluster obtained by grouping all the leaf nodes reachable from the orange star, and so on.

A commonly used unsupervised algorithm is **hierarchical clustering**, which, in addition to grouping similar samples together, also builds a hierarchy among those clusters. This hierarchy can be represented as a tree in which samples appear in the leaves and similar samples will be close to each other in the tree. Let us consider an arbitrary node in the tree and follow all the paths that go from that node to leaves. The samples that are in the leaves that are reachable from that node form a cluster. If the sample clustering worked well, then the samples within this cluster will be more similar to each other than to samples that are not in the cluster. This shows that it is not necessary to know a priori how many clusters one wants to split the samples into, because it is possible to build a hierarchical clustering tree in which each sample is its own leaf. Figure 9B represents an example of hierarchical clustering. This is commonly performed when analyzing gene expression data, and intermediate nodes in the tree are then used to define broad categories of similar samples or genes. Hierarchical clustering builds this tree in a greedy manner, meaning that at each step of the algorithm, it makes the optimal decision for this step alone, even though this succession of locally optimal decisions might not lead to the best final solution.

The clustering algorithm can work in a divisive or in an agglomerative way. **Divisive clustering** starts by having a single cluster (corresponding to the root node of the tree) that is then split into the two most different subgroups of samples, forming two clusters, which are then each split into two again, and so on. This continues until each cluster consists of a single sample. **Agglomerative clustering** starts by having one cluster per node and grouping the two most similar clusters, thus forming a node in the clustering tree, and then repeating this grouping until all clusters are connected in the tree. Hierarchical clustering algorithms rely on two metrics: the distance between two samples and the distance between two clusters of samples, also called the **linkage criteria**. Choosing the right distance metric based on the properties of the features is important to obtain good clusters. Commonly used metrics include the Euclidean and Manhattan distances defined in the section on k-nearest neighbor classifiers above. Commonly used linkage criteria between two samples include using the maximum distance between any sample in one cluster and any sample in the other cluster, using the minimum distance, or using the average distance over all pairs with one sample in each cluster.

As we have seen, hierarchical clustering builds a hierarchy of clusters and does so in a greedy way. There are settings in which one is only interested in splitting the samples into k clusters (where k is known) and to do so in a globally optimal way. An intuitive algorithm to do so is ***k*-means clustering**. When the samples are grouped into k clusters, then each cluster has a mean (a point in the feature space). It is therefore possible to compute the distance between each sample and the mean of the cluster to which the sample belongs. The idea behind k-means clustering is to find the assignment of samples to clusters that minimizes the total sum of these distances. However, there is no fast algorithm to find this optimal assignment, and thus, doing so is only possible for small data sets. In practice, there is an efficient heuristic algorithm that results in good clusters, even though there is no guarantee of obtaining the best clustering. The algorithm starts by choosing a random mean for each of the k clusters and then performs two steps repetitively. In the first step, it assigns each sample to the cluster whose mean is closest to the sample. In the second step, it recomputes the mean of each cluster based on the samples in the cluster. Although the initial means are random, repeating these two steps will lead the algorithm to converge to a cluster assignment that is usually a good approximation of the optimal k-means clustering. In practice, the algorithm can be repeated several times with different initial clusters so as to avoid the situations in which an unfortunate initial set of cluster means leads to poor performance. Furthermore, if the number of clusters is unknown, then the algorithm can be repeated with various values for k.

Unsupervised learning is not limited to clustering. A different approach is to go from a very large feature space to a much smaller space that still captures most of the variation in the data, a process called **dimensionality reduction**. For example, let us come back to the case in which we want to understand whether there are unknown subtypes of cancer based on gene expression data. Because there are several thousand genes, it is virtually impossible for a human being to get an intuitive overview of how close samples are to each other in this space. Hierarchical clustering would produce a tree that captures distances between samples, but one would then have to believe that the clustering algorithm made the right choices. What would be really useful is a 2D representation of the data that captures as much information as possible. This can be performed using **principal component analysis** (PCA). Although, in practice, most features are somewhat correlated, PCA finds a set of vectors in the feature space called principal components that are orthogonal (not correlated). The first principal component corresponds to the axis in the high-dimensional space that explains most of the variability in the data, the second principal component to the axis that explains the highest amount of variability once the variability explained by the first principal component has been removed, and so on. In order to obtain a 2D view of the data, one would find the first two principal components, which will become the two axes of this view. How this is performed using linear algebra is beyond the scope of this chapter. Each sample is then projected onto these two axes, resulting in a 2D representation of the data that represents as much variation as possible. This representation can then be interpreted much more easily than the raw data. A particularly spectacular application of PCA is the work of Novembre et al. (2008) in population genetics. A data set of thousands of individuals from across Europe were genotyped at several hundreds of thousands of SNPs, and then PCA was used to reduce this very large feature space to a 2D plot. The first two principal components mimic the north/south and east/west geographic axes particularly well, thus illustrating how well genetic information can be linked to geographic origin.

MACHINE LEARNING RESOURCES

In this final section, we briefly present some popular resources that allow you to learn more regarding machine learning and to use some of the algorithms presented in this chapter. One of the most commonly used machine learning applications is **Weka** (Hall et al. 2009), a free program that provides both an implementation of the most commonly used learning algorithms in Java and an intuitive graphical user interface. Weka also provides useful tools for data preprocessing and for visualizing data and parameters (e.g., a visual view of a decision tree). The Weka website (http://www.cs.waikato.ac.nz/ml/weka/) provides many resources in addition to the software itself, such as documentation, example data, and some tutorials. The creators of the Weka software have also published a book, currently in its third edition, that provides a thorough introduction to many data mining and machine learning approaches and explains how to use Weka in practice (Witten et al. 2011). Weka is probably the best starting point if you are interested in trying out some machine learning algorithms or if you want to quickly evaluate the performance of several approaches using your data. Another approach is to use a data analysis framework that you may already be familiar with, such as R or Matlab, together with a toolbox that adds machine learning algorithms to those programs. This approach lets you easily integrate machine learning into an existing workflow and gives access to simple ways for data input and output, and to statistical functions that are often useful to preprocess the data, generate features, and evaluate performance. A good starting point, if using R, is the Machine Learning & Statistical Learning page of the official R website (http://cran.r-project.org/web/views/MachineLearning.html), which provides a list of the most commonly used machine learning packages. Implementations of the most commonly used machine learning algorithms can also be found for Matlab. The SHOGUN Machine Learning Toolbox (http://www.shogun-toolbox.org/) (Sonnenburg et al. 2010) provides interfaces between state of the art support vector machine algorithms and frameworks such as Matlab, R, and Octave. For Python users, Scikit-learn (http://scikit-learn.org/stable/) integrates machine learning algorithms with other commonly used scientific packages. An alternative to using such a framework are stand-alone implementations of individual machine learning algorithms. In this case, you would have to generate input files that match the format expected by the tool, for example, using a scripting language like Python.

If you are interested in learning more regarding the details of how machine learning algorithms work, including their mathematical and statistical foundation, we recommend a very comprehensive resource, the book *Pattern Recognition and Machine Learning* by Christopher M. Bishop (Bishop 2006). *The Elements of Statistical Learning: Data Mining, Inference, and Prediction* by Trevor Hastie, Robert Tibshirani, and Jerome Friedman (Hastie et al. 2009) presents machine learning from a more statistical perspective. This book also contains several examples of applications of machine learning to biological data. The full text of the book can be downloaded from the authors' website (http://www-stat.stanford.edu/~tibs/ElemStatLearn/). A reference on probabilistic models is *Probabilistic Graphical Models: Principles and Techniques* by Daphne Koller and Nir Friedman (Koller and Friedman 2009).

Thanks to the recent development of interactive online higher education platforms, some of the most popular graduate-level machine learning classes are now freely available online. Andrew Ng's Stanford Machine Learning class and Daphne Koller's Probabilistic Models class can be taken online at Coursera (http://www.coursera.org). The Introduction to Artificial Intelligence course taught at UC Berkeley by Dan Klein and Pieter Abbeel can be found online at EdX (https://www.edx.org/). Andrew Moore at Carnegie Mellon University maintains a web page (http://www.autonlab.org/tutorials/) with very clear tutorial slides on a wide variety of machine learning algorithms.

REFERENCES

Bishop CM. 2006. *Pattern recognition and machine learning*, 1st ed. (corr. 2nd ed. printing, October 1, 2007). Springer, New York.

Breiman L. 1996. Bagging predictors. *Mach Learn* **24:** 123–140.

Breiman L, Friedman J, Stone CJ, Olshen RA. 1984. *Classification and regression trees.* Wadsworth International Group, Belmont, CA.

Coates A, Abbeel P, Ng AY. 2008. Learning for control from multiple demonstrations. In *Proceedings of the 25th International Conference on Machine Learning*, July 5–9, Helsinki, Finland. International Machine Learning Society, ACM, New York.

Golub TR, Slonim DK, Tamayo P, Huard C, Gaasenbeek M, Mesirov JP, Coller H, Loh ML, Downing JR, Caligiuri MA, et al. 1999. Molecular classification of cancer: Class discovery and class prediction by gene expression monitoring. *Science* **286:** 531–537.

Hall M, Frank E, Holmes G, Pfahringer B, Reutemann P, Witten IH. 2009. The WEKA data mining software: An update. *ACM SIGKDD Explorations Newslett* **11:** 10–18.

Hastie T, Tibshirani R, Friedman J. 2009. *The elements of statistical learning: Data mining, inference, and prediction*, 2nd ed. Springer, New York.

Hinds D, Do C, Wu S. 2012. Studies of extreme longevity extremely challenging. http://blog.23andme.com/news/studies-of-extreme-longevity-extremely-challenging/.

Hull JJ. 1994. A database for handwritten text recognition research. *IEEE Pattern Anal Mach Intell* **16:** 550–554. http://doi.ieeecomputersociety.org/10.1109/34.291440.

Koller D, Friedman N. 2009. *Probabilistic graphical models: Principles and techniques.* MIT Press, Cambridge, MA.

Novembre J, Johnson T, Bryc K, Kutalik Z, Boyko AR, Auton A, Indap A, King KS, Bergmann S, Nelson MR, et al. 2008. Genes mirror geography within Europe. *Nature* 456: 98–101.

Pearl J. 1985. Bayesian networks: A model of self-activated memory for evidential reasoning. In *Proceedings of the 7th Conference of the Cognitive Science Society* August 15–17, University of California, Irvine, pp. 329–334. Cognitive Science Society, Austin, TX.

Ramaswamy S, Tamayo P, Rifkin R, Mukherjee S, Yeang C-H, Angelo M, Ladd C, Reich M, Latulippe E, Mesirov JP, et al. 2001. Multiclass cancer diagnosis using tumor gene expression signatures. *Proc Natl Acad Sci* **98:** 15149–15154.

Schapire RE. 1990. The strength of weak learnability. *Mach Learn* 5: 197–227.

Schapire RE, Freund Y. 2012. *Boosting: Foundations and algorithms.* MIT Press, Cambridge, MA.

Sonnenburg S, Raetsch G, Henschel S, Widmer C, Behr J, Zien A, de Bona F, Binder A, Gehl C, Franc V. 2010. The SHOGUN machine learning toolbox. *J Mach Learn Res* **11:** 1799–1802.

Thrun S, Montemerlo M, Dahlkamp H, Stavens D, Aron A, Diebel J, Fong P, Gale J, Halpenny M, Hoffmann G, et al. 2006. Stanley: The robot that won the DARPA Grand Challenge. *J Field Robotics* **23:** 661–692.

Witten IH, Frank E, Hall MA, Kaufmann M. 2011. *Data mining: Practical machine learning tools and techniques*, 3rd ed. Morgan Kaufmann, Burlington, MA.

WWW RESOURCES

http://www.autonlab.org/tutorials Statistical Data Mining Tutorials, with tutorial slides by Andrew Moore (Carnegie Melon University)

http://www.coursera.org Machine Learning, Andrew Ng, Stanford University

http://www.coursera.org Probabilistic Models, Daphne Koller, Stanford University

http://www.cs.waikato.ac.nz/ml/weka WEKA 3: Data Mining Software in Java

http://cran.r-project.org/web/views/MachineLearning.html CRAN Task View: Machine Learning & Statistical Learning

https://www.edx.org Introduction to Artificial Intellegence, Dan Klein and Pieter Abbeel, University of California, Berkeley

http://scikit-learn.org/stable Scikit-learn. Machine learning in Python. Tools for data mining and data analysis

http://www.shogun-toolbox.org Shogun 2.0—Machine learning open source software

http://www-stat.stanford.edu/~tibs/ElemStatLearn Hastie T, Tibshirani R, Friedman J. 2009. *The elements of statistical learning: Data mining, inference, and prediction*, 2nd ed. Springer, New York

5

Image Analysis

Marina Sirota, Sarah J. Aerni, Tiffany Liu, and Guanglei Xiong

Stanford University School of Medicine, Biomedical Informatics Training Program, Stanford, California 94305

IMAGING BASICS

What is a digital image? To our eyes, the world appears as a continuous space of color and brightness. When we observe a scene, we are able to move around within it freely and focus on different elements. The human eye can distinguish 10 million colors, has a dynamic contrast ratio of 1 million to 1, and can focus at distances from a fraction of an inch to hundreds of feet away. However, when we capture a scene in an image such as a digital photograph, we are forced to choose a limited selection of what we can observe with our eyes. We are forced to choose a focal plane as well as specific ranges of colors and light intensities to capture. (Of course, most modern cameras have an "automatic" setting that will make these choices for us.)

For example, imagine that you are visiting Stanford and wish to take a picture of the Hoover Tower (Fig. 1A). There are a few variables that you must consider when capturing the scene photographically. First, what is the focus of your photograph? If you want to capture the details of the Hoover Tower, you would have to zoom in and focus on it, leaving the tree in front out of focus (Fig. 1B). On the other hand, if you want to capture the detail of the palm tree, you would lose the architectural details of the Tower (Fig 1C). You also must consider the ambient lighting. Otherwise, the architectural details could be lost in a shadow or overexposed in the sun.

You snap a photo and end up with a beautiful digital image of the Hoover Tower that you can open on your computer. At low magnification, the image on your screen looks like a true-to-life reproduction of the scene that you saw on that day. However, the image is actually a collection of colored squares, called pixels, arranged in a two-dimensional grid of fixed size. To understand the "digital" part of digital photography, imagine that you held a grid in front of you on the day that you took your picture. Each square in the grid recorded the average color and intensity of the light falling within that square. If the grid was a single square, you would end up with an average

FIGURE 1. Hoover Tower images showing focus. (*A*) The original image, (*B*) an image with the focus on the Hoover Tower, and (*C*) the image with the focus on the palm tree. (Reprinted, with permission, from King of Hearts/Wikimedia Commons/CC-BY-SA-3.0.)

of the light color and intensity for the whole scene and lose all visual detail. As you increase the number of squares, more details begin to emerge, and at some point the image transforms into what seems to be a continuous scene (Fig. 2), just as your eye would see it. But if you zoom in on the image closely enough, you can see that all of the little squares are still there (Fig. 3).

A **pixel** is the smallest unit of information in a digital image. Literally, the word pixel means "picture element" and is based on the contraction of pix (for "picture") and el (for "element"). It represents the intensity of a given color at any specific coordinate in an image. The color and brightness of each pixel are typically represented by several components. A **color space** is defined by a set of possible colors that can be created by combining different amounts of three primary colors. You might remember from art class that mixing primary colors lets you expand your palette from red, blue, and yellow to purples, greens, oranges, and of all the colors in between. A primary color in an image is referred to as a **channel**; an image from a standard digital camera has red, green, and blue channels.

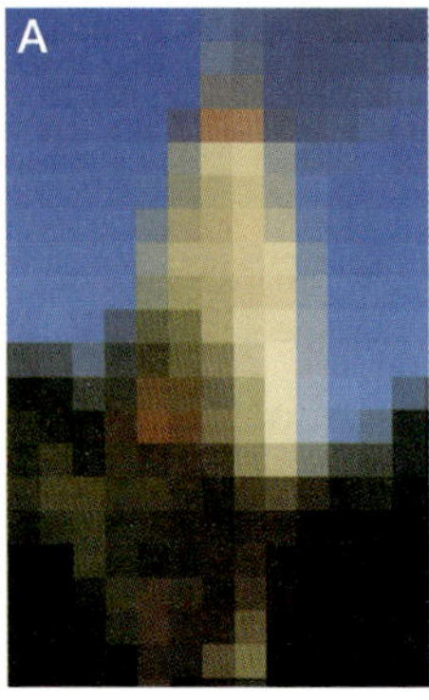

FIGURE 2. Resolution. (*A*)–(*D*) The image in increasing resolution. (*A*) Has the lowest resolution to (*D*), the highest. (*B,C*) Intermediate.

FIGURE 3. Pixelation. (*A*) The original image and (*B*) a zoomed-in version of the palm tree. As we zoom in, we see that the image consists of individual pixels.

The number of distinct colors that can be represented by a pixel depends on the number of bits per pixel (bpp). A 1-bpp image uses one bit for each pixel, so each pixel can be either on or off and can represent, for example, only the extremes of black or white. Each additional bit doubles the number of colors that can be represented, so a 2-bpp image can have four colors, a 3-bpp image can have eight colors, etc. (see Box 1).

An **RGB** (red, green, and blue) image has three channels: red, green, and blue. In a 24-bit RGB image, each of the three channels has eight bits for red, green, and blue. The image is composed of three respective matrices, where each can store discrete pixels with conventional brightness intensities between 0 and 255. The RGB color model uses **additive color mixing** (Fig. 4A) as it describes the kind of light that must be added to black to create a color. If all colors are added in equal parts, the color produced is a shade of gray, and as their intensity values increase, the resulting color approaches white. RGB channels correspond roughly to the three colors recognized by the three classes of cones (color photoreceptors) in the human eye and are used in computer displays and image scanners.

A **CMYK** image has four channels: cyan, magenta, yellow, and black. A 32-bit CMYK image is made of four 8-bit channels, one for cyan, one for magenta, one for yellow, and one for black. CMYK is used in the printing process. It uses **subtractive color mixing** (Fig. 4B), which describes what combinations of ink colors need to be applied so the light reflected from the substrate and through the ink produces a given color. One starts with a white substrate (canvas, page, etc.) and uses ink to subtract color from white to create an image. CYMK is therefore the standard for print.

Color space conversion is the translation of the representation of a color from one basis to another. This typically occurs in the context of converting an image that is represented in one color space to another color space. The goal of the process is to make the translated image look as similar as possible to the original. This happens when you print an image from your computer. On your screen, the image is represented in RGB. When printed, these colors must be appropriately converted so that the printed photograph looks like the image on your screen.

BOX 1. Bits in Images

In computer science, bits are used to perform computations and store information. They are the fundamental building blocks of all that we represent digitally. A bit is a binary entity—that is, it can assume two values, 0 or 1. However, you can string together many such bits to represent larger and more complex values, as discussed in the introductory chapter. Now, let us try to understand what color space means in terms of bits. Recall that an image is a number of pixels set up in a matrix. Each pixel represents the colors captured from the perspective of the lens at that location. If we were able to represent infinitely many colors at each pixel, we would be able to have an image that is very true to life. Unfortunately, we are constrained by computer memory and must find a representation that is appropriate, both from the perspective of the computer hardware and from capturing all of the relevant data in the image.

Let us look at the images in Box 1, Figure 1. Each row is 256 pixels wide and 20 pixels tall (5120 pixels). In row one of the figure, each pixel is represented by exactly one bit, making the size of the image in that row a total of 640 bytes (recall our discussion, 8 bits = 1 byte). We are able to represent exactly two colors at each pixel using one bit, and if our goal were to create an image of a gradient from dark to light, we would end up with the black and white image in row 1 of the figure. If we allowed two bits per image (row 2), we would be able to achieve four colors (recall that each bit contributes two possible values, and there are four possible combination: 00, 01, 10, and 11). In the second gradient, we see that it is only slightly improved. The image, however, has grown from 640 bytes to a 1280-byte image (or 1.28 kilobytes). As we increase the number of bits per image, the memory usage grows linearly as the number of colors grows exponentially. That is, with three bits, we have a total of 2*2*2 = 8 colors (row 3, Box 1, Fig. 1), with four bits we have 16, etc. In the final image, we have a total of eight bits per pixel, resulting in 256 colors represented and a 163.84-kilobyte image. The image looks quite smooth, but we see that already with six bits, we achieve a relatively smooth gradient using only 40.96 kilobytes. It is important to choose the right color representation that enables proper processing. However, if memory is a limiting factor, it is also important to consider when compressed representations are appropriate.

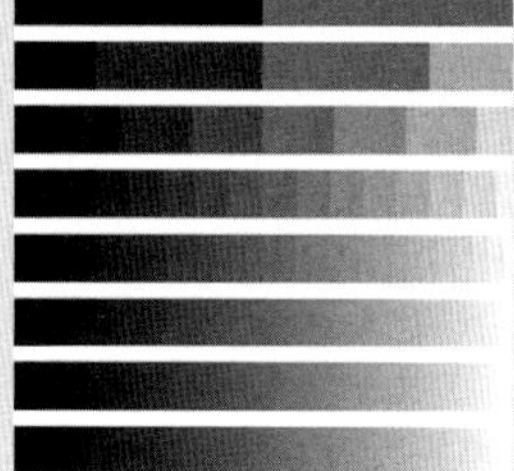

BOX 1, FIGURE 1. Increasing bits, showing a gradient from dark to light with increased bits.

A **grayscale** digital image (Fig. 5) is an image in which each pixel has a single value that carries only intensity information. Images of this sort are composed exclusively of shades of gray, varying from black at the weakest intensity to white at the strongest. Radiological and electron microscopy (EM) images are often grayscale.

As part of an analysis, scientists often may wish to convert an image from color to grayscale. To convert any color to grayscale, one must first obtain the values of its

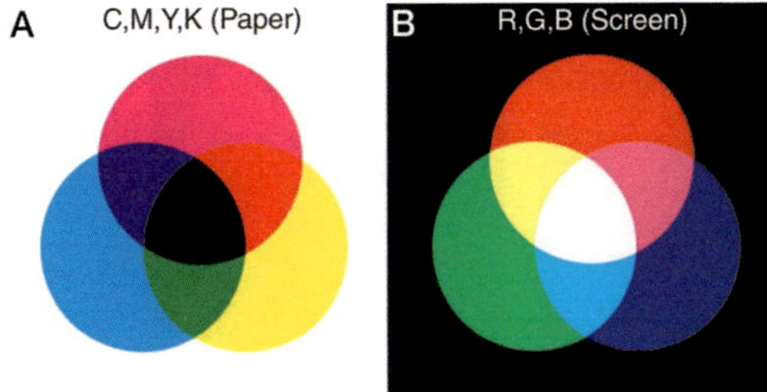

FIGURE 4. RGB and CMYK color models. (*A*) shows the CMYK color model, a subtractive color model consisting of four channels: cyan, magenta, yellow, and black; (*B*) shows the RGB, an additive color model, consisting of red, green, and blue channels.

RGB channels. Then, the values of the red (Fig. 5A), green (Fig. 5B), and blue (Fig. 5C) intensities are added together proportionally to obtain a value that corresponds to a grayscale representation (Fig. 5D). Thus, a grayscale image has just one channel.

BIOMEDICAL IMAGES

In this section, we describe different kinds of images widely used in biomedical research. On the basis of the size of the specimen being imaged, biomedical images generally fall into one of two categories: (1) microscope images that capture small objects, such as cells and molecules, and (2) radiological images that capture larger structures, such as tissue and organs of the body (Fig. 6).

Microscope Images

There are many types of microscopes, but some of the most widely used are the **light microscope, fluorescence microscope,** and **electron microscope**. These differ in the kinds of illumination that they use to make the specimen visible to the lens.

FIGURE 5. Grayscale image. (*A*), (*B*), and (*C*) show the red, green, and blue channels of the original image, respectively. (*D*) The combined intensities that produce the final grayscale image.

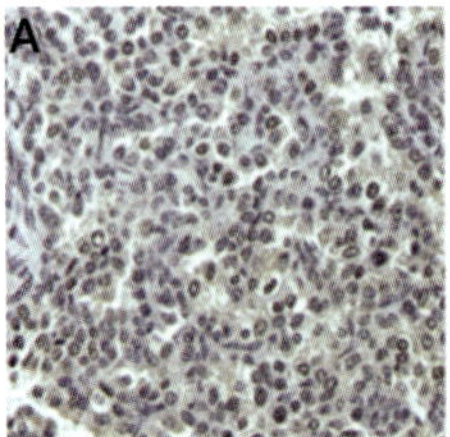

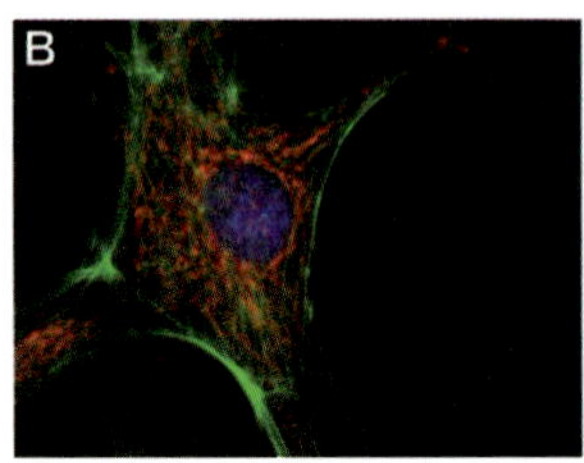

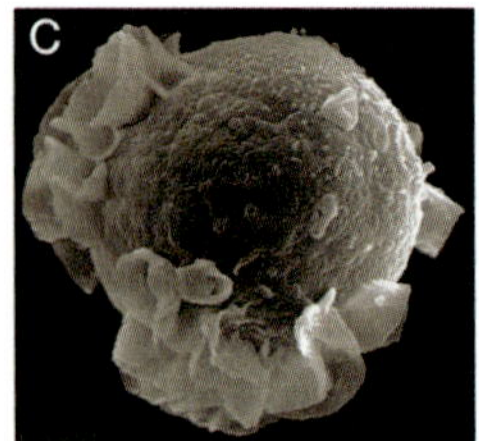

FIGURE 6. Microscope images with increasing resolution. (*A*) Partial view of an immunohistochemistry image of a lung nodule of a mouse using a light microscope with resolution on the order of micrometers. (*B*) Fluorescence. (*C*) B-cell budding virus shown in an electron microscope with resolution on the order of nanometers. (*B*, From http://www.phy.cam.ac.uk/research/research-groups-images/bss/images/fluorescence.jpg/view; *C*, reprinted, with permission, from Analytical Imaging Facility at the Albert Einstein College of Medicine, http://cammer.net/historical/aif/gallery/sem/sem.htm.)

The light microscope is the oldest of the three technologies. In light microscopy, a beam of full-spectrum visible light is passed through the object of interest to create an image (Fig. 6A). Fluorescence microscopy uses fluorescent molecules intrinsic to (or introduced into) the biological subject to create an image, rather than illuminating the target externally as in light microscopy. Fluorescence images are widely used to interrogate cellular biology (Fig. 6B). The electron microscope works by a mechanism similar to that of the light microscope, but it uses beams of electrons rather than photons (light) to illuminate the target. It has a much higher resolution than the other two imaging technologies because electron beams are much smaller in width than light beams. Thus, proteins and other large molecules can be visualized in an electron microscope image (Fig. 6C).

CT

Computed tomography (CT), short for computed axial tomography (CAT), was developed based on conventional X-ray techniques. The CT scanner has arrays of X-ray tubes and electronic X-ray detectors arranged around an examination table. The X-ray tubes rotate around the body and deliver beams as the examination table moves through the X-ray beams, resulting in multiple X-ray paths through the body. The absorption of X-ray beams by different body parts varies depending on tissue composition and is measured by electronic X-ray detectors. The data are processed by a computer program and converted to a series of sectional images (tomograms) through the body. Denser materials generally appear lighter on a CT scan; thus, bones appear to be white, soft tissues shades of gray, and air black. CT can be used to image bone, soft tissue, and blood vessels simultaneously (Fig. 7A) but provides particularly good clarity for hard tissues such as bone. Because CT uses ionizing radiation, it is associated with a slightly increased risk of cancer. In most cases, however, the diagnostic benefits of CT outweigh this risk.

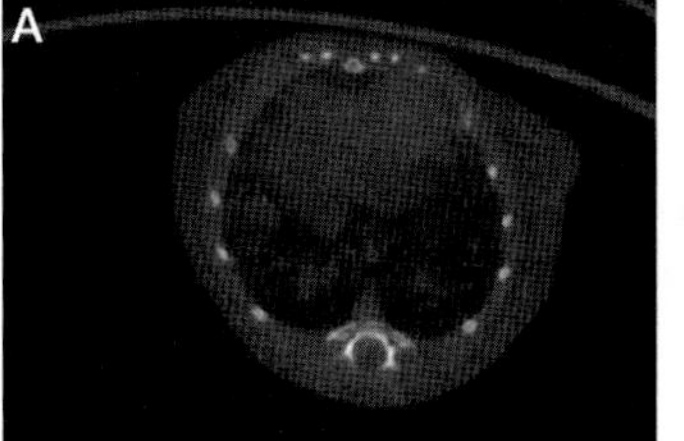

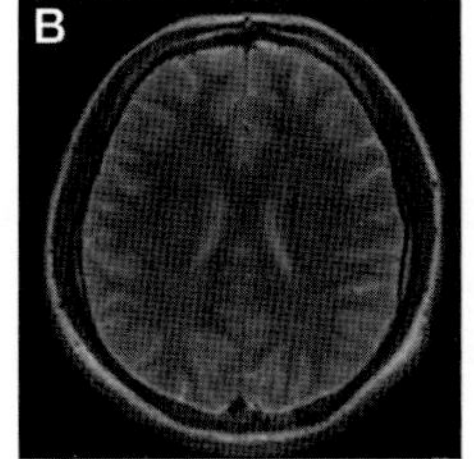

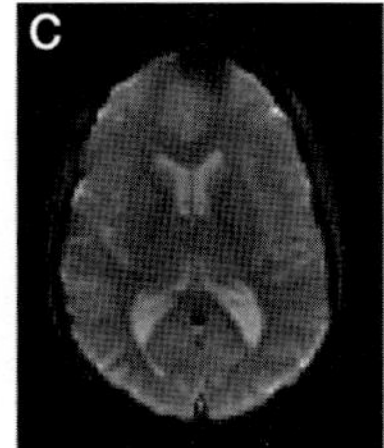

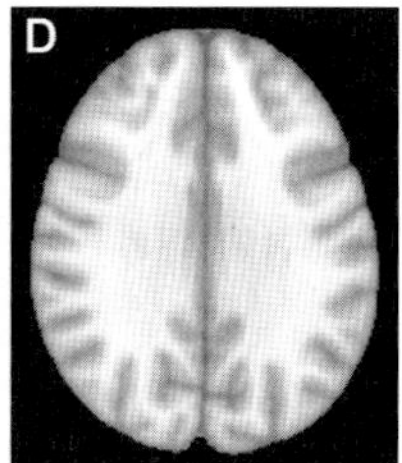

FIGURE 7. Radiographic images on the visible scale. (*A*) CT image of the lung of a mouse. (*B*) Anatomical MRI image of a cross section of the brain with resolution on the order of millimeters. (*C*) Perfusion-based MR image of a cross section of the brain. (*D*) A slice of MR brain image clearly showing the anatomy of the white matter and gray matter structures.

MRI

Magnetic resonance imaging (MRI) can generate high-contrast images of soft tissue, making it a particularly valuable tool for clinicians and researchers interested in brain imaging (Fig. 7B). MRI uses powerful magnetic fields to stimulate nuclei of atoms, such as hydrogen that is charged (called protons), within the body. Because protons are essentially like small magnets, the electromagnetic energy around protons can be detected after these protons are excited by an external rotating electromagnetic field. MRI uses the interaction of external electromagnetic fields and the magnetic moment of protons in the body to create detailed cross-sectional images of the body. The protons—abundant in the body as the essential element of molecules of water, protein, lipids, or carbohydrates—vary in concentration in different tissue types and pathologies and thus behave differently in the external magnetic field, which enables image contrast and recognition of the different tissues in the image.

Traditional MRI provides images only of body structure (and is therefore also called "**structural MRI**"). Other MRI techniques have been developed that enable visualization of physiological processes as well as anatomy. For example, **functional MRI** (fMRI) can be used to visualize changes in brain activity after subjects are presented with different stimuli or asked to engage in specific kinds of tasks. fMRI works by detecting regional changes in brain blood flow and oxygenation that are associated with the activation of large groups of neurons, and it is used primarily as a basic research tool. **Perfusion MRI** enables visualization of dynamic changes in blood flow through tissues. In dynamic susceptibility contrast perfusion imaging (DSC-MRI), a contrast agent is injected into the body before MRI imaging to amplify signal intensity. The signal intensity in perfusion MR images is used to derive a set of blood perfusion parameters, called hemodynamic parameters, that can be used in diagnosis (Fig. 7C). We discuss perfusion-based MR images again in the Analysis of Brain Images section below. Often, to extract hemodynamic parameters from a specific anatomical structure, the anatomical with the perfusion image must be aligned in a process called registration that is discussed later in the chapter. Before registration, a skull-stripping step is often performed to isolate the brain parenchymal (Fig. 7D).

PET

Positron emission tomography (PET) makes use of radiolabeled tracers to image biological processes within the body. A biologically active compound (for example, a glucose analog) is labeled with a positron-emitting radionuclide. Gamma emission from the radioactive tracer is collected by scintillation detectors within the PET scanner and is analyzed to produce two-dimensional (2D) or three-dimensional (3D) images of the distribution of the tracer through the body. By using different radioactive tracers, researchers can visualize different kinds of physiological processes, including blood flow, glucose uptake, neurotransmitter receptor binding, and drug metabolism. PET is used both in basic research and as a clinical tool. Brain pathologies, such as strokes, tumors, and neurodegenerative diseases, are often accompanied by changes in brain metabolism or other biochemical changes (e.g., the accumulation of amyloid in Alzheimer's disease) that can be visualized by PET and used for clinical diagnosis and risk assessment.

GENERATING IMAGES FOR ANALYSIS

Image acquisition is a critical step in image analysis. When capturing an image, one must think about those aspects that one would wish to process computationally, so that it is possible to obtain the necessary data at the correct level of detail. Some variables to consider follow.

Resolution refers to the number of pixels in an image. The more pixels used to represent an image, the more closely the image will resemble the original. High-resolution images contain more information, and they also take more computational power to process and analyze. The resolution of a biomedical image ranges from nanometer (nm) to millimeter (mm), depending on the imaging instrument used to capture the image. The imaging medium is chosen based on the resolution needed to properly image the target object. Imaging techniques such as light microscopy and electron microscopy are widely used in biological research. They have the appropriate resolution in the nanometer to 0.1-mm range, because objects of interest that may be at the molecular, cellular, or tissue level (Fig. 6) are quite small. Anatomic images such as MRI and CT images have resolutions on the order of millimeter or larger and are appropriate for visualizing organs or entire organisms (Fig. 7).

Indexed color is a technique used to efficiently manage colors in a digital image. To save a computer's memory and file storage, color information can be stored in a separate data structure called a **palette**, an array of color elements in which every color is indexed by its position within the array (Fig. 8). This way, each pixel does not have to contain the full specification of its color, only its index into the palette (Fig. 8).

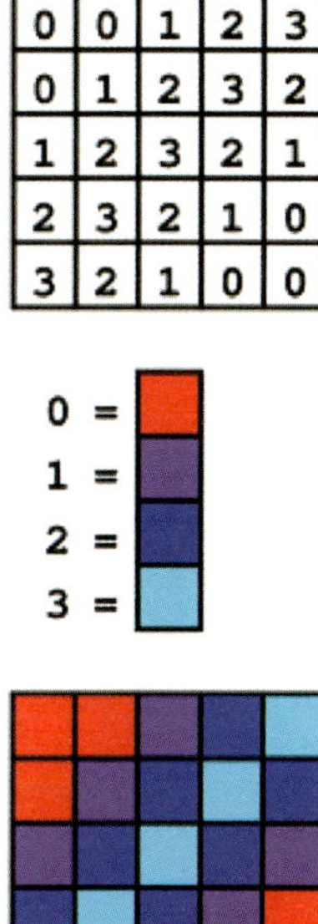

FIGURE 8. Example of palette indexing. Bottom image is indexed using the code in the middle. The table at the top shows that every color is indexed by its position within the array.

Focus, as we previously mentioned, is an important aspect to consider when choosing the parameters for image acquisition. As shown in Figure 1, the captured image might be quite different depending on the focus of the image.

Contrast is the difference in color and/or brightness that makes an object in an image distinguishable from other objects and the background (Fig. 9). Before image analysis, one might want to consider ways to enhance image contrast for better results. Figure 9 shows an example in which the contrast on the right half of the photo has been increased relative to the left half, and you can see (for example) that the buildings stand out more clearly against the sky in the right half.

Dimensionality. A voxel (volumetric pixel), the 3D analog of a 2D pixel, is a volume element, representing a value on a regular grid in 3D space. 3D voxels are

FIGURE 9. Contrast. This figure demonstrates the same image with higher (*right*) and lower (*left*) contrast.

commonly discussed in radiographic images, where each slice represents a section of anatomical structures with a certain thickness. A generalization of a voxel is the doxel (dynamic voxel). This is used in the case of a four-dimensional (4D) data set, for example, an image sequence that represents 3D space together with another dimension such as time.

COMPUTATIONAL IMAGE ANALYSIS

Image of a Cell

Here, we present an example of how a biologist might use computational techniques in image analysis to automate a process that would otherwise have to be performed by manual curation.

The field of image analysis is growing rapidly, and many complex methods are being applied to difficult problems. However, you may be surprised to find how valuable a few basic techniques might be in analyzing images from your own research.

As we described earlier, a 2D image is represented by pixels arranged in a matrix of rows and columns that allows many operations to be implemented by uniformly applying the same operation to each pixel independently. At each of these pixels are one or more numbers representing a color at that location. Let us start our discussion with the 8-bit grayscale image shown in Figure 10. If you recall from our discussion, each pixel in such an image is a number between 0 and 255 that represents a shade between black and white. If we look at Figure 10, our eye can easily pick out cell shapes in the image. In fact, if we asked for a boundary of each cell, you could draw a line around it. In the work from which this image is taken, the scientists were interested in the shapes of the bacteria in the image. More specifically, they wanted to count the ones that were star-shaped. That task could be done by hand, but if we could tell a computer how to pick out the star-shaped bacteria, the computer could quickly count them in thousands of different cultures and greatly speed our research.

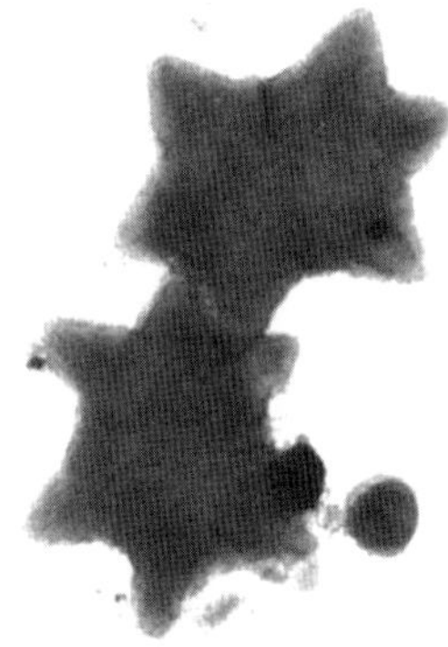

FIGURE 10. Grayscale image of a cell. We use this image of a star-shaped cell for several examples of image processing techniques. (Modified from Staley 1968; Staley et al. 2007.)

To a computer, the image in Fig. 10 is just a set of numbers in a matrix. How do we tell the computer how to pick out the bacteria? In image processing, this task is known as **segmentation**. It results in the partitioning of a digital image into two or more components. The goal of segmentation is to simplify an image so that a computer can detect and quantify objects in it. Several techniques can be used to do this.

Intuitively, we see that the bacteria are darker than the background in the image. If we simply pick an intensity (a number between 0 and 255) to use as a cutoff, we can see how many bacteria the computer picks out at different intensities (see Fig. 11). Once we choose an intensity cut-off value, we can draw a new image where all pixels that are above this value are black and all others are red. Looking at Figure 11, we can see several red and black (**binary**) images of the bacteria at generated different intensity cutoffs.

When we compare the images generated at the different intensity cutoffs, we find one (Fig. 11D) in which all of the bacteria are well separated from the background. We also see a segment of the image that is an artifact and not a bacterium (the round object on the bottom right). To exclude artifacts, we could add variables to the analysis, such as a size cutoff (if we know the approximate size of the bacteria in question).

If we look at multiple images, we may find that the range of intensities among them is inconsistent. Some images may be dimmer than others, and what works

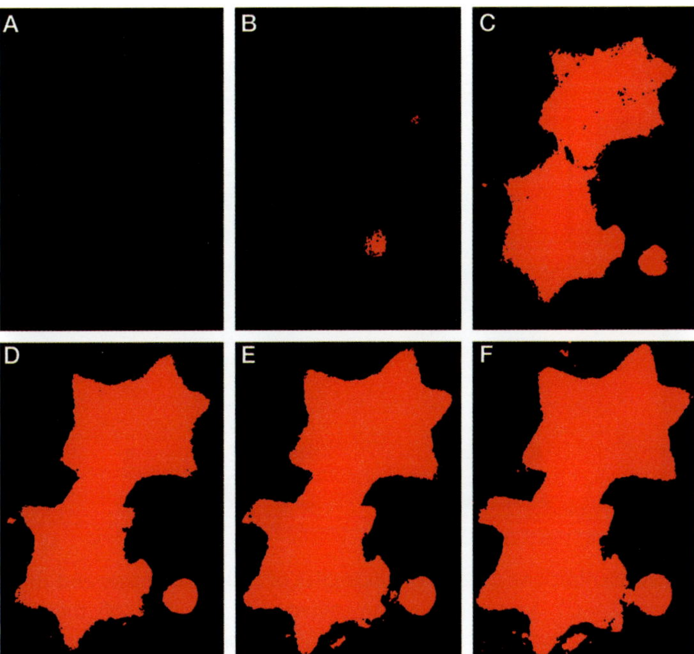

FIGURE 11. Varying intensity levels. (*A–F*) Applications of increasing intensity cutoffs are used here to generate a binary image to separate the bacteria from the background. (Modified from Staley 1968; Staley et al. 2007.)

as an intensity cutoff for one image may not work for the next. Picking the right intensity cutoff for each image could be done by human intervention, but it would be better if we could set a unique cutoff for each image automatically. In fact, methods exist that do just this. For example, **Otsu's method** enables the computer to choose the optimal cutoff to separate the image into two segments, the foreground and background.

Otsu's Method for Image Segmentation

This technique works by using the statistics of the distribution of intensities in an image to separate the image into two components with optimal variance between the sets of pixels in each. Every image can be represented by a histogram that displays the number of pixels at each intensity level. Such a histogram can give you a general sense of the image's overall brightness. For example, we show five images and their corresponding histograms in Figure 12. We see that dark images have histograms that are skewed to the left and bright images have histograms that are skewed to the right. In the image of the star-shaped bacteria, we see two peaks (Fig. 12D). If we pick the center point between those two peaks as the intensity threshold for segmentation, we achieve a good separation between the bacteria and background (Fig. 13).

That is the essence of Otsu's method. Let us describe the technique in a bit more detail. We start by choosing an arbitrary intensity threshold and create two distinct

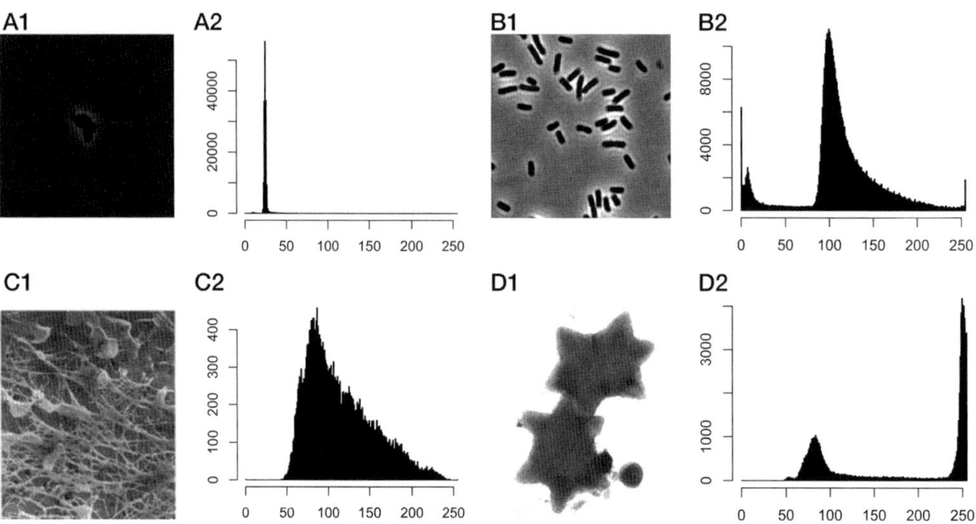

FIGURE 12. Intensity distributions. (*A–D*) Four examples of biological images and their corresponding intensity histograms. (*A–C*) Scanning electron micrograph of cells of the hyperthermophilic crenarchaeum, Pyrodictium occultum. (*A,B*, Reprinted, with permission, from Carolina Tropini and Kerwyn Casey Huang, Biophysics Program, Stanford University; *C*, courtesy of Gertraud Rieger and Reinhard Rachel, University of Regensburg, Regensburg, Germany, published online at http://microbewiki.kenyon.edu/index.php/Pyrodictium; *D*, star-shaped bacterium, reprinted, with permission, from Staley 1968; Staley et al. 2007.)

FIGURE 13. Image segmentation at optimal threshold. Results of applying image segmentation at the optimal threshold. (Modified from Staley 1968; Staley et al. 2007.)

groups of pixels: pixels with intensities at or greater than the cutoff and those with lower intensities. For each group of pixels, we compute the variance of all intensity values within that group. Our goal is to create two groups that are very different from each other (maximizing interclass variance). Mathematically, it turns out that minimizing the variance within each of the groups (intraclass variance) is equivalent to maximizing the variance between the two groups (interclass variance), and the former value can be calculated more efficiently. By calculating intraclass variances for every potential segmentation threshold and choosing the threshold that generates the lowest intraclass variance, we determine the optimal threshold computationally. The resulting segmented image at the optimal threshold for our example is shown in Fig. 13.

Otsu's method is effective in a variety of situations. Look at the different examples online and convince yourself that this method works well in cases of images in which two groups can be nicely separated at a single intensity. Of course, there are many cases in which this is a difficult task. Take an image where the acquisition caused problems, so that the image is dim at the edges and bright in the center. In this case, a group of lower pixel intensities at the center of the image could represent a cell as in our previous example. However, lower pixel intensities at the edges of the image are due to an artifact of image capture, not the presence of bacteria. This situation will cause errors in segmentation.

Histogram-based methods, such as Otsu's method, are very efficient compared to other image segmentation methods because they typically require only one pass through the pixels. One disadvantage of the histogram-seeking method is that significant peaks and valleys may be difficult to identify (Fig. 12C).

k-*Means Clustering Algorithm*

Clustering algorithms provide an alternate approach to tackle the problem of segmentation. The ***k*-means clustering** algorithm is an iterative technique that partitions an image into *k* clusters. We start out by picking centers of each *k* cluster, either randomly or based on some initial data. Each pixel in the image is then assigned to the cluster that minimizes the variance between the pixel and the cluster center. The cluster centers are then recomputed by averaging all of the pixels in the cluster. This process is repeated until none of the pixels change clusters and convergence is achieved. This algorithm requires the number of clusters or segments to be known in advance. For example, if we are trying to locate cells in an image versus the background, k will be set to 2: a "cell" cluster and a "background" cluster.

Now, let us return to the image of two star-shaped cells in Figure 10, and use the k-means method to pick out the cells from the background. The "cell" cluster is coded in dark gray and the "background" cluster is coded in white (Fig. 14). One drawback of this approach is that we need to specify the number of clusters or segments k in advance of running the k-means algorithm.

Analysis of Brain Images

Next, let us go back to the anatomical MRI image of the brain in Figure 7D and examine it more closely. We mentioned earlier that conventional MRI measures tissue density and is particularly good for visualizing soft tissues. The brain is composed of gray matter and white matter. White matter is myelinated, which means it has a high lipid content and is therefore less dense than gray matter. In Figure 7D, the outermost light-gray region of the brain is the gray matter and the darker-gray region inside is the white matter. In images of normal brain, the white and gray matter compartments appear to be fairly homogeneous, as in Figure 7D. By segmenting out the gray matter and white matter compartments, we can quantify the areas of different brain regions, see how the areas of those regions change under different experimental or pathological conditions, and detect abnormal features such as hemorrhages or brain tumors.

Perfusion-based MRI images, like that shown in Figure 7C, measure brain blood flow and are used for evaluation of cerebrovascular conditions. Certain kinds of brain abnormalities show specific features in perfusion-based MRI images. For example,

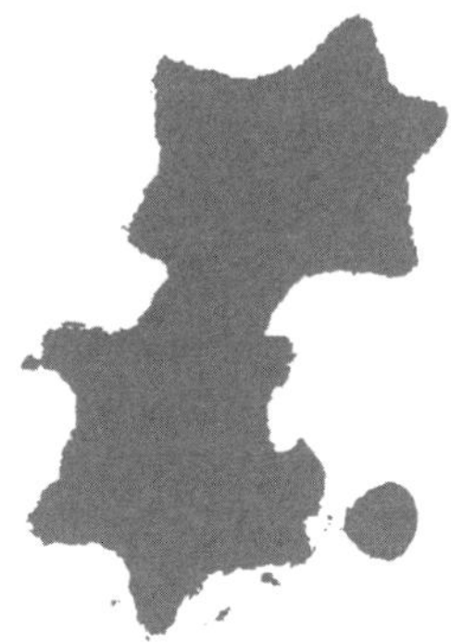

FIGURE 14. k-means clustering. Results of the k-means segmentation method on the cell segmentation task, where k is chosen to be 2 (cell and background).

brain tumors may appear darker than surrounding normal tissue, whereas hemorrhages appear lighter. Radiologists in clinics often use simple visual inspection of these images in making diagnoses. However, perfusion MRI images can also be used to generate quantitative hemodynamic parameters, including cerebral blood volume (CBV) and cerebral blood flow (CBF). The hemodynamic parameters can be used for more definitive diagnosis. For example, lesions such as hemorrhages and tumors have different CBV ranges and thus can be identified on that basis.

k *Means*

The *k*-means algorithm can be applied relatively simply and so is used widely in analyzing brain MRI images. Now, we will use the *k*-means algorithm to segment the structural MRI image of the brain in Figure 7D. As previously mentioned, a major drawback of the *k*-means algorithm is that it requires the value of *k* to be specified in advance. However, standard brain MRI images have three colors: black (the background), pale gray (gray matter), and dark gray (white matter). Therefore, *k* is typically 3 for such images. This segmentation process essentially corresponds to assigning each pixel to one of the three colors (Fig. 15).

Active Contour Algorithm

We can also use the active contour method to segment the MRI brain image. Active contour, or so-called snakes, is a well-studied technique to segment an object in 2D or recently extended to work in 3D. The active contour algorithm is essentially an energy minimization problem consisting of mainly two components of energy: an external edge-based energy that should be lowest at a boundary and an internal energy that should be minimal for the shape desired. For example, if we know the segmentation results should be round and smooth, elongated contours with high curvature would be penalized by assigning high energy.

The minimization problem can be solved by converting the image to a partial differential equation, where internal and external energy are translated to internal and external force terms. There are many variants of the basic algorithm, mostly

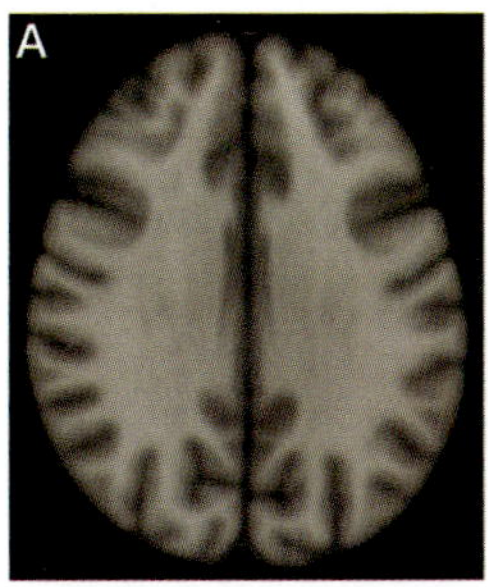

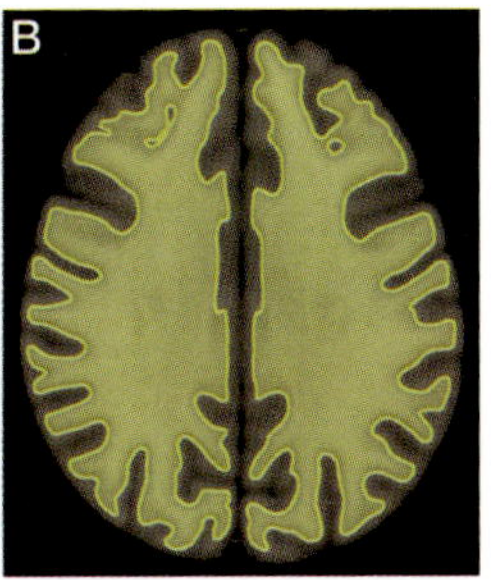

FIGURE 15. MRI brain images segmented by the *k*-means method. The original brain image (Fig. 7D) (*A*) has been segmented using *k* means ($k = 3$) into three distinct compartments: background (black), gray matter, and white matter (yellow), resulting in the segmented image (*B*).

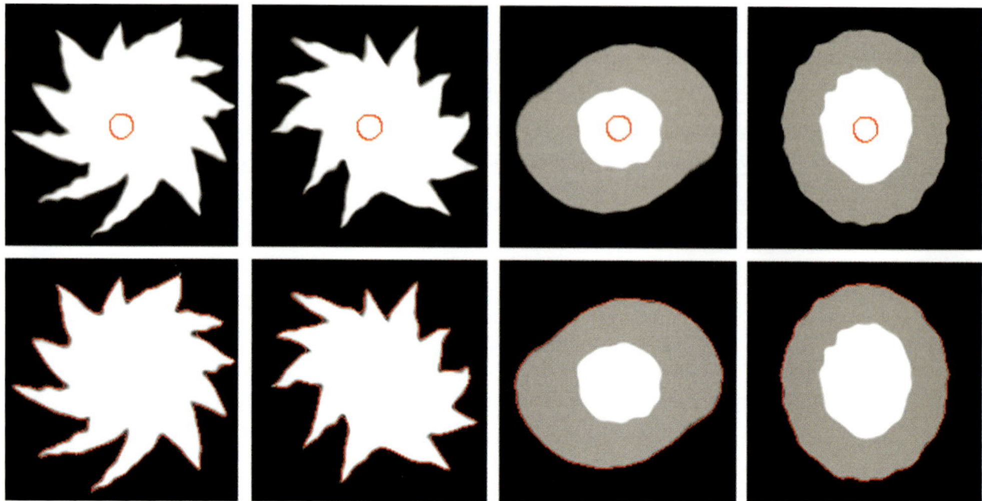

FIGURE 16. Active contour algorithm. The panels Illustrate the initial, intermediate, and final contours. (Reprinted, with permission, from Xiong et al. 2006.)

differing in how and which type of forces are applied. For example, the external forces can be a combination of edge- and region-based features, and the internal force can be often taken from the local curvature of the contour. Figure 16 illustrates the initial, intermediate, and final contours. Figure 17 shows the segmentation result on the brain MR image after applying the active contour algorithm.

Level Set Algorithm

Another popular segmentation method is the level set algorithm that implicitly represents the interface between the objects and the background. Unlike the active contour method, which explicitly delineates the shape, in the level set algorithm the interface is implicitly defined to be the zero level of a level set function. In general, the level set function is negative in the region corresponding to the object and positive for the background. Due to the nature of implicity, level set is well suited to analyzing images in 2D, 3D, or even more dimensions. Similar to the active contour, the evolution of level set function, and thus the segmentation process, is controlled by a

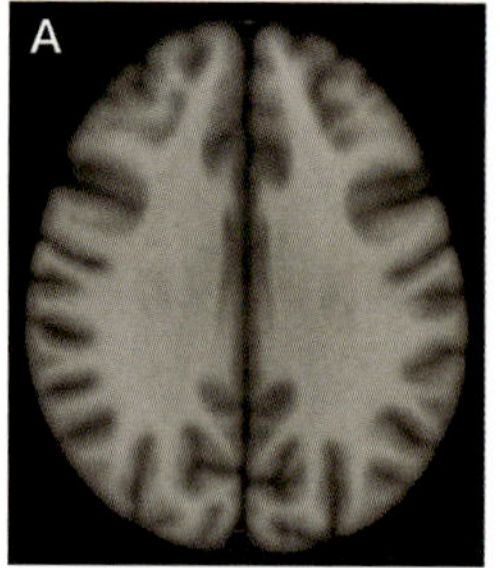

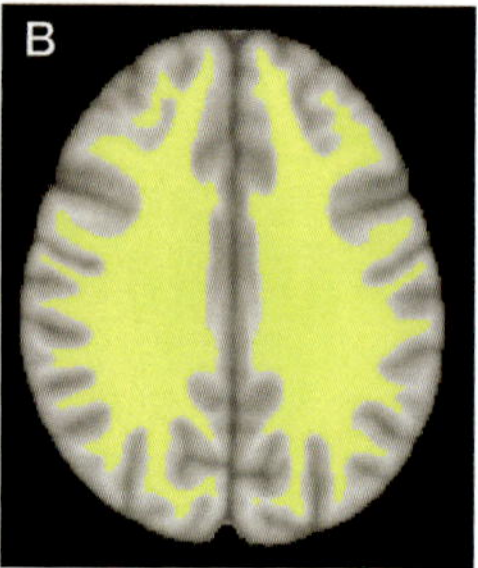

FIGURE 17. Active contour algorithm. Segmentation result on the original brain MR image (Fig. 7D) (*A*) after applying the active contour algorithm to identify the white matter in the brain image (*B*).

partial differential equation that has multiple terms for image features and smoothing constraints. Advantages of the level set method include automatic adaption of topology changes and rapid computation. The former is attractive when multiple objects (e.g., a set of cells) are simultaneously segmented when initialized with multiple seeds. The latter makes the level set method as efficient as active contour because the level set function is actually solved only in a narrowly banded region. Finally, an extra procedure to convert the implicit representation to an explicit one (e.g., marching cubes) is necessary if the actual surface is desired. Figure 18 is an illustration of segmenting multiple cells using level set. Figure 19 shows the segmentation result on the brain MR image using the level set algorithm.

Image Registration

Multiple Images

Multiple images acquired via the same imaging modality are called intramodality images, whereas those acquired via different modalities are called intermodality

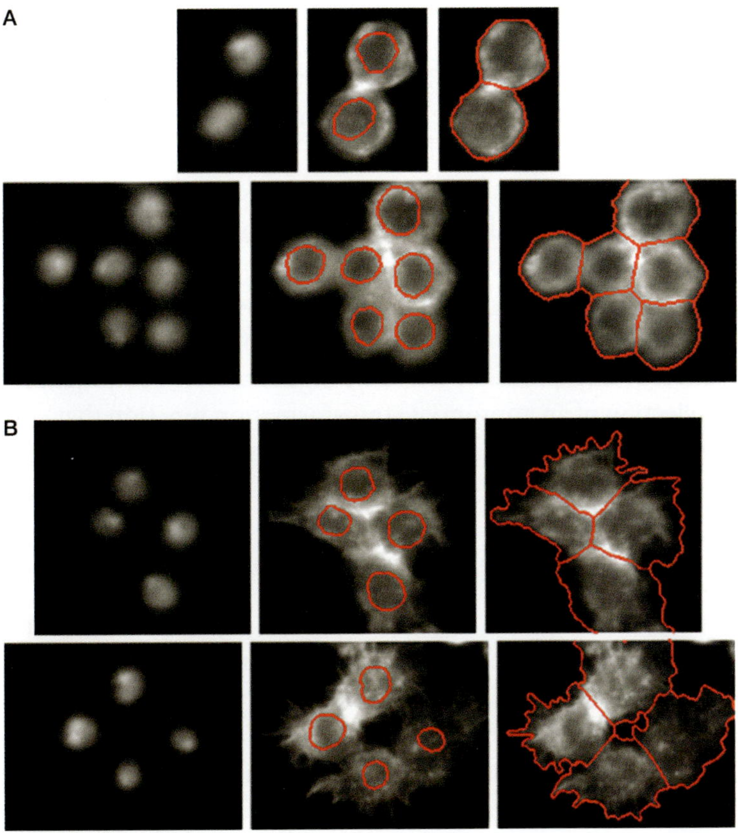

FIGURE 18. Level set algorithm. (*A,B*) Illustration of segmenting multiple cells using level set method—intermediate steps shown. (Reprinted, with permission, from Xiong et al. 2006.)

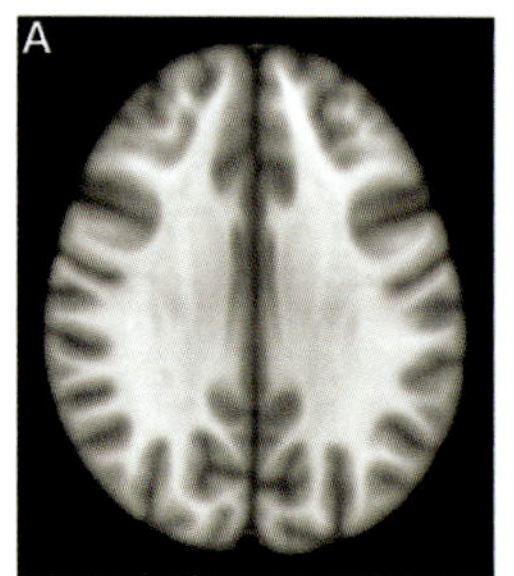

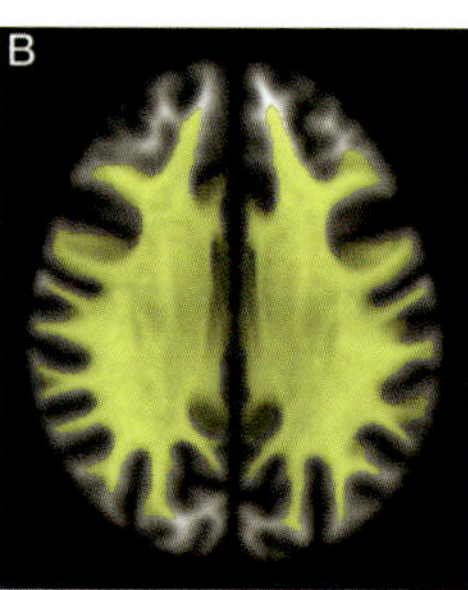

FIGURE 19. Level set algorithm. Segmentation result on the brain MR image (Fig. 7D) (*A*) using level set algorithm (*B*).

images. For example, two MR imaging series are considered intramodality images, whereas MR and CT images of the same patient acquired from a CT/MR combined system are intermodality images. Multiple images may also be taken of the same subject at different points in time. Such images can be categorized as *dynamic* images (acquired over a short period of seconds to minutes) or *serial* images (acquired over the time course of months to years). The temporal information is embedded in these images as the third dimension; therefore they are called 3D images.

Image registration is the alignment of images for the purpose of comparing or integrating them. This process involves mapping points in one image to corresponding points in another image via a transformation function, which can be applied in 2D to 2D, 3D to 3D, or 2D to 3D. Registration has a wide range of applications in medical imaging, including image-guided therapy, integration of information from multiple imaging modalities, and monitoring changes in the sizes or intensity of specific features over time that are of biological or clinical significance. One of the applications of image registration is the annotation of structures in a new image based on a reference atlas image. In neuroimaging, for example, an anatomical brain atlas may be used to annotate specific brain structures within an MRI image.

The task of registration is relatively straightforward and robust for images acquired at different time points by the same modality. The computation of registration requires selection of an appropriate transformation function that creates a mapping of points within the original image space S to the transformed image space S'.

Spatial Transformation

There are three main categories of transformation: rigid, affine, or nonrigid. Rigid transformation assumes that the structure of interest behaves as a rigid body with no distortion in spatial orientation or position. Rigid transformation includes translation and rotation, whereas affine transformation allows additional skew and scaling. Rigid and affine transformations are appropriate for images acquired using the same imaging modality. Nonrigid transformation, also called elastic transformation, is more effective for modeling changes in the size or shape of a structure. It is used for monitoring changes in soft tissue that have been imaged over long intervals of time.

Suppose that we want to identify various structures in a new brain image. Registration of this image with a brain atlas image can be well approximated using a rigid transformation, because the shape of the skull changes minimally. Thus, rigid transformation is the first step in brain imaging processing and is usually adequate to produce excellent results.

There are a number of algorithms that can be used for image registration. Of these, the two most commonly used classes are intensity based and feature based. These algorithms generally boil down to a generic framework that involves a cost function calculated based on intensity and/or spatial information, an optimization method, and a transformation function.

Algorithms

Intensity-Based Registration

The intensity-based registration uses both spatial transformation and intensity transformation, whereas feature-based registration uses only spatial information. Intensity-based registration involves defining a cost function that measures similarities between intensity patterns of two images. The optimal mapping between the two images can then be found by solving the cost function using an optimization method. The optimization method minimizes the difference in intensity patterns between the two images and builds a global relationship between pixel intensities in one image and those in another image. Ultimately, the relationship is a mapping that can be a combination of rigid, affine, and nonrigid transformation.

The simplest and most commonly used cost function is the sum of squared differences (SSD) between intensities of two images S and S' (Equation 1).

$$\mathrm{SSD} = \frac{1}{n}\sum_{i}^{n}\left(S_i'' - S_i'\right)^2, \tag{1}$$

where S_i and S_i' denote intensity of the ith pixel in the original and transformed image, respectively.

This cost function assumes that the difference between the two images is only noise, and thus minimizing the difference increases accuracy of registration. Another popular cost function is correlation coefficient (CC; Equation 2):

$$\mathrm{CC} = \frac{\sum_i (S_i - \overline{S}) \cdot (S_i' - \overline{S}')}{\sqrt{\sum_i (S_i - \overline{S})^2 \cdot \sum_i (S_i' - \overline{S}')^2}}, \tag{2}$$

where S_i and S_i' are corresponding points on images S and S'.

The CC method assumes a linear relationship between the two images. This less strict assumption allows the cost function to be used over both the spatial domain and the intensity domain, whereas the SSD method can only be used in the intensity domain.

Mutual Information Theory

The above two cost functions discussed are commonly used with images acquired from the same modality. Next, we introduce a method called **mutual information theoretic technique**. This method, of which the cost function is defined by a measure of the mutual information content in the two images, is one of the earliest and still the most successful registration methods for mapping images acquired both from the same modality and from different modalities. Mutual information is a measure of information redundancy or dependency formulated by a joint probability function. The intuition behind this idea is that if two images are independent, the mutual information defined by a joint probability function is 0. On the other hand, the mutual information is highest when the joint probability function is 1, where information in one image completely determines that of the other image.

Let us first define an important concept in mutual information theory. Entropy is a measure of uncertainty associated with a random variable. For example, given a random variable X, entropy $H(X)$ is defined as $H(X) = \Sigma p(X = \mathrm{x}) \log p(X = \mathrm{x})$, where $p(X = x)$ is the probability that outcome x occurs for the random variable X. The mutual information between X and Y is defined as

$$I(X, Y) = H(X) - H(X \mid Y) = H(X) + H(Y) - H(X, Y),$$

where $H(X \mid Y)$ is the conditional entropy of X given Y. Thus, mutual information can be interpreted as the probability that the uncertainty in X is reduced when Y is known. Imaging registration is equivalent to maximizing $I(X, Y)$ or minimizing $H(X, Y)$ in order to optimize the mutual information content between the two aligned images.

Feature-Based Algorithm

Iterative closest point (ICP). Another type of algorithm in image registration attempts to align corresponding features from two images. Examples of these features can be anatomical landmarks or salient edges that are identified manually or preferably automatically. If the correspondence of point-based features between both images is known, the rigid transform can be analytically solved that aligns these features optimally in a least-squares sense. Otherwise, an iterated approach, namely, the well-known ICP algorithm, is used. The basic idea is to alternate the step of searching for corresponding closest points and the step of solving for the optimal transform, until a tolerant minimal change is reached. Many variants of the original ICP algorithm exist, mainly to improve for speed and stability, as well as to allow for more general transforms, such as affine transform. Recall, as described earlier, that affine transformation encompasses not only rigid transformation, but also scaling and shearing. This algorithm is straightforward and can be run in real time. The algorithm is subject to converging to local minimal. To overcome this challenge, refinements such as stochastic ICP and simulated annealing can be applied.

Edge Detection

Edge detection is the task of identifying points in a digital image at which the image brightness changes sharply. Many edge detection algorithms apply this first principle and produce robust results. Edge detection is an essential step in many applications of image analysis, including feature extraction in computer vision, cell detection in biological images, etc. It filters out useless information and preserves important structural information, thus reducing the amount of data that must be processed subsequently. Edge detection is closely related to the edge-based segmentation technique we mentioned previously. However, edge detection does not necessarily detect closed regions; instead, it simply detects whether a window of pixels contains a change in intensities resembling an edge. We present an example of an edge detection method called **Canny edge detection** in Box 2.

CONCLUDING COMMENTS

From pathology to basic biology, the use of digital images is growing increasingly prevalent in the biomedical field. In this chapter, we introduce the basic concepts of image processing that form the foundation of the complex field of computer vision that is being developed and applied to analyze both laboratory and medical data. We introduce several algorithms that enable the reader to understand how images are represented and the mathematical foundation behind computational problems such as shape recognition and edge detection.

As the volume of biomedical digital images grows, it will become increasingly unfeasible for individuals to manually analyze and curate their data, imposing the need to turn to automated tools. The approaches described here arm the reader with some basic tools to allow them to identify the computational challenges in processing their images, and examine and choose among the tools available in the literature. In making more educated decisions, we hope to allow the reader to harness the full power yielded by the images that they produce, to gain more meaningful insight into their field of study.

BOX 2. Canny Edge Detection

We describe the concept of edge detection using the **Canny edge detection** algorithm. It is claimed to be one of the most robust edge detection algorithms by many. The seminal paper by John F. Canny in 1986 entitled "A Computational Approach to Edge Detection" introduced the algorithm as a multistep process. Because an image may contain many spurious edges, it is important to follow a multistep image processing pipeline that (1) first smooths the image to reduce noise that may resemble an edge, (2) identifies all gradients in an image, and (3) performs a postprocess to identify a clean set of edges.

Step 1. Gaussian Smoothing to Reduce Noise

The Canny edge detection algorithm first applies noise filtering using a Gaussian mask, known as Gaussian smoothing, which is a type of linear smoothing filter. Gaussian smoothing uses the process of image convolution. The convolution process consists of sliding a mask over the image, applying the defined transformation over a square of pixels of the size of the mask. Thus, the mask transforms the intensity value of each pixel to be in close proximity to the values of its surrounding neighbors. Thus, the effect of a single pixel with significantly high or low intensity will be reduced, resulting in a blurred version across the pixel and its surrounding region. This step is called **Gaussian smoothing** because the mask used is of Gaussian distribution. An example of a mask is shown in Equation 2.1.

Step 2. Computation of Image Intensity Gradient

The next step is to compute the intensity gradient of the image to determine the edge strength. Because an edge may point to different directions, the gradient is decomposed into the *x* and *y* directions. Mathematically, this is reduced to the task of taking the derivatives of the image in the *x* and *y* directions, which can be estimated using a filter such as Sobel, Roberts, and Prewitt. For example, the Sobel filter in the *x* and *y* directions is shown in Equation 2.2. Then, the edge strength and direction can be determined using the gradient in the *x* and *y* directions (Equations 2.2 and 2.3). In Box 2, Figure 1 shows an example of the output of the algorithm. Figure 1B demonstrates that the edge gradient in the *x* direction detects vertical edges and edges with a small angle to the vertical. Figure 1C shows the edge gradient in the *y* direction, similarly showing edges close to horizontal as the most salient. Figure 1D is the edge gradient combined with information from both *x* and *y* directions.

$$G_x - \begin{bmatrix} -1 & 0 & 1 \\ -2 & 0 & 2 \\ -1 & 0 & 1 \end{bmatrix} G_y - \begin{bmatrix} 1 & 2 & 1 \\ 0 & 0 & 0 \\ -1 & -2 & -1 \end{bmatrix}, \tag{2.1}$$

$$|G| = \sqrt{(Gx)^2 + (Gy)^2}, \tag{2.2}$$

$$\theta'' = \arctan\left(\frac{Gy}{Gy}\right). \tag{2.3}$$

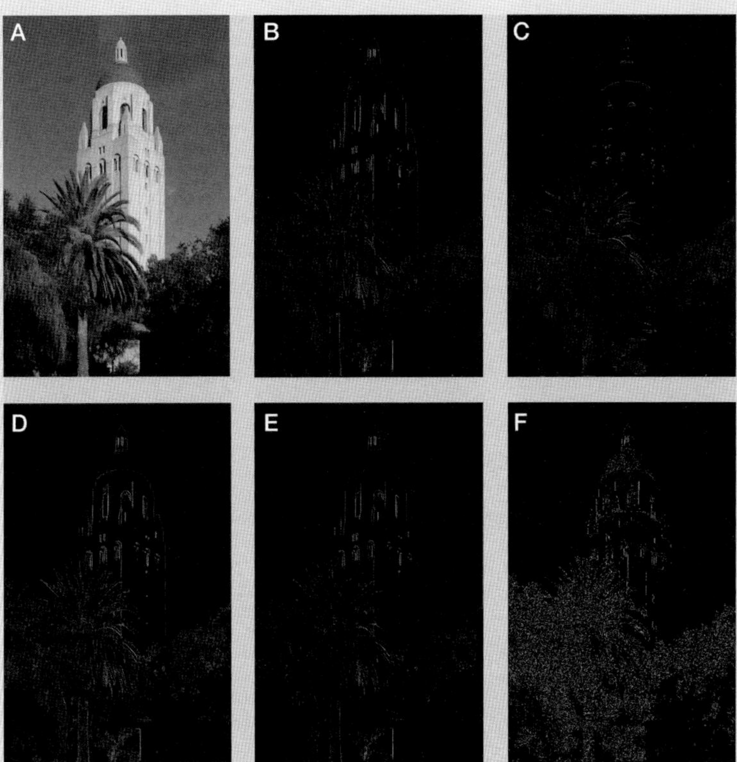

BOX 2, FIGURE 1. (*A–F*) Canny edge detection algorithm. Iterative application of the Canny edge detection algorithm to the original image (Fig. 1A).

Step 3. Postprocessing

The final step of the procedure attempts to reduce to the false edges using nonmaximum suppression and hysteresis thresholding.

Nonmaximum suppression attempts to guarantee that the edge is one pixel wide. To do this, at each pixel the algorithm examines the neighboring pixels and suppresses the other nonedge pixels.

To further reduce the false-positive edges, a threshold could be used to eliminate edges with a small gradient, under the generally good assumption that true edge pixels should have a large gradient. However, a single threshold may result in a set of disjoint segments of an edge when regions fall below that threshold. The Canny algorithm uses a more sophisticated thresholding method, called **hysteresis thresholding** to overcome this problem. Two thresholds are used: one high (T_{high}) and one low (T_{low}). Pixels with a gradient less than T_{low} are removed immediately, whereas pixels with a gradient greater than T_{high} are kept as edges. Pixels with gradients between T_{high} and T_{low} are kept only if their neighboring points are edge pixels, therefore preserving continuous edges. Figure 1F shows the final result of the Canny edge detection algorithm.

REFERENCES

Staley JT. 1968. Prosthecomicrobium and Ancalomicrobium: New prosthecate freshwater bacteria. *J Bacteriol* **95:** 1921–1942.

Staley JT, Gunsalus R, Lory S, Perry J. 2007. *Microbial life*, 2nd ed, Fig 19.12. Sinauer Press, Sunderland, MA.

Xiong G, Zhou X, Ji L. 2006. Automated segmentation of *Drosophila* RNAi fluorescence cellular images using deformable models. *IEEE Trans Circuits Syst Regular Papers* **53:** 2415–2424.

WWW RESOURCES

http://cammer.net/historical/aif/gallery/sem/sem.htm Analytical Imaging Facility at the Albert Einstein College of Medicine.

http://microbewiki.kenyon.edu/index.php/Pyrodictium Microbewiki, the student-edited microbiology resource, Kenyon College.

http://www.phy.cam.ac.uk/research/research-groups-images/bss/images/fluorescence.jpg/view University of Cambridge, Department of Physics, Cavendish Laboratory, UK.

6

Expression Data

David Ruau

Stanford University School of Medicine, Stanford, California 94305

INTRODUCTION TO MICROARRAY TECHNOLOGY

As the Central Dogma (DNA→mRNA→protein) has become more complex over the past decade, especially with the discovery of functional noncoding RNAs, increasingly sophisticated technologies are required to identify, track, and compare various aspects of gene expression levels across the genome. Gene expression microarrays are robust and reliable platforms for measuring mRNA abundance on a genomic scale. Since their conception in 1995, microarrays have developed into one of the hallmarks of genomic tools used in biomedical research. These classic expression arrays continue to be a tool of choice to assess the transcriptome of living cells. There are several reasons for that. Next-generation sequencing (NGS; discussed in Chapter 8), the main alternative to gene expression microarrays today, remains expensive and has not yet been adopted as a mainstream gene expression technique. More importantly, microarray technology has evolved beyond its initial objective of simply measuring mRNA expression to other types of genomic assays.

In particular, expression array technologies have progressed from their primary goal of measuring relative mRNA abundance to a more specialized application aimed at competing with DNA sequencing. The ever-increasing density of the probes made available on the array surface offers the ability to test a portion of the genome to genotype organisms for single-nucleotide polymorphisms (SNPs) or structural DNA variations such as inversions, deletions, insertions, and copy number. Tiling microarrays used for this purpose may cover parts of specific chromosomes or the entire genome, depending on the level of focus needed. Tiling arrays are conceived by splitting the entire genome into blocks of 35–75 nucleotides spaced by 10–100 nucleotides, representing a set of seven to 10 microarrays depending on the manufacturer and the organism. An additional popular application of DNA microarrays is chromatin immunoprecipitation on chip (ChIP-on-chip), used as an alternative to sequencing to map epigenetic modifications or protein–DNA binding sites. Clinical applications of microarrays have also served as tests to improve the accuracy of a diagnosis (mainly for cancers) or to genotype specific enzymes important for metabolizing certain classes of drugs (Glas et al. 2006; Heller et al. 2006).

Expression arrays are typically used to compare the expression levels of mRNA isolated under two or more conditions, such as in diseased and in healthy tissues. From a technical point of view, expression microarrays use glass or plastic supports for DNA probes representing baits for complementary DNA (cDNA) sequences reverse transcribed from mRNA. The two types of microarrays—one color and two color—are both designed to test differential gene expression levels between two samples (noted as "control" and "condition" in this chapter). There is, however, a fundamental difference in their utilization. One-color arrays use only a single fluorescently labeled sample that is hybridized to the array, producing an "absolute" fluorescence intensity value that reflects abundance. The consequence of such an approach is that two arrays must be run and compared, one for the control sample and one for the condition sample. In contrast, two-color microarrays rely on the use of competitive hybridization involving two nucleic acid samples on a single array, in which each sample is labeled with a different fluorophore (e.g., one sample with red, the other with a green fluorophore). In this hybridization setup, the differentially labeled samples compete for the same complementary sequences (probes) on the same array. Following hybridization, the array is scanned with lasers to detect the labels. The readout provides a ratio of fluorescence intensity (red/green) that is interpreted as the fold change in mRNA abundance between the conditions being compared.

An important notion here is the relativity of the measurement yielded by microarrays that, per design, do not produce quantitative measurements. This characteristic is obvious for two-color microarrays, where competitive hybridization shows a fold change in mRNA abundance between samples. We may ask why intensity is not a quantitative measure of abundance. The main explanation is that the specificity and sensitivity of each probe for its cDNA sequence are unique. This has been attributed, in part, to the proportion of GC content in the probe sequence that introduces nonspecific hybridization. In addition, because the quantity of probe sequences on the array is unknown, it is difficult to estimate the mRNA abundance required for signal saturation. Hence, probes only allow for comparison of the variation in expression for the same probe between arrays/conditions, not between probes on the same array. Because of these limitations, it is not uncommon that key genes found to be significantly regulated are often additionally validated by quantitative real-time–polymerase chain reaction (qRT-PCR).

In this chapter, we present the different microarray technologies and contrast the benefits regarding their experimental design requirements. We introduce the methods used for preprocessing and normalizing raw expression values for two- and one-color approaches. Different data mining tools for exploring gene expression profiles will be presented in a logical order for understanding the global structure of the data and then extracting relevant lists of up- and down-regulated genes. Finally, statistical concepts, such as multiple hypothesis testing correction, are introduced.

TWO-COLOR MICROARRAYS

Overview

Two-color microarrays are performed by hybridizing the two targets of interest (e.g., control and condition), mixed in equal proportion, to the same array. The hybridization is performed in large excess of the fixed probe relative to the labeled target so that competitive hybridization is not a factor (Duggan et al. 1999). The technique relies on using different fluorophores, generally cyanine 5 (Cy5; red) and cyanine 3 (Cy3; green), for labeling the two RNA samples to be hybridized. The hybridization is performed under stringent conditions, allowing only complementary sequences to hybridize to the probes. As noted above, two-color microarrays yield gene expression ratios that are a direct measure of relative mRNA abundance between samples. The number of microarrays required varies according to the experimental design chosen. However, because Cy5 and Cy3 dyes are incorporated into the DNA at different rates, they introduce sequence-specific noise in the measured fluorescence intensity. In addition, the green dye may provide a "brighter" or more intense signal than the red dye, when detected at the same sensitivity. This dye bias can result in a high false-positive rate, especially for genes with low fold change or low overall expression value. Thus, it has been suggested as good practice to perform dye-swap experiments, in which the same samples, now labeled with the alternative dyes, are measured again, and the fold changes between the replicates are averaged. However, this method doubles the number of arrays needed and the costs associated with this (along with the amount of reagent needed) may be a problem for some laboratories (thus, alternative approaches are discussed further below in the text). For example, consider the hypothetical experiment with matched control and condition samples with four replicates for each. Control Sample 1 will be compared with condition Sample 1 and so on for Samples 2, 3, and 4. The number of arrays needed to perform the dye swap will be eight.

In microarray experimental design, hybridized samples are also called "blocks," and the example above represents a balanced, complete block design. Such a design, the favored design in two-color microarray experiments, can be achieved when only two factors are compared (control vs. condition). When additional factors are of interest, for example, evaluating the effect of sex, balanced incomplete block design should be considered (Wit et al. 2005; Nguyen and Williams 2006; Knapen et al. 2009). Such complex designs have led to the creation of simpler so-called reference design, consisting of using a common cDNA reference on every array. Classically, it is an internal control to the experiment such as an untreated sample or an unrelated, commercially available "standard" sample made of a mixture of RNA from different tissues or cell lines. If we assume continuing availability of the product, the advantages of using a commercial reference RNA standard are (1) the possibility to compare results from different experiments within the same laboratory, because all would be hybridized against the same control; and (2) to improve the reproducibility of the results between laboratories and the potential for data reuse by the community upon publication of the results.

In addition to dye-swap experiments, another approach has been developed consisting of performing self–self hybridization (SSH), in which the same sample, labeled with Cy5 and Cy3, is hybridized to the same single array. The expression ratios obtained represent a direct measure of the dye bias and, theoretically, should all be equal to 1. In their study, Fang et al. (2007) suggested the use of SSH for the reference sample only in the context of a reference design. The SSH of the reference sample (reference vs. reference) is then used as the true control in a two-sample *t*-test by comparing ratios obtained in the experiment (treated sample vs. reference) against the ratios of the SSH. This method effectively reduces the number of arrays needed (Fang et al. 2007).

Overall, the experimental design of two-color microarrays can be harder to plan for than their one-color counterpart. Nevertheless, results from two- and one-color microarray platforms have shown high correlation and concordance of differentially expressed genes. The choice between the two platform types is influenced by cost, experimental design complexity, and personal expertise with a particular platform (Patterson et al. 2006).

Preprocessing and Normalization

After hybridization, the microarray chips are scanned, and image analysis software extracts the fluorescence intensity values of the spotted probes. First, a background correction is performed during the image analysis step, which determines if the spotted probes will obtain a positive or negative intensity value. Negative values are probes having a lower fluorescent intensity than their local background and are handled differently according to the platform (see the *MA* plot in Fig. 1A, which is discussed further in the text below). Gene expression ratio data are then processed for an additional background correction and normalized within and between arrays. Several normalization methods exist. Traditional approaches consist in either (1) dividing the expression ratio of every probe by a constant such that the array mean of the ratio is equal to 1, or (2) scaling the expression ratio to an equal mean or median within or across arrays (Quackenbush 2002). An alternative approach that is commonly used is the Lowess normalization method (Lowess for "locally weighted scatterplot smoothing"). This method is popular for two-color microarrays, because it is able to correct the global, but not the gene-specific, dye bias (see the dye-swap technique description above).We use as an example the publicly available data set GSE16026 from GEO. In this study, the researchers investigated anticancer and anti-inflammatory drug actions on a leukemia cell line using the two-color Agilent Whole

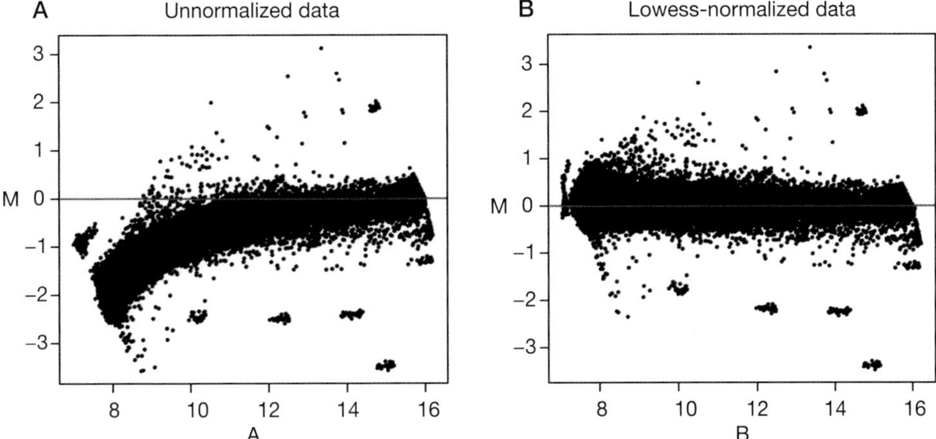

FIGURE 1. MA scatterplots for the array GSM401041 (replicate 1 of TNF-α-treated vs. untreated cells from data set GSE16026 referenced in the text) available in the Gene Expression Omnibus. (*A*) Unnormalized data display a strong bias in fold change in the green channel. (*B*) Lowess-normalized array. Outlier groups are "spike-in" controls.

Human Genome Microarray 4x44 K (Schumacher et al. 2010). *MA* plots are used to present the intensity dependence of the ratio of the variables by displaying the log-ratio (M) between green and red fluorescence on the *y*-axis and the average log intensities (A) on the *x*-axis (see Box 1). Figure 1 compares unnormalized data with data that is normalized by using the Lowess method.

In Figure 1A, we see that unnormalized data show a consistent bias in fluorescence intensity for low intensity. Such bias is induced by a systematic detection deficiency between the Cy3 and Cy5 dyes as well as a sequence-specific dye integration bias. Figure 1B shows that the Lowess normalization method detects deviation in the *MA* plot, applies a local weighted linear regression, and subtracts the calculated best-fit average $\log_2$ of the ratio from the experimentally observed ratio for each data point.

BOX 1. *MA* Plots

MA plots are used to study the dependence between the log ratio of *two* variables (in our discussion, the control and the condition under study) and the mean values of the two variables. Here, the log ratio (*M*) and the average log intensities (*A*) are defined as follows (where *R* = fluorescence intensity of the red dye and *G* represents fluorescence intensity of the green dye):

$$M = \log(R/G) = \log_2(R) - \log_2(G);$$

$$A = \frac{1}{2}\log(RG) = \frac{1}{2}[\log(R) + \log_2(G)].$$

ONE-COLOR MICROARRAYS

Overview

One-color technology simplifies the on-chip experimental design required for two-color arrays. Samples are hybridized separately from each other on different arrays. mRNA abundance is measured in absolute fluorescence intensity because no concurrent hybridization is performed.

Preprocessing and Normalization

Multiple commercial platforms exist, and each possesses its own technique aiming at the most accurate measurement possible. Depending on the manufacturer, probes are of different lengths (60 nucleotides for Agilent and 25 nucleotides for Affymetrix). To control for inherent nonspecific hybridization, multiple probes are usually used to assess different segments of the same gene or are present on the array multiple times. In addition, Affymetrix, Agilent, and Illumina have specific probes designed to measure nonspecific hybridization. For example, Affymetrix probes are organized as follows: All of the probes for a particular gene are grouped together into a probe set that is divided into perfect-match (PM) and mismatch (MM) probes. There is one MM probe for each PM probe. The MM probe sequences are the same as the PM probes save for one nucleotide that has been mutated in an attempt to control for unspecific hybridizations.

A consequence of using multiple probes to test a single gene is that measurements must be summarized to a single gene expression value. For Affymetrix, the original proposed summarization model, MAS 4.0, summarized gene expression after preprocessing for background correction is

$$AvDiff = \frac{1}{|A|}\sum_{j\in A}(PM_j - MM_j). \tag{1}$$

AvDiff is the expression value for the probeset and *A* the number of pairs of MM and PM probes in the probeset. As an internal quality control to their technology, Affymetrix excluded the probe from the probeset summary where the difference $PM-MM$ was superior to three times the standard deviation from the mean $PM-MM$ of the rest of the probeset. However, the MAS 4.0 model did not hold when the MM probe value was higher than the PM, leading to aberrant negative expression values. Consequently, Affymetrix produced a new version of their summary method, MAS 5.0 (Equation 2).

$$signal = Tukey\ Biweight\{\log(PM_j - CT_j)\}. \tag{2}$$

In this model, the probe intensities are preprocessed for global background correction, and intensity differences are log-transformed before signal extraction. Briefly, the *CT* value is the *MM* value as long as $MM < PM$ and is replaced by the robust average of the *MM* values of the entire probeset otherwise. Thus, CT_j is never greater than PM_j.

Following the release of MAS 5.0, several studies proposed alternative preprocessing methods that outperformed the manufacturer's method (Li and Wong 2001; Naef et al. 2002). The most widely used of these is the Robust Multiarray Analysis (RMA) model, which does not directly use the *MM* values (Irizarry et al. 2003). In the proposed model, the *PM* value is decomposed into a background signal plus the true signal of interest.

$$PM_{ij} = \beta_i + s_{ij}. \tag{3}$$

In this model, β_i is the global background estimate for the microarray i and s_{ij} is the true signal for the probe j. The background signal β_i is estimated as the mode of the *MM* value. Once the true signal is estimated, quantile normalization (discussed below) and median Polish summarization steps are performed. RMA was proved to outperform MAS 5.0 in precision at high concentration, thus avoiding attenuation of the signal due to the *MM* subtraction. This also produces less noise at low concentrations and thus offers a better detection of differentially expressed genes.

Building on the RMA method, Wu et al. published GC-RMA, which improved RMA by more accurately modeling the background noise using the GC dinucleotide content of the probe sequences (Naef and Magnasco 2003; Wu et al. 2004). The implementations of MAS 5.0, RMA, and GC-RMA algorithms are available in the R software (http://www.r-project.org) through the Bioconductor project (Gentleman et al. 2004).

Other one-color microarrays from Agilent and Illumina manufacturers provide their own background correction method similarly to Affymetrix. The Bioconductor project provides several packages (limma, Agi4x44PreProcess, and lumi), allowing users to apply the preprocessing workflow recommended by the manufacturer as well as alternatives developed by the bioinformatics community.

From a general perspective, a greater diversity of methods exists for background correction methods than for normalization and summarization procedures, maybe because it has been observed to have the greatest influence on the signal estimation. Normalization procedures for one-color arrays tend to follow the consensus method of quantile normalization. Quantile normalization aims at making the distribution of the probe intensities for each array the same. This is achieved by making the highest value on all of the arrays identical, the second highest value identical, and so on. Consequently, when arrays are quantile-normalized, the gene expression patterns have the same distribution (Bolstad et al. 2003). Of note, the median Polish summarization method used by RMA has recently been found to produce artifacts for odd numbers of replicates. The tRMA method has been suggested as a replacement (Giorgi et al. 2010).

ANALYZING GENE EXPRESSION VALUES

After preprocessing the raw gene expression data, one generally obtains a matrix of gene expression values presenting the samples in columns and genes in rows. In this section, we present the available tools for finding which genes or samples are significantly changed by the experiment.

Metrics

Metrics are measurements used to determine how similar or dissimilar genes or samples are. Downstream data mining performed on microarray data is as robust as the choice of a relevant metric in the first step. In general, metrics can be subdivided into three categories: distances, parametric correlation, and nonparametric correlation. Here, we will try to contrast these three categories sequentially. Among distances, the Euclidean and Manhattan distances are the most widely used. Formulas for both of these distances can be derived from the Minkowski equation (Equation 4) with $p = 1$ and $p = 2$ for Manhattan and Euclidean distances, respectively.

$$D(x, y) = \sqrt[p]{\sum_{i=0}^{n} |x_i - y_i|^p}. \tag{4}$$

Distance can be defined using the following conditions, for all objects i, j, and h:

(a) $d(i,j) \geq 0$;

(b) $d(i,i) = 0$;

(c) $d(i,j) = d(j,i)$;

(d) $d(i,j) \leq d(i,h) + d(h,j)$.

Condition (a) states that distances are always nonnegative. Condition (b) states that the distance between an object and itself is zero. Condition (c) states the symmetry of the distance function. Condition (d) defines the triangular inequality that allows geometrical interpretation of a distance between two objects. These definitions allow us to contrast distances with correlation similarity measures such as the Pearson correlation and its nonparametric counterpart, the Spearman rank correlation.

$$D(x, y) = \frac{\sum_{i=1}^{n} (x_i - \bar{x}_i)(y_i - \bar{y}_i)}{\sqrt{\sum_{i=1}^{n} (x_i - \bar{x}_i)^2 \sum_{i=0}^{n} (y_i - \bar{y}_i)^2}}. \tag{5}$$

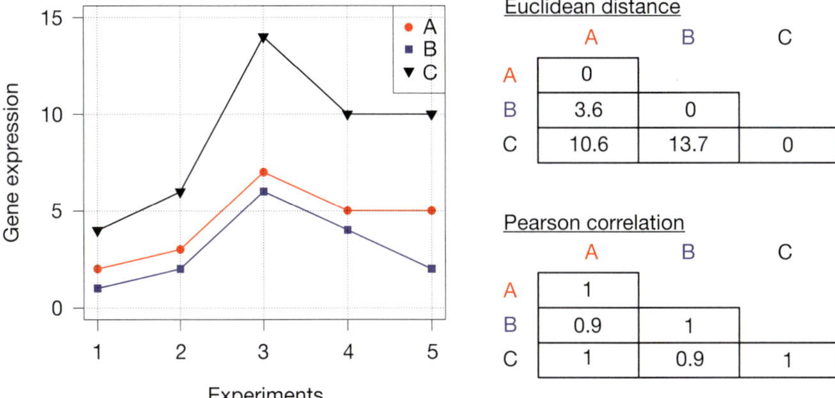

Euclidean distance

	A	B	C
A	0		
B	3.6	0	
C	10.6	13.7	0

Pearson correlation

	A	B	C
A	1		
B	0.9	1	
C	1	0.9	1

FIGURE 2. Gene expression pattern for hypothetical genes A (red), B (blue), and C (black) with their Euclidean distances and Pearson correlations.

Correlations usually respect conditions (b) and (c) but can adopt a negative score if the triangular inequality does not hold for any of them. To illustrate the difference between distances and correlations, let us consider the hypothetical gene expression profiles A, B, and C presented in Figure 2. If we ask which gene expression pattern is similar to A and use the Euclidean distance to answer the question, B would be considered the closest and, by extension, the most similar to A (Euclidean distance table in Fig. 2). However, using the Pearson or the Spearman correlation, the gene expression pattern that most correlated to A would be C followed by B (Pearson correlation table in Fig. 2).

The Pearson and Spearman rank correlations look at the similarity in the expression profile, whereas the Euclidean distance is, in fact, the distance between them. In other words, the Pearson and Spearman correlations can be considered scale-independent similarity measures in which the physical distance between the objects is irrelevant and only the trend is important. The difference between Pearson and Spearman is that one looks for linear and the other for monotone relations, respectively. Choosing between distance and correlation is a matter of estimating the importance of a change in gene expression. If an increase in gene expression value from 200 to 300 is equivalent to an increase from 1000 to 1100, then the Euclidean distance would be a good choice. However, microarray gene expressions are non-quantitative, as explained above, and, consequently, only the fold change is relevant. In the example above, a variation of 50% from 200 to 300 should be more or less equivalent to an increase from 1000 to 1500 for another gene.

Unsupervised Clustering Methods

Unsupervised clustering methods are first line data exploratory methods applied to microarray data. These methods are able to find groups (a.k.a. clusters) in data without prior knowledge regarding the data such as the number of groups that are expected to be found. Hierarchical clustering is an unsupervised clustering approach widely used to take a first look at the internal structure of data sets. It was first applied to gene expression data by Eisen et al. (1998). Hierarchical classification methods are intuitive to understand because they reproduce, in a binary manner, how our mind constantly classifies its surroundings into categories. Hierarchical clusters are built iteratively starting from a matrix of distance or correlation between all of the genes considered. In the first round, the two genes detected to be the most similar will be merged into the first node of the tree (Fig. 3). If the average linkage method is used, this node will be considered as a virtual gene having the average expression profile of both genes composing it. The same operation is repeated looking through the distance or correlation matrix for the next two most similar objects. The two objects are then merged into a new node and so on, until all of the genes are grouped into one node, the root of the tree.

Because the starting similarity matrix is determined, the solution of the hierarchical clustering is reproducible. However, when the hierarchical clusters are represented graphically, the software can rotate the branches of the dendrogram around their nodes freely, and the graph can sometimes look different. In the example presented in Figure 4, a hierarchical clustering was performed on the genes found differentially expressed between neurosphere cells left untreated or treated with two drugs. Neurosphere cells are a multipotent neuronal cell type. Here, the hierarchical cluster is presented on the left of the heatmap as a dendrogram. Heatmaps are regularly associated to hierarchical clusters in gene expression analysis to display, in a clear manner, up- or down-regulation of gene expression profiles across samples (Eisen et al. 1998; Wilkinson and Friendly 2009). However, hierarchical clusters have some disadvantages. First, they display a static representation of the data without proposing a solution to the actual number of clusters in the data. It is left to the user to interpret

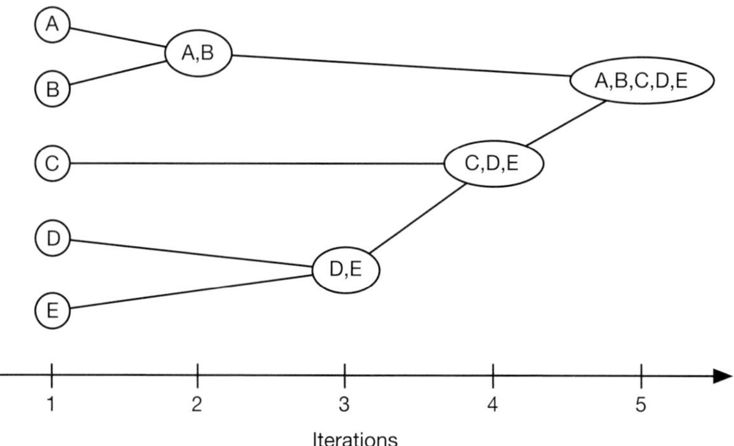

FIGURE 3. Hierarchical clustering iterative building process. (Adapted, with permission, from Kaufman and Rousseeuw 1990; © John Wiley & Sons Inc.)

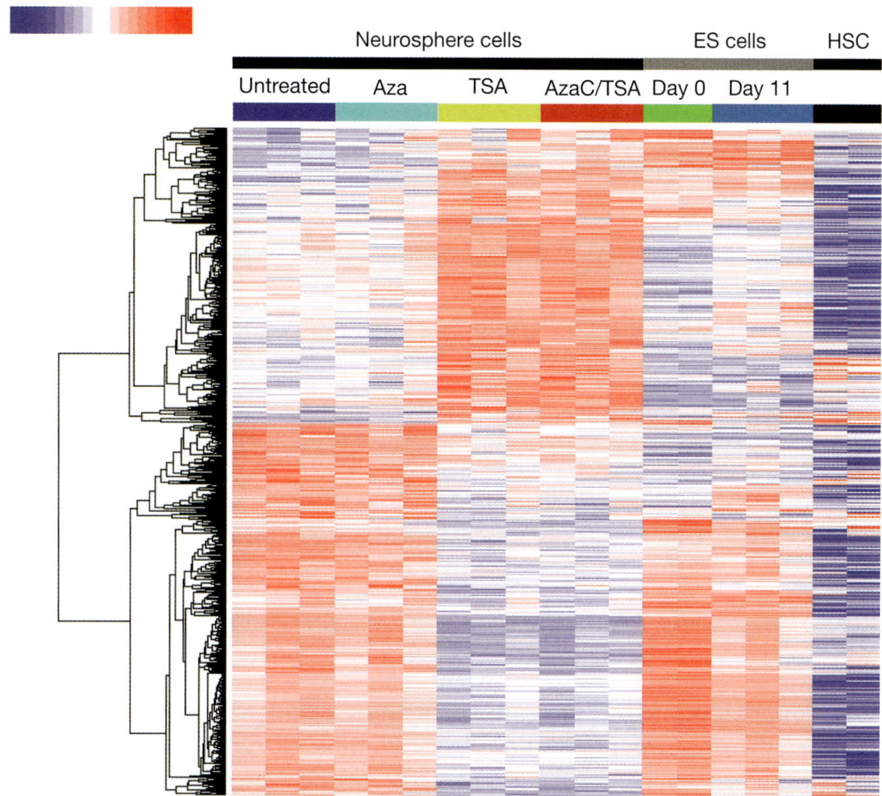

FIGURE 4. Hierarchical clustering and heatmap representation of gene expression (Ruau et al. 2008). Each row represent a gene colored according to its fold change across the conditions. The color scale indicates gene expression change: (blue) decreased expression; (white) no change; (red) increased expression. Three cell types under different conditions are represented in the columns. Neurosphere cells were treated with one or more chromatin-modifying agents: azacytidine (Aza), Trichostatin A (TSA), or both, and compared with (1) undifferentiated embryonic stem (ES) cells undifferentiated (Day 0), (2) ES cells differentiated toward neuronal lineage (Day 11), and (3) hematopoietic stem cells (HSC) that are blood cell type precursors. Gene Expression Omnibus accession number: GSE2375. (Adapted, with permission, from Ruau et al. 2008; © John Wiley & Sons Inc.)

the dendrogram and decide to which cluster each object belongs. Second, for large data sets, generating the dendrogram is computationally intensive and offers little help in understanding the data when too many objects are displayed. For these reasons, it is sometimes recommended to use semisupervised clustering methods.

Semisupervised Clustering

The main difference between unsupervised and semisupervised clustering methods is that, with the latter, the user must provide input on how to group the data. Partitioning algorithms are a perfect example of semisupervised clustering methods, and here we present the *k*-means algorithm. *k* means aims at forming *k* clusters using an iterative process that searches to optimize high intracluster similarity and low intercluster similarity. This is generally evaluated using the sum of squares error criterion, E, between the object in the cluster and the cluster center.

$$E = \sum_{i=1}^{k} \sum_{p \in C_i} |p - m_i|^2, \tag{6}$$

where p is a point in the cluster C_i, and m_i is the cluster C_i center or mean.

The algorithm works as follows.

1. Select k random objects, each representing a cluster center.
2. Assign all of the objects to the cluster center to which they are the most similar or closest (depending if one uses correlation or distance as a metric).
3. Update the cluster center by computing the mean of all of the objects in the cluster.
4. Repeat Steps 2 and 3 until the square errors criteria do not change or a preset maximum number of iterations is reached.

A quick search of the literature reveals many variants of the k-means algorithm that differ by the selection method of the initial random objects such as the cluster center, how the dissimilarity between objects is calculated, or how the cluster means are interpreted to calculate the similarity criterion.

Partitioning algorithms are fast but have two major drawbacks. First, as mentioned above, the user has to specify the number of clusters beforehand, and this can be challenging when the natural grouping is not obvious. To guess the correct number of clusters present in the data, one can use unsupervised hierarchical clustering to explore the data. At least one solution has been proposed that combines both clustering techniques into one algorithm and suppresses the need for specifying the desired number of clusters (Chen et al. 2005). This hybrid clustering technique was shown to perform better than the traditional k mean, mainly because of its capacity to handle outliers. A traditional approach to select the k number of clusters is to use silhouette plots to evaluate the global structure of the data (Kaufman and Rousseeuw 1990). A silhouette plot is an exploratory graphical representation of the clusters found using a partitioning algorithm. The silhouette algorithm works as follows: For each object in each cluster, a similarity score is computed to evaluate how well the object fits into its attributed cluster. Then, for each cluster, the similarity scores of their objects are ordered from highest to lowest. The resulting ordered scores are then plotted as a bar graph, thus displaying the coherence of the objects within their cluster. A silhouette coefficient is also computed for each cluster as the average of the similarity score, and a global average silhouette coefficient for the entire cluster solution is computed. The optimal number of clusters k is found when the silhouette

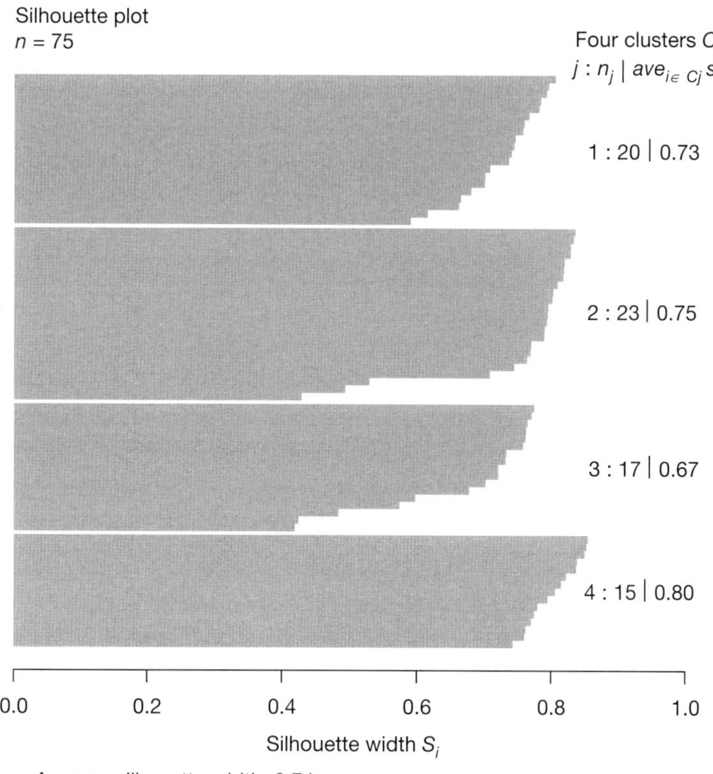

FIGURE 5. Silhouette plot of a partitioning clustering solution for $k = 4$ clusters produced by the silhouette function in R. The numbers on the side of each cluster indicate the number of objects in the cluster followed by the silhouette coefficient. The global average silhouette coefficient is indicated as average silhouette width.

coefficient is maximal. Figure 5 shows an example of a silhouette plot for a cluster solution with $k = 4$ clusters.

The second drawback is that partitioning algorithms returns "hard clusters," meaning that objects can belong only to one cluster at a time. Sometimes no ideal solution exists to a particular clustering problem, and, in this case, obtaining information regarding the degree of membership of each object to each cluster can provide a better understanding of the data. Based on this idea of partial membership of objects to multiple clusters, the expectation–maximization (EM) algorithm extends the *k*-means idea and represents the data as a mixture of underlying probability distributions (Dempster et al. 1977). The EM algorithm assigns to the objects the weighted probability of membership in clusters. However, cluster solutions generated with EM are hard to interpret on large data sets, and hard-cluster interpretations of the results are often made in which objects are assigned to the cluster to which they belong with the highest probability. EM is often used in conjunction with *k* means to gain additional information.

Self-organizing maps (SOMs) is a neural network method that uses an iterative process (similar to k means) to attribute objects to a predefined k number of clusters. The difference is that SOMs try to cluster high-dimensional objects in a lower-dimensional space, usually two dimensional (2D) or three dimensional (3D), allowing a graphical representation of all of the clusters together to represent their similarity visually.

Statistical Approaches to Gene Expression Data Interpretation

In this section, we describe some of the most popular statistical methods used to derive lists of genes that are significantly differentially expressed. Multiple hypothesis-testing correction methods will be explained in the context of fold change and p-value traditional filtering to find genes that are significantly regulated. We then introduce the highly popular and superior approaches of significance analysis of microarray (SAM) and rank product methods.

When the microarray became popular 15 years ago, fold change was the classic intuitive analysis that biologists looked at to determined up- and down-regulation. However, fold change alone has two major problems. First, the cutoff value is difficult to justify (twofold or 1.5-fold?), and it can be calculated in two different manners, either as a ratio (a/b) or a difference ($a - b$), leading to unreliable results. Second, there is no measure of significance. To evaluate the confidence of a fold-change, a simple t-test can be used to help justify the minimal fold change observable.

$$\frac{\bar{x}_i - \bar{x}_j}{\sqrt{\frac{s_i^2}{n_i} + \frac{s_j^2}{n_j}}}, \tag{7}$$

where $\bar{x}_i$ and $\bar{y}_i$ refer to the means for sample groups i and j and s refers to the standard deviation.

When performing a t-test to evaluate the significance of a gene expression change, a test hypothesis is implicitly formulated, called the null hypothesis (H_0). In the context of microarrays, the null hypothesis is often defined as "the gene under consideration is not significantly regulated." Thus, rejecting H_0 means that the gene is significantly regulated with an error rate below α (usually $\alpha = 0.05$). However, performing a t-test to accept or reject H_0 for each of the 30,000 genes on a microarray with a confidence level of 95% for each test may introduce many false positives where we incorrectly rejected H_0. Thus, it is important to correct for multiple hypothesis testing by evaluating the false-positive probability defined in statistics as the type I error rate (Table 1).

To calculate the probability of type I errors (i.e., rejecting the null hypothesis when it is true), a variety of tests are available such as the family-wise error rate (FWER) procedures or the false discovery rate (FDR).

TABLE 1. Type I error rates

Number of	Accepted H_0 (declared insignificant)	Rejected H_0 (declared significant)	
True null hypotheses	U	V	m_0
False null hypotheses	T	S	m_1
	$m - R$	R	m

m is the total number of tests performed against the null hypothesis $H_0, j = 1, ..., m$. U are the true negatives, T is the theoretical number of false negatives or type II errors, S are the true positives, V is the theoretical number of false positives, or type I errors, and R is the total number of rejected null hypotheses (genes declared nondifferentially expressed).

The FWER can be written as $\Pr(V \geq 1) \leq \alpha$, meaning that the probability of a gene to be false positive is inferior to the user-defined threshold α (usually 0.05). The FWER can be applied using the Bonferroni procedure that guarantees the $\Pr(V \geq 1) \leq \alpha$ condition by calling all of the features (genes) with p value $\leq \alpha/m$ significant. This is equivalent to: adjusted p value = raw p value $\times$ m. Because m (the total number of genes) in microarrays can easily be very large, most of the time the Bonferroni procedure results in corrected p values close to 1 for all of the genes. Thus, since the Bonferroni procedure is too conservative, alternative methods such as the Bejamini–Hochberg's FDR are sometimes preferred (Benjamini and Hochberg 1995). The FDR approach evaluates the proportion of false positives as the random variable $Q = V/(V + S)$. Q is random or unknown because the true values of V (number of false positives) and S (number of true positives) are unobservable. However, it is possible to calculate an expected Q value, $E[Q]$, and use it to establish the adjusted p value for each gene (Benjamini and Hochberg 1995). FDR, Bonferroni, and several other multiple hypothesis correction procedures can be easily computed using the *mt.rawp2adjp* function from the *multtest* R package.

J.D. Storey developed an improvement to the FDR in 2001 called the pFDR (Storey 2002; Storey and Tibshirani 2003). The pFDR was motivated by the too-conservative methodology of FDR for genomic applications with a high number of features (i.e., genes). The pFDR produces q values for each gene using a sample label permutation approach. Briefly, the t-test is recomputed for every gene on the array after the sample labels are randomly shuffled. The expected effect is to obtain unrelated t-tests that are due to chance, forming what is called a null distribution. The q value is the expected proportion of false positives for a gene with a specified p value when using the null distribution as control.

The SAM method is a compelling method for extracting differentially expressed genes that merge the benefit of the t-test, fold-change, and multiple hypothesis testing methods (Tusher et al. 2001). It is based on a modified t-test:

$$t_i' = \frac{\overline{x_i} - \overline{y_i}}{sd_i + s_0}, \tag{8}$$

where s_0 is a small positive constant that minimizes the coefficient of variation of t'. The SAM method is combined with an estimation of the FDR for multiple testing, calculated by repeated permutation of the data, thus giving a directly adjusted p value. SAM software is available through R (Ihaka and Gentleman 1996) (www.r-project.org) and also as a Microsoft Excel plug-in (www-stat.stanford.edu/~tibs/SAM/).

Finally, Rank Product (RP) is a statistical approach for finding differentially expressed genes based on fold change (Breitling et al. 2004). The RP method first ranks each gene according to the fold change obtained when considering two conditions. For two-color arrays, a gene g will be described with as many ranks as replicate k of the experiment were performed. For one-color arrays, the ranks are calculated for all of the possible pairwise comparisons k possible. The RP algorithm then proceeds to calculate the geometric mean of the rank of gene g in the different replicates k:

$$RP_g^{\mathrm{up}} = \left(\prod_i^k r_{i,g}^{\mathrm{up}}\right)^{1/k}, \tag{9}$$

where r^{up} is the rank of gene g in replicate i. An important point to consider before applying RP is to check if the assumptions hold over the data. The creators of RP assume that (1) relevant expression change affects only a minority of genes, (2) measurements are independent between replicate arrays, (3) most changes are independent of each other, and (4) measurement variance is approximately equal for all genes. The RP score has an associated significance value obtained by multiplying RP by a factor F (the number of different possible product permutations of the k replicate). RP has been shown to outperform SAM for small sample group size (Jeffery et al. 2006). RP implementation is available as an R package, RankProd, through Bioconductor (Gentleman et al. 2004).

CONCLUSIONS

In this chapter, we discuss the microarray technology and computational tools needed to analyze expression data. We talk about the differences between one- and two-color arrays, various normalization approaches, and different techniques needed to study differential gene expression. Finally, we discuss the application of hierarchical clustering to gene expression data. Many of these concepts and methods may also be applied to other data types of interest.

REFERENCES

Benjamini Y, Hochberg Y. 1995. Controlling the false discovery rate: A practical and powerful approach to multiple testing. *J R Stat Soc Series B Stat Methodol.* **57:** 289–300.

Bolstad BM, Irizarry RA, Astrand M, Speed TP. 2003. A comparison of normalization methods for high density oligonucleotide array data based on variance and bias. *Bioinformatics* **19:** 185–193.

Breitling R, Armengaud P, Amtmann A, Herzyk P. 2004. Rank products: A simple, yet powerful, new method to detect differentially regulated genes in replicated microarray experiments. *FEBS Lett* **573:** 83–92.

Chen B, Tai PC, Harrison R, Pan Y. 2005. Novel hybrid hierarchical-*K*-means clustering method (H-K-means) for microarray analysis. *Proc IEEE Comput Syst Bioinform Conf* **2005:** 105–108.

Dempster AP, Laird NM, Rubin DB. 1977. Maximum likelihood from incomplete data via the EM algorithm. *J R Stat Soc Series B Stat Methodol.* **39:** 1–38.

Duggan DJ, Bittner M, Chen Y, Meltzer P, Trent JM. 1999. Expression profiling using cDNA microarrays. *Nature genetics* **21:** 10–14.

Eisen MB, Spellman PT, Brown PO, Botstein D. 1998. Cluster analysis and display of genome-wide expression patterns. *Proc Natl Acad Sci* **95:** 14863–14868.

Fang H, Fan X, Guo L, Shi L, Perkins R, Ge W, Dragan YP, Tong W. 2007. Self–self hybridization as an alternative experiment design to dye swap for two-color microarrays. *OMICS* **11:** 14–24.

Gentleman RC, Carey VJ, Bates DM, Bolstad B, Dettling M, Dudoit S, Ellis B, Gautier L, Ge Y, Gentry J, et al. 2004. Bioconductor: Open software development for computational biology and bioinformatics. *Genome Biol* **5:** R80.

Giorgi FM, Bolger AM, Lohse M, Usadel B. 2010. Algorithm-driven artifacts in median Polish summarization of microarray data. *BMC Bioinformatics* **11:** 553.

Glas AM, Floore A, Delahaye LJ, Witteveen AT, Pover RC, Bakx N, Lahti-Domenici JS, Bruinsma TJ, Warmoes MO, Bernards R, et al. 2006. Converting a breast cancer microarray signature into a high-throughput diagnostic test. *BMC Genomics* **7:** 278.

Heller T, Kirchheiner J, Armstrong VW, Luthe H, Tzvetkov M, Brockmoller J, Oellerich M. 2006. AmpliChip CYP450 GeneChip: A new gene chip that allows rapid and accurate CYP2D6 genotyping. *Ther Drug Monit* **28:** 673–677.

Ihaka R, Gentleman R. 1996. R: A language for data analysis and graphics. *J Comput Graph Stat.* **5:** 299–314.

Irizarry RA, Hobbs B, Collin F, Beazer-Barclay YD, Antonellis KJ, Scherf U, Speed TP. 2003. Exploration, normalization, and summaries of high density oligonucleotide array probe level data. *Biostatistics* **4:** 249–264.

Jeffery IB, Higgins DG, Culhane AC. 2006. Comparison and evaluation of methods for generating differentially expressed gene lists from microarray data. *BMC Bioinformatics* **7:** 359.

Kaufmann L, Rousseeuw PJ. 1990. *Finding groups in data: An introduction to cluster analysis.* Wiley, New York.

Knapen D, Vergauwen L, Laukens K, Blust R. 2009. Best practices for hybridization design in two-colour microarray analysis. *Trends Biotechnol* **27:** 406–414.

Li C, Wong WH. 2001. Model-based analysis of oligonucleotide arrays: Expression index computation and outlier detection. *Proc Natl Acad Sci* **98:** 31–36.

Naef F, Magnasco MO. 2003. Solving the riddle of the bright mismatches: Labeling and effective binding in oligonucleotide arrays. *Phys Rev E Stat Nonlin Soft Matter Phys.* **68:** 011906.

Naef F, Lim DA, Patil N, Magnasco M. 2002. DNA hybridization to mismatched templates: A ChIP study. *Phys Rev E Stat Nonlin Soft Matter Phys.* **65:** 040902.

Nguyen NK, Williams ER. 2006. Experimental designs for 2-colour cDNA microarray experiments. *Appl Stochastic Models Bus Ind.* **22:** 631–638.

Patterson TA, Lobenhofer EK, Fulmer-Smentek SB, Collins PJ, Chu TM, Bao W, Fang H, Kawasaki ES, Hager J, Tikhonova IR, et al. 2006. Performance comparison of one-color and two-color platforms within the MicroArray Quality Control (MAQC) project. *Nat Biotechnol* **24:** 1140–1150.

Quackenbush J. 2002. Microarray data normalization and transformation. *Nat Genet* **32:** 496–501.

Ruau D, Ensenat-Waser R, Dinger TC, Vallabhapurapu DS, Rolletschek A, Hacker C, Hieronymus T, Wobus AM, Muller AM, Zenke M. 2008. Pluripotency associated genes are reactivated by chromatin-modifying agents in neurosphere cells. *Stem Cells* **26:** 920–926.

Schumacher M, Cerella C, Eifes S, Chateauvieux S, Morceau F, Jaspars M, Dicato M, Diederich M. 2010. Heteronemin, a spongean sesterterpene, inhibits TNFα-induced NF-κB activation through proteasome inhibition and induces apoptotic cell death. *Biochem Pharmacol* **79:** 610–622. Erratum: **79:** 1837.

Storey JD. 2002. A direct approach to false discovery rates. *J R Stat Soc Series B Stat Methodol* **64:** 479–498.

Storey JD, Tibshirani R. 2003. Statistical significance for genomewide studies. *Proc Natl Acad Sci* **100:** 9440–9445.

Tusher VG, Tibshirani R, Chu G. 2001. Significance analysis of microarrays applied to the ionizing radiation response. *Proc Natl Acad Sci* **98:** 5116–5121.

Wilkinson L, Friendly M. 2009. The history of the cluster heat map. *Am Stat* **63:** 179–184.

Wit E, Nobile A, Khanin R. 2005. Near-optimal designs for dual channel microarray studies. *Applied Statist* **54:** 817–830.

Wu Z, Irizarry R, Gentleman R, Martinez-Murillo F, Spencer F. 2004. A model-based background adjustment for oligonucleotide expression arrays. *J Am Stat Assoc* **99:** 909–917.

WWW RESOURCES

http://cran.r-project.org/web/packages/samr Significance Analysis of Microarrays (SAM), R version

http://www.stat.stanford.edu/~tibs/SAMR SAM Microsoft Excel plug-in

7

A Gentle Introduction to Genome-Wide Association Studies

Chuong B. Do,[1] Marc A. Schaub,[2] Marina Sirota,[3] and Karen Lee[4]

[1,2]Stanford University, Computer Science, Stanford, California 94305

[3]Stanford University School of Medicine, Biomedical Informatics Training Program, Stanford, California 94305

[4]Stanford University, Stanford, California 94305

During the last decade, genome-wide association studies (GWASs) have emerged as an exceptionally powerful tool for the study of human genetics. What are GWASs and how do they work? In this chapter, we provide a high-level overview of this fascinating area of research, with an emphasis on applying GWASs to the genetics of complex human traits.

This chapter covers four main topics.

1. What is a GWAS?
2. What can we learn from GWASs?
3. How does one conduct a GWAS?
4. How does one interpret the results of a GWAS?

Conducting a proper GWAS requires careful attention to technical detail, from data quality control to analysis and reporting of results. Numerous tutorials and reviews are already available that describe best technical practices for conducting GWASs (Carlson et al. 2004; Hirschhorn and Daly 2005; Wang et al. 2005; Balding 2006; Donnelly 2008; Kruglyak 2008; McCarthy et al. 2008; Pearson and Manolio 2008; Cichon et al. 2009). In this chapter, we offer a "gentle" introduction to the basic concepts underlying GWASs. Our goal is to provide a nonexpert with the knowledge needed to read and interpret a standard scientific paper describing a GWAS.

FUNDAMENTAL CONCEPTS

The goal of a GWAS is to identify **associations** among genetic variants in a human population (i.e., **genotypes**) and specific biological characteristics of individuals in that population (i.e., **phenotypes**). To understand this definition, we start by taking a closer look at the meaning of its three key components: genotypes, phenotypes, and associations.

What Is a Genotype?

The complete DNA sequence describing every human being is more than 6 billion nucleotides long and is organized into 46 **chromosomes** (22 pairs of **autosomes** and two **sex chromosomes**). Each individual inherits half of these chromosomes from his or her mother and the other half from his or her father. Although the genome is long, the nucleotides at the vast majority of positions (>99%) within the genome are identical from one human being to the next. Studies of genetic variation, therefore, focus on the 1% of the genomic sequence that makes each human unique.

Most of the genetic variation among individuals in a particular population consists of single-nucleotide sequence variations, known as **single-nucleotide polymorphisms** (SNPs), which are present at specific locations in the genome. Generally speaking, the term SNP is reserved for sequence variations for which the less common variant occurs in at least 1% of the population. SNPs occur frequently throughout the genome—once every 300 positions on average—although this number varies depending on the population considered.

At any given SNP location (or **locus**), the different variants occurring in a population are known as **alleles**. Most SNPs are **biallelic**, meaning that only two different variants occur with an appreciable frequency. The more common variant in a particular population is known as the **major allele** and the less common variant as the **minor allele**. For all chromosomes except for sex chromosomes, an individual has two alleles at each SNP—one inherited from the mother and one inherited from the father. These two alleles (known as the **maternal** and **paternal** alleles, respectively) together comprise the individual's genotype for the given SNP.

Current technologies for determining the alleles at a given SNP are generally unable to distinguish the parent of origin for each allele. Therefore, genotypes are often considered to result from one of three possible outcomes: two copies of the major allele, one copy each of the major and minor alleles, or two copies of the minor allele. These three outcomes are often conveniently represented using the integers 0, 1, or 2 (the numbers corresponding to the number of copies of the minor allele). An example of an SNP that is associated with eye color is presented in Figure 1.

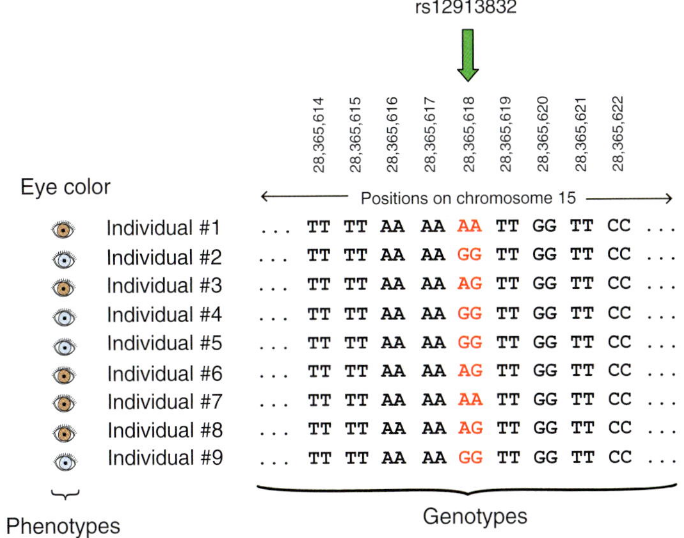

FIGURE 1. rs12913832, a single-nucleotide polymorphism on chromosome 15. The figure shows genotypes from a region of chromosome 15 from base positions 28,365,614 through 28,365,622 (according to build 37.3 of the NCBI reference human genome) in nine European individuals. As shown in the diagram, the nucleotides at most positions within this region do not vary among individuals. The variation at position 28,365,618, however, displays an interesting association with eye color: Individuals with at least one copy of the A allele (instead of the G allele) at this position tend to have brown eyes instead of blue eyes. This position is known as rs12913832 according to dbSNP build 135, a public repository of SNPs that have been identified to date (Sherry et al. 2001). In European populations, the A allele occurs at ~20% frequency and is therefore considered to be the minor allele; conversely, the G allele is considered to be the major allele.

What Is a Phenotype?

A phenotype is any physical or biological characteristic of an individual that can be observed and measured. In many GWASs, phenotypes are quantified as the simple presence or absence of a trait (e.g., curly hair) or condition (e.g., asthma). In other cases, phenotyping may involve classifying individuals into one of several categories (e.g., different eye colors) or the use of a continuous measurement (e.g., height). The range of possible phenotypes that can be studied by genome-wide association is virtually limitless.

The choices of what phenotype to study and how to measure it can be crucial to the success of a GWAS. A phenotype too broadly defined may capture a spectrum of conditions that seem to be related but that have distinct underlying genetic etiologies, a situation known as **genetic heterogeneity**. Heterogeneity dilutes the strength of each true association signal in the data, thus reducing the probability that a study will detect any single association individually. Conversely, a phenotype defined too narrowly may make it difficult for a researcher to find a sufficient number of individuals to conduct a GWAS. A small sample size, in turn, will reduce the **statistical power** of the study (i.e., the likelihood that the study will be able to detect a significant association). Both extremes, of defining phenotypes too broadly or narrowly, can be fatal to the success of a study. Thus, careful attention must be paid to defining phenotypes that cover a sufficiently broad set of cases and also plausibly share a common genetic mechanism (see Fig. 2).

What Is an Association?

In everyday usage, the word **correlation** typically refers to a relationship in which two quantities vary together. For example, the statement "increases in poverty are positively correlated with rises in crime rates" means that changes in poverty levels tend to occur in the same direction as changes in crime rates. That is, when poverty levels go up, crime rates also go up, and when poverty levels go down, so do crime rates.

In statistics, the term correlation has a narrower meaning: Correlation between two variables implies a linear relationship between them. For example, a statement of statistical correlation might be that "every 1% increase in the percentage of Americans living in extreme poverty results in a 2% increase in the frequency of crimes on average." For statisticians, associations refer to the more general concept that levels of one variable predict levels of another variable (even in situations in which the relationship between the two variables is not necessarily linear). A **genetic association** (sometimes referred to as a **genotype–phenotype association**) occurs whenever a particular SNP genotype is associated with a specific phenotype in a given population.

To make this concept clear, we consider throughout this chapter a hypothetical GWAS investigation whose goal is to understand the genetics of telepathy, the ability to read minds. Suppose we have somehow managed to assemble a collection of 950 individuals with demonstrated telepathic ability and a separate collection of 1050 individuals who lack this ability. We refer to the individuals in the first set as **cases** and individuals in the second set as **controls**. Suppose also that for each individual,

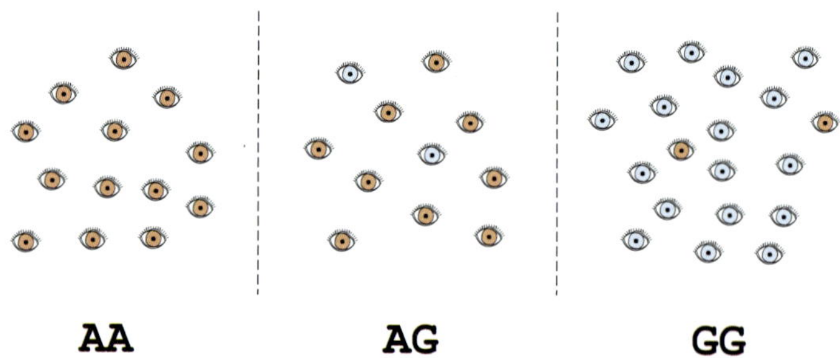

FIGURE 2. A genetic association for eye color. Eye color in European individuals is related to their genotype at position 28,365,618 on chromosome 15 (i.e., rs12913832). Individuals with at least one copy of the A allele tend to have brown eyes, whereas individuals with two copies of the G allele tend to have blue eyes. In a GWAS, genotypes and phenotypes for large sets of individuals from a population (typically thousands) are used to identify genetic associations such as the one shown here.

we have determined the genotypes at 1 million different SNPs in the genome. For one of these SNPs, we show below the 2 × 3 **contingency table** that indicates the number of cases and controls with each of the three possible genotypes (AA, AG, or GG) for that particular SNP.

	AA	AG	GG
Telepathy	38	310	602
No telepathy	12	400	638

Among the participants in the study, individuals with the AA genotype appear to have a higher probability of showing telepathy (i.e., 38/[38 + 12] = 76%) than those with the GG genotype (i.e., 602/[602 + 638] ≈ 48.5%). A standard way of summarizing these probability differences in a GWAS is the **genotypic odds ratio** for telepathy in the AA genotype relative to the GG genotype, mathematically defined as

$$
\begin{aligned}
\text{odds ratio}_{\text{AA}} &= \frac{P(\text{telepathy} \mid \text{AA})}{P(\text{no telepathy} \mid \text{AA})} \bigg/ \frac{P(\text{telepathy} \mid \text{GG})}{P(\text{no telepathy} \mid \text{GG})} \\
&= \frac{38/(38+12)}{12/(38+12)} \bigg/ \frac{602/(602+638)}{638/(602+638)} \\
&= \frac{38 \cdot 638}{12 \cdot 602} \approx 3.4.
\end{aligned}
$$

Alternative genotypes with odds ratios that are >1 relative to a reference genotype are said to confer increased risk of having the phenotype. Conversely, alternative genotypes with odds ratios that are <1 relative to the reference genotype confer decreased risk of having the phenotype.

In some situations, it can be appropriate to estimate an **allelic odds ratio**, which measures the effect of each copy of the alternative allele on disease risk. For the SNP above, the allelic odds ratio is calculated by reducing the 2 × 3 contingency table to a 2 × 2 table showing the phenotypic counts per allele.

$$
\begin{array}{l|ll}
 & \text{A} & \text{G} \\
\hline
\text{Telepathy} & 2 \cdot 38 + 310 & 2 \cdot 602 + 310 \\
\text{No telepathy} & 2 \cdot 12 + 400 & 2 \cdot 638 + 400
\end{array}
=
\begin{array}{l|ll}
 & \text{A} & \text{G} \\
\hline
\text{telepathy} & 386 & 1514 \\
\text{no telepathy} & 424 & 1676.
\end{array}
$$

The odds ratio computed in this case is known as an allelic odds ratio for telepathy of the alternative allele (A) relative to the reference allele (G).

$$
\text{odds ratio}_{\text{A}} = \frac{P(\text{telepathy} \mid \text{A})}{P(\text{no telepathy} \mid \text{A})} \bigg/ \frac{P(\text{telepathy} \mid \text{G})}{P(\text{no telepathy} \mid \text{G})} = \frac{386 \cdot 1676}{424 \cdot 1514} \approx 1.01.
$$

Both genotypic and allelic odds ratios are measures of the statistical effect size that a particular genotype has on disease risk.

When should we consider an odds ratio different from 1 to be "interesting," and when might it simply reflect sampling error (i.e., an error that results when the characteristics of the subset of individuals used in a study do not match those of the whole population)? The answer to this question depends on a combination of factors, including the size of the effect (i.e., the magnitude of the odds ratio), the direction of the effect (i.e., does the alternative genotype increase or decrease risk?), the frequency of the alternative genotype (i.e., what proportion of the population has modified risk?), and the **statistical significance** of the result (i.e., how likely is it that this result would have arisen by chance?). We return to the topic of assessing significance below. For now, suffice it to say that if a person's genotype at this SNP is indeed predictive of his or her telepathic ability, then we can say that this SNP is associated with that phenotype.

WHY ARE GENOME-WIDE ASSOCIATION STUDIES PERFORMED?

In this section, we discuss the scientific rationale for GWASs and the reasons for their current popularity as a technical approach. We also compare GWASs with linkage studies, another approach used to study complex traits in human genetics research.

What Can We Learn from a Genome-Wide Association Study?

As described in the previous section, GWASs are large-scale investigations that seek to identify statistical associations between genotypes (typically SNPs) and phenotypes. But why are such statistical associations interesting in the first place?

These associations are interesting because they can help to elucidate the biological pathways underlying the development of a disease. Given that there are 20,000 known human genes, identifying specific genes involved in the etiology of a particular disease is often very challenging. Nonrandom association of a SNP with disease status can be a powerful indicator that a gene close to that SNP in the genome is part of a biological pathway whose disruption leads to the disease phenotype. Genes identified in this way can often suggest potential therapeutic targets: Drugs might be designed (or may already exist) that can specifically affect the behavior of the proteins encoded by these genes in vivo.

In some cases, statistical associations enable predictions of phenotypes based on genotypes. Among European individuals, for example, possession of either one or two copies of the A allele for the SNP rs12913832 is strongly associated with having brown eyes instead of blue eyes (Eiberg et al. 2008; Kayser et al. 2008; Sturm et al. 2008). Not many phenotypes are so strongly associated with a single SNP, however. The vast majority of complex human diseases (type 2 diabetes, prostate cancer, etc.)

are not wholly genetic in origin but, rather, arise from the interplay of many genetic and environmental risk factors, and the associations discovered through GWASs typically provide only very modest predictive ability because of their small effect sizes (odds ratios between 0.7 and 1.5).

The small effect sizes of most SNPs likely reflect the fact that genetic variants that strongly increase risk for a disease would typically be subject to strong negative selection during evolution and thus tend to be rare. Exceptions to this rule are seen for late-onset diseases, such as Alzheimer's, or age-related macular degeneration, where certain common genetic variants are known to convey significantly increased risk. Because of their late onset, these conditions have little effect on reproductive success, and their associated genes are not particularly affected by selective evolutionary pressures.

Other important exceptions to the rule of small SNP effect sizes are seen in **pharmacogenetics**, the study of how genetic variation affects individual responses to different drugs. By elucidating why some individuals respond favorably to certain drugs and less favorably to others based on their genetics, pharmacogenetic research promises to enable the coming era of *personalized medicine*, a model of healthcare in which medical treatments are highly tailored to the individual being treated. The effect sizes for SNPs involved in drug responses are often larger than those involved in disease pathways. The likely reason for this difference is that most drugs used in modern medicine were developed relatively recently and have had little opportunity to impact human genetic evolution.

The Development of Genome-Wide Association Studies as a Research Tool

The concept of the GWAS was anticipated as early as 1996 (Risch and Merikangas 1996), 7 years before the sequencing of the human genome was complete. However, the first GWASs were not published until 2006 (Dewan et al. 2006), and standard protocols for conducting and analyzing GWASs were first established in 2007, with the publication of a seminal study of seven complex diseases by the Wellcome Trust Case-Control Consortium (WTCCC) (Burton et al. 2007). Since then, GWASs have come into widespread use, and more than 2000 significant associations between SNP genotypes and complex human traits have been identified to date (Hindorff et al. 2009).

Why did it take so long for GWASs to come into general use? Two major challenges had to be met to make this development possible. First, a comprehensive map of the SNPs present in modern human populations was needed. Second, the cost of SNP genotyping had to become sufficiently low to enable researchers to undertake studies of sufficiently large size to discover novel associations.

The first challenge was addressed by a large collaborative effort known as the International HapMap Project (International HapMap Consortium 2005; Frazer et al. 2007). In this project, researchers resequenced biological samples from individuals in three different populations from around the world (specifically, 30 parent–child Yoruban trios from Ibadan, Nigeria; 30 trios of U.S. residents from Utah with northern and western European ancestry; 45 unrelated individuals from Tokyo, Japan; and 45 unrelated Han Chinese from Beijing, China). The purpose of the resequencing efforts was to identify genetic variants whose minor allele occurred at >1% frequency in one of the HapMap populations. Once a sufficient number of these SNPs had been identified, genotypes were determined at each known SNP for all individuals from the three HapMap populations. The key outcome of the HapMap project was a comprehensive map of allele frequencies and correlations among alleles in each population.

The challenge of lowering the costs of GWASs was met by rapid advances in the development of **high-density DNA microarrays** for measuring SNP genotypes. As a result of those advances, the cost of SNP genotyping has fallen dramatically during the last decade. Although SNP genotyping may be replaced by exome and whole-genome sequencing technologies in the next few years, the current low cost of SNP genotyping ($200 or less per sample) makes this approach still very attractive for studies involving thousands of samples.

How Can Genome-Wide Association Studies Work When Only a Small Subset of Known SNPs Is Measured?

As of this writing, several million SNPs have been identified in the human genome. However, the genotyping technologies currently used in GWASs typically assess SNP variation at only a small fraction of known polymorphic loci. For example, the early WTCCC studies used a genotyping platform that assessed variation at only ~500,000 SNPs, a practice that is still fairly common today. If one assumes that the total number of SNPs in the human genome is roughly 10 million, then a panel with 500,000 markers covers only 5% of all SNPs. Given this situation, it may seem odd that GWASs have been so successful at identifying disease-predisposing genetic variations!

In practice, however, the alleles at specific SNPs in a particular population may be correlated with one another. In particular, if the allele of one SNP for a given individual in that population is known, the alleles of nearby SNPs can often be predicted. These patterns of nonrandom association in large populations are referred to as **linkage disequilibrium** (LD). Informally, when the allele present at one SNP is associated with the allele present at another SNP in a particular population, we say that the two SNPs are "in LD" with each other.

There are multiple reasons why SNPs may occur in LD in a given population. For example, close proximity of two SNPs on a chromosome can lead to correlation because those two SNPs are more likely to stay together on the same chromosome during the process of meiosis than are two SNPs that lie far apart. As you may recall, the genetic contributions from each parent are kept physically separate as distinct chromosomes within each cell of an individual. When DNA is replicated in germ cells during meiosis, crossing-over and exchange of genetic material occur among homologous chromosomes, so that the new chromosomes are produced as mosaics consisting of long, alternating chunks from each of the parental chromosomes. The specific locations in a germline chromosome where there is a change in the parent of origin are known as **recombinations**. Typically, only 20–50 recombinations occur across all chromosomes per meiosis. Hence, in the short term, statistical associations among physically proximate SNPs will remain strong, causing LD. Over generations, these associations generally fade as more recombination events take place (a process known as **linkage equilibrium**).

Why is LD important? Even though several million SNPs have been identified to date, many SNPs carry redundant information because of the strong patterns of LD between them. Therefore, in GWASs, the SNPs included on a genotyping panel are usually limited to a set of a **tag SNPs**, which are specifically selected because they are good predictors of most of the other SNPs in the genome. For example, the Illumina 610 k genotyping chip assays only 610,000 markers but provides good coverage ($r^2 \geq 0.8$, a measure of correlation) of 87% of all SNPs in the European HapMap population (Spencer et al. 2009). Given any true association between a phenotype and an untyped SNP (i.e., one that is missing from a genotyping panel), if there exists a tag SNP on the panel in sufficiently high LD with the untyped marker, chances are good that the tag SNP will also be associated with the phenotype. Linkage disequilbrium, therefore, enables association studies to capture most of the information content in the genome using only a small sample of the SNPs identified to date (see Fig. 3).

How Do Genome-Wide Association Studies Differ from Linkage Studies?

Before GWASs, **linkage analysis** was the most widely used statistical technique for identifying disease-causing genes in humans. Many different types of linkage analysis methods have been used, but all involve typing genetic markers from individuals belonging to one or more family pedigrees in which multiple individuals are affected by a disease and identifying genomic regions whose pattern of inheritance (as inferred from the typed genetic markers) appears to correlate with the pattern of disease status in the pedigree.

Linkage studies are based on exploiting linkage—the phenomenon in which genetic loci located close to one another on the same chromosome tend to be inherited together from one generation to the next. In the context of a linkage study, regions whose patterns of inheritance closely correlate with patterns of disease transmission in a pedigree are more likely to be located close to the genes responsible for disease.

Linkage studies and association studies are similar in that both look at statistical associations between phenotypes and genotypes in a collection of individuals. However, the two approaches differ in several key ways.

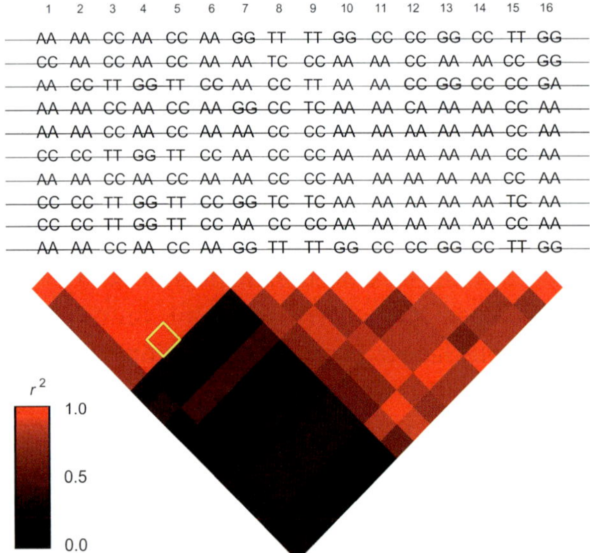

FIGURE 3. Illustration of linkage disequilibrium (LD). In the diagram, each row of the table represents the genotype calls for 10 individuals (labeled A–J) at 16 consecutive SNPs in the genome. Beneath the table is a heat map depicting the pattern of correlations (i.e., LD) in genotype calls between all pairs of SNPs in this region. The box outlined in yellow, for example, represents the correlation between SNP 3 and SNP 6. Based on the pattern of LD in this region, if one knew an individual's genotype at SNP 3, one could impute (i.e., infer) the genotype of that individual at SNP 6 by noticing that individuals with CC at SNP 3 tend to have AA at SNP 6, and individuals with TT at SNP 3 tend to have CC at SNP 6. In practice, genotyping platforms often take advantage of LD structure by selecting tag SNPs with good correlation with as much of the genome as possible, given a constraint on the number of SNPs allowed on the platform. In the example shown, the SNPs shown appear to be divided into roughly two LD blocks with strong correlations between SNPs 1–6 and SNPs 7–16. Hence, a tagging algorithm might select one SNP from each block as an approximate proxy for the other SNPs in that block.

1. Linkage analyses focus on shared genomic regions in closely related individuals, whereas GWASs look at specific sequence differences among unrelated individuals.
2. Because recombination events are so infrequent, disease-associated regions found in a linkage analysis tend to be extremely large, spanning several million positions in the genome. In contrast, LD diminishes rapidly with physical distance and the number of generations since the disease-associated (or protective) allele was introduced.
3. For linkage studies involving multiple separate pedigrees, sharing of alleles is assessed only within pedigrees and not across different pedigrees. As a result, linkage studies can find associations even in cases in which no single allele shows a consistent direction of effect in all pedigrees (consistent with the hypothesis that there exists a causal SNP that appears on different genetic backgrounds within each pedigree). In contrast, GWASs test for only one direction of

allele effect, so that any heterogeneity of effect across individuals impairs the method's ability to identify the association.

Today, GWASs have largely supplanted linkage studies as the method of choice for studying complex traits and diseases. This development can be attributed to several factors. First, linkage studies could be accomplished using relatively low-density marker sets, whereas GWASs became possible only with the availability of high-density genotyping panels. Second, GWASs usually provide much-higher-resolution localization of association signals, owing to the limited extent of linkage disequilibrium on a population level. Third, linkage studies typically provide good power for detecting disease-causing genes only of very strong effect, whereas similarly sized GWASs can often detect SNP associations of substantially smaller effect (Risch and Merikangas 1996).

It should be pointed out that GWASs are not the right choice of study design for every situation. For example, diseases known to be caused by rare mutations are very difficult to study by genome-wide association because of the low frequency of the disease-causing mutations in the general population. In such cases, a linkage study is more likely to be successful because a rare mutation is more likely to be found in families in which the disease occurs in an inherited fashion than it would be in the general population.

HOW ARE GENOME-WIDE ASSOCIATION STUDIES CONDUCTED?

In this section, we provide a brief description of the main steps in a GWAS, from measuring genotypes to conducting tests of association.

Genotype Calling

Compared with many other types of biological assays, genotyping technologies are remarkably consistent in their measurements. Modern genotyping technologies routinely quote accuracy statistics in excess of 99.9% in determining (or "calling") sequence information. However, even a 0.1% error rate can translate to thousands of incorrectly measured SNPs when one is simultaneously assessing millions of SNPs at a time.

To understand how errors arise, we take a brief look at how SNP genotyping technology works. DNA microarrays are high-throughput biochemical assays using millions of single-stranded oligonucleotide **probes** attached to a glass slide (i.e., the genotyping chip). Each of the probes on a standard genotyping chip is designed to bind specifically (through complementary base pairing) to a unique sequence within the genome. Different SNP genotyping platforms differ in the manner by which the probe identifies the target SNP. One popular approach (used in the popular Illumina platform) is **primer extension**, in which the probe is designed to bind to DNA upstream of the SNP of interest. A minisequencing reaction using nucleotides labeled with different fluorescent dyes is then used to determine the identity of the allele (i.e., major or minor) at the SNP itself (Oliphant et al. 2002). Each allele binds a different complementary nucleotide and thus fluoresces a different color, and the relative intensities of the two colors can be used to identify the allele present at any given genomic location.

Graphically, the fluorescence intensities for a particular SNP across a large collection of individuals can be depicted on a two-dimensional (2D) plot known as a **cluster** or **intensity plot**, as shown in Figure 4. In this plot, each dot corresponds to a single individual, and the *x* and *y* positions of each dot correspond to the fluorescence intensities for the major and minor alleles. For well-behaving DNA probes, intensity plots typically contain points grouped into three tight clusters: one with signal concentrated along the major allele axis (indicating **homozygous** individuals with two copies of the major allele), one with signal concentrated along the minor allele axis (indicating homozygous individuals with two copies of the minor allele), and one with signal along both axes (indicating **heterozygous** individuals with one copy of each allele). For less common SNPs (or SNPs for which the minor allele has a lethal phenotype), the homozygous cluster for the minor allele may be absent.

Sometimes, however, clusters for each genotype are not well defined. This situation can occur for many reasons, including small sample size (e.g., in the case of extremely rare minor alleles), problems with the chemistry of a particular probe (such as unintended RNA secondary-structure formation that prevents proper binding of the probe to the DNA target), or nonspecific probe binding (i.e., the probe binds to locations in the genome other than the intended target SNP).

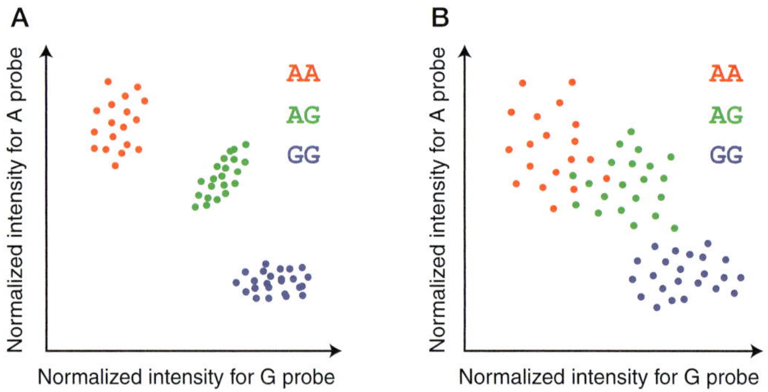

FIGURE 4. Cluster plots. (*A*) The example cluster plot on the *left* shows a well-behaving SNP with tight clusters. Each point corresponds to the genotype call for a single individual; individuals with AA, AG, and GG genotypes, respectively, form three visually separate clusters. (*B*) In contrast, the clusters on the *right* are poorly defined and there exists significant overlap among clusters. Genotype calls assigned on the basis of point location in the *right* plot, therefore, can be expected to be far less reliable than those based on the *left* plot.

Sometimes these problems can be diagnosed visually by identifying overlapping clusters in the intensity plots. In other situations (as when a probe binds to a different SNP than it was targeted to), confirmatory sequencing of the larger genomic region surrounding the SNP must be performed to ensure that the correct SNP is being identified.

Practically speaking, however, one cannot expect to manually inspect genotyping quality for every SNP in a modern panel. To address the potential concerns of call quality, most GWASs rely on various heuristics to analyze SNPs as a preprocessing step before performing a GWAS. This approach will reduce the number of cluster plots that must be examined after association tests are completed. These techniques include exclusion of SNPs with low call rates (which may indicate that not all measurements fall within well-defined clusters) or low minor allele frequencies (clusters for SNPs with low minor allele frequencies tend to include very few points and hence may not be well defined either).

Another commonly used quality filter involves looking for SNPs whose allele distributions are inconsistent with the assumption of random mating (also known as **Hardy–Weinberg equilibrium**). Specifically, for a SNP whose minor allele occurs with frequency p in a population, then the expected number of individuals (from a sample of size n) with 0, 1, or 2 copies of the minor allele would be $n(1-p)^2$, $2np(1-p)$, and np^2 in populations in which pairing of alleles occurs randomly. The random mating assumption does not hold for all populations—there are many cases in which **assortative mating** occurs (i.e., individuals mate preferentially based on specific genetic characteristics). However, very significant departures from Hardy–Weinberg are usually a good indication that a technical failure has occurred during the genotype calling process (Xu et al. 2002). None of these filter-based methods are foolproof, and they may often remove SNPs that are actually correctly called; nonetheless, they greatly reduce the number of false positives in an association study (Weale 2010).

Association Testing

Once genotypes and phenotypes are available, the mechanics of association testing are fairly straightforward. In the simplest cases, the strength of an association can be measured using any standard statistical test of independence for contingency tables. Conceptually, statistical association tests involve three key steps.

1. Compute a test statistic T based on the data.
2. Determine the distribution of T under a null hypothesis of no association.
3. Compare the value of T determined from the data with the null distribution of T.

We discuss each of these steps in the context of the telepathy example given above; each of these steps is also illustrated in Figure 5.

A Observed counts

	AA	AG	GG	Total	Proportion
Cases	38	310	602	950	0.475
Controls	12	400	638	1050	0.525
Total	50	710	1240	2000	1.000
Proportion	0.025	0.355	0.620	1.000	

B Expected counts

	AA	AG	GG
Cases	2000(0.475)(0.025) = 23.75	2000(0.475)(0.355) = 337.25	2000(0.475)(0.620) = 589.00
Controls	2000(0.525)(0.025) = 26.25	2000(0.525)(0.355) = 372.75	2000(0.525)(0.620) = 651.00

C Computing the test statistic

$$T = \sum_{i=1}^{r}\sum_{j=1}^{c}\frac{(O_{i,j} - E_{i,j})^2}{E_{i,j}}$$

$$= \frac{(38.000 - 23.750)^2}{23.750} + \frac{(310.000 - 337.250)^2}{337.250} + \frac{(602.000 - 589.000)^2}{589.000} + \frac{(12.000 - 26.250)^2}{26.250} + \frac{(400.000 - 372.750)^2}{372.750} + \frac{(638.000 - 651.000)^2}{651.000}$$

$$\approx 8.55 + 2.20 + 0.29 + 7.74 + 1.99 + 0.26$$

$$\approx 21.03.$$

D Distribution of test statistic under the null hypothesis

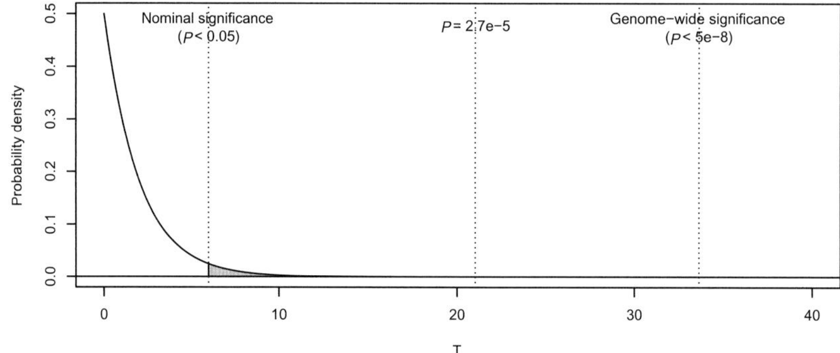

FIGURE 5. Example of statistical calculation. This figure outlines the steps involved in a basic X^2 test of association of a common variant with a binary phenotype. (*A,B*) Illustration of how one can use the observed row sums and column sums of the 2 × 3 table of genotype counts to compute the counts that would be expected under a null hypothesis of no association. These values are used to compute a test statistic (*C*) that is compared against a critical value (i.e., 5.99 for nominal significance or 33.62 for genome-wide significance in this example). Here, the association observed appears to be nominally significant (i.e., it would be considered significant if this were the only hypothesis being tested) but fails to meet genome-wide significance (i.e., after adjusting for the large number of simultaneous hypotheses being tested in a typical GWAS). (*D*) This step also illustrates the common step of converting test statistics into *p* values. Generally, *p* values simplify the process of significance assessment because their critical values do not depend on the particular statistical test being performed. In practice, genome-wide significance generally corresponds to a fixed *p*-value threshold of 5×10^{-8}, although the critical values needed for genome-wide significance vary depending on the exact properties of the test statistic used.

1. **Compute a test statistic *T* based on the data.** In our telepathy example, we use the 3×3 contingency table indicating the frequencies of different combinations of genotypes and phenotype to compute a quantity, known as the **test statistic**, that provides a measure of the strength of the association represented by the observed counts in the table. In the popular X^2 (chi squared) **test of statistical independence**, the formula for T as a function of the counts n_{ij} in the ith row and jth column of the contingency table can be explicitly written as

$$T = \sum_{i=1}^{2} \sum_{j=1}^{3} \frac{(O_{ij} - E_{ij})^2}{E_{ij}},$$

where E_{ij} is the expected number of counts for the cell in the ith row and jth column in a model where genotype and phenotype are assumed to be unrelated or **statistically independent**. The calculation of T is illustrated for our telepathy contingency table in Figure 5C.

2. **Determine the distribution of *T* under a null hypothesis of no association.** Now, suppose we examined contingency tables for a large collection of SNPs that are known to have no true association with telepathy. If we were to restrict our attention to SNPs with genotype frequencies similar to the one in Figure 5, we might see contingency tables such as the following:

	AA	AC	CC	Total	Proportion
Cases	23	330	597	950	0.475
Controls	25	360	665	1050	0.525
Total	48	690	1262	2000	1.000
Proportion	0.024	0.345	0.631	1.000	

or

	GG	GT	TT	Total	Proportion
Cases	20	340	590	950	0.475
Controls	30	370	650	1050	0.525
Total	50	710	1240	2000	1.000
Proportion	0.025	0.355	0.620	1.000	

For each of these random contingency tables, we can use our formula for T to compute a test statistic as well ($T \sim 0.05$ and $T \sim 1.17$ for the two tables shown above).

More generally, it turns out that under certain mathematical assumptions, one can show that if one looks at sufficiently many contingency tables representing nonassociated SNPs with genotype frequencies similar to that of the SNP being tested, the distribution of T tends to look like a χ^2 random variable with two degrees of freedom.

3. **Compare the value of T determined from the data with the null distribution of T.** If the observed test statistic differs sufficiently from the typical values calculated for the random contingency tables, one may conclude that the observed test statistic is highly unlikely to be consistent with a model in which genotypes and phenotypes are independent—that is, that one can reject the null hypothesis.

At first glance, one can see that the value of the test statistic T that we computed for our original SNP ($T \approx 21.03$) is much larger than the values for the two random tables above. But how can we quantify this difference and determine if it is large enough to be potentially meaningful?

One way to accomplish this task is to ask: In the distribution of T values for the collection of SNPs known to have no association (i.e., in the null hypothesis, Case 2 above), how often does one observe a T value greater than the one seen for the observed data? This key quantity is known as a p value, and the smaller the p value for a statistical test, the stronger the evidence of an association. Figure 5D depicts this graphically by showing the distribution of T values under the null hypothesis for our telepathy example (i.e., a χ^2 distribution with two degrees of freedom). It should be evident that the fraction of times a T statistic >21.03 would occur by chance is very small ($p \approx 2.7 \times 10^{-5}$, corresponding to roughly one in every $1/p \sim 37{,}000$ random contingency tables).

How small of a p value is usually considered to be significant? The standard convention in most scientific investigations is that a p value of 0.05 or less should be deemed significant. For genetic association studies, a threshold of $p < 0.05$ is often referred to as a **nominal significance threshold**. A threshold of $p < 0.05$ means that the probability that any particular nonassociated SNP is declared significant (i.e., a **type 1 error**) is at most 5%.

In a GWAS, however, often hundreds of thousands or even millions of markers are being tested simultaneously. Even for diseases where there exist no true associations for any SNP marker, roughly 5% of all SNPs will have contingency tables whose test statistics generate p values below 0.05. This is a huge problem for GWASs, because $(0.05) \times (1{,}000{,}000) = (50{,}000)$ spurious associations per study is far too many to be useful for any biologist! Clearly then, nominal significance is not a strict enough criterion to prevent an abundance of false-positive results in GWASs. This situation is known in statistics as the **multiple hypothesis testing** (or **multiple testing**) problem.

The most common solution to the multiple testing problem is to require an even higher threshold for **genome-wide significance** than one would use to test the association of a single SNP. In particular, for most GWASs, $p < 5 \times 10^{-8}$ is commonly accepted as a reasonable choice. Roughly speaking, such a threshold means that in any study where roughly 1 million independent SNPs are being tested, one would expect to see $(5 \times 10^{-8}) \times (1{,}000{,}000) = 0.05$ significant associations per study (or conversely, roughly one false association out of every 20 studies). Altering the significance threshold in this manner is referred to as applying a Bonferroni correction (in this case, "a Bonferroni correction assuming 1 million tests"). Such a significance threshold may seem to be quite austere, but such strict requirements are, in fact, necessary to counteract the great potential for false associations.

Dealing with the multiple testing problem is an active area of statistical research, and several alternative approaches have been suggested (Storey and Tibshirani 2003; Manly et al. 2004). In general, however, the extremely stringent significance levels needed for genome-wide significance mean that, quite often, sample sizes involving thousands of individuals are needed to detect associations of moderate effect.

Improving Statistical Power

The probability that a GWAS will identify a true association is known as the **statistical power** of the study. Formal power calculations are the process of analyzing a study design to assess its statistical power as it relates to several factors, including the number of individual samples included in the study, the expected differences in genotype frequencies between cases and controls (i.e., the odds ratio), the frequency of the minor allele, and phenotyping error rates (i.e., the frequency with which cases and controls are incorrectly labeled). Power calculations can be laborious, but are necessary to ensure that a GWAS is worth conducting before committing significant resources to it and for setting expectations about the results to be expected.

A number of different techniques are used to ensure that the most statistical power possible is extracted from a given genotype and phenotype data set. These techniques include the following.

1. **Alternate tests of association**. Although the χ^2 test of association described above is fairly straightforward, it is not the only way of determining statistical association. In fact, the development of tests for association is an active area of research in the statistical community. Other well-known tests include the **Cochran-Armitage trend test**, the **likelihood ratio test**, **Fisher's exact test**, and **permutation tests**.

   The differences in how each of these tests define the test statistic T can sometimes have a profound effect on the properties of the resulting statistical test. For example, certain choices of T may be quick and easy to compute and thus may be advantageous in situations in which limited computing resources are available. Other choices may yield statistical tests that are much more likely to distinguish true associations from false associations. Finally, other tests may allow for more flexibility in the choice of T but may require computationally intensive sampling schemes to evaluate the null distribution of T, due to the lack of known **closed-form expressions**. Closed-form expressions are explicit formulas for describing the shape of the distribution. Choosing between different statistical tests is a complex endeavor and should generally be done in consultation with a trained statistician.

2. **Testing for associations with alternate types of variation.** Although SNPs are by far the most common genetic variation studied in GWASs, they are not the only kind amenable to this type of analysis. Others include insertions and deletions (known collectively as **indels**) (Mills et al. 2006) and sequence repeats for which the number of copies of the repeat differs among individuals (known as **copy-number variants** or CNVs) (McCarroll and Altshuler 2007). These genetic variations can often be inferred to some extent from the output of a genotyping chip. For example, deletions can sometimes be detected by observing the intensity plots of consecutive SNPs, and copy-number variants show up as alterations in the magnitude of signal intensity.

   For any given SNP genotype, identifying the parent-of-origin for each allele is generally difficult with current genotyping technologies. In some cases, however, computational algorithms can be used to determine whether specific alleles from groups of SNPs lying close to one another are likely to have been inherited from the same parent (even though it may not be possible to determine which parent was the source of these alleles) (Browning and Browning 2011). Regions of SNP alleles that are inherited in aggregate from the same parent are known as **haplotypes**, and the problem of determining which alleles belong to the same haplotype is known as **haplotype phase inference**. **Haplotype blocks**, which are long stretches of DNA containing SNP alleles that tend to be inherited together, may also be analyzed as individual units in an association study (Daly et al. 2001; Gabriel et al. 2002).

3. **Genotype imputation**. During the last several years, drastic reductions in the cost of technologies for sequencing **whole genomes** (or at least the protein-coding components, known as **exomes**) have led to massive increases in the availability of sequence data for various reference populations. The best known of these data catalogs is the 1000 Genomes Project (Altshuler et al. 2010), an international effort to produce publicly available sequence data for thousands of individuals from multiple populations across the world. Although phenotype data are not generally available for the individuals in the 1000 Genomes Project, the sequence data are nonetheless valuable for augmenting existing GWASs through the use of **genotype imputation** (Li et al. 2009; Marchini and Howie 2010).

   Genotype imputation is a process in which data from individuals typed on a small SNP genotyping panel (i.e., the **study sample**, such as the individuals in a GWAS) are compared against data from an independent set of individuals for whom full sequences or high-density SNP genotypes are available (i.e., the **reference panel**, such as the 1000 Genomes data set). Two types of SNPs are of interest in the problem of genotype imputation: SNPs that are *typed* in both the study panel and the reference panel, and SNPs that are *untyped* in the study panel but typed in the reference panel. By studying the statistical correlations between these two sets of SNPs in the reference panel, one can build predictive models that allow one to statistically infer the most likely genotype for each untyped SNP in the study panel based on the typed SNPs that are present. The end result of imputation is a set of very-high-density SNP genotype data for individuals in the study panel, inferred using the statistical correlations among SNP loci seen in the reference panel (see Fig. 6).

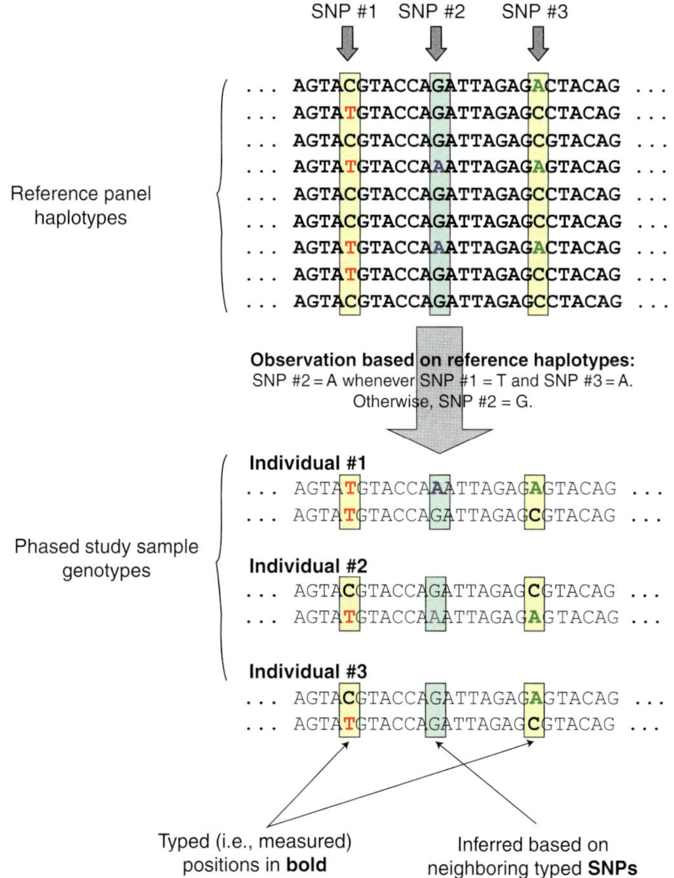

FIGURE 6. Illustration of genotype imputation. (*Top*) Haplotypes from a reference panel are used to infer a rule for predicting the allele at SNP #2 based on the alleles at SNPs #1 and #3. This rule is then applied to the phased study sample genotypes (i.e., genotypes to which a computational algorithm has been applied infer the most likely haplotypes that gave rise to the observed genotypes). In the example shown, the study sample has only been typed at the positions shown in bold (i.e., SNPs #1 and #3), and therefore the rule inferred above must be used to predict the allele at the remaining SNPs (i.e., SNP #2).

In general, genotype imputation techniques cannot predict genotypes for untyped markers with perfect accuracy, but provided that false genotypes are uncorrelated with the phenotype of interest, this method should not generate systematically false-positive results. The great advantage of imputation is that it provides a much higher-resolution picture of association based on existing typed data and, in some cases, may reveal the presence of an untyped variant with a much stronger association effect size than any typed variant on the particular genotyping array used in the study.

4. **Meta-analysis**. One of the biggest barriers to obtaining significant results is accruing cohorts that are sufficiently large to detect SNPs of small or moderate effect. In a recent trend toward realizing this goal, geneticists in multiple disease communities have begun to pool resources to conduct large-scale genetic analyses that combine information from multiple separate data sets. Studies of this kind, which involve aggregating resources from multiple separate GWASs, are known as **meta-analyses** (de Bakker et al. 2008).

   Meta-analysis can often lead to dramatic increases in statistical power; however, it can also give rise to many complications. For example, genotype data from the cohorts involved may be acquired on different genotyping platforms. Differences in populations or subtle variations in phenotype inclusion and exclusion criteria may also give rise to heterogeneity of SNP effects across different cohorts (because each cohort may be revealing the effect on an SNP on a distinct subphenotype). Data-sharing constraints imposed on the genotype data from individual cohorts (as a result of ethical or privacy-related concerns) can also be an important factor. In general, however, the advantages of performing meta-analyses are so great that research consortia find ways to resolve these challenges. For example, genotype imputation can be used to resolve issues with differing genotyping platforms, and statistical techniques for dealing with heterogeneity and combining aggregated test statistics (without combining the underlying individual-level genotype data) have become popular techniques for conducting meta-analyses.

Interpreting Genetic Associations

What does a genetic association mean? As a purely statistical matter, a genetic association is a significant difference in the genotype frequencies for an SNP between case and control individuals in a GWAS. Identifying associations, although mathematically involved, is a straightforward application of statistical techniques to a large-scale data set. But what does one then do with the list of associations identified in a GWAS?

The usual motivation for running an association study is to identify genetic differences related to the etiology of a disease or that provide insight into a specific biological pathway. The SNPs associated with a particular phenotype are often good candidates for being biologically interesting, but it is important to realize that there are a plethora of other reasons that an SNP genotype may be statistically associated with a phenotype. Unfortunately, in most of these cases, the reason has little to do with the biological question that motivated the study. In this section, we discuss several scenarios in which a noninteresting association may arise.

Bad Luck

Sometimes, spurious genetic associations appear by chance. This should not happen often if proper statistical standards are in place to deal with multiple hypothesis testing. In fact, Lander and Kruglyak (1995) established guidelines for reporting results for linkage studies that are still commonly used today in the context of GWASs. Of particular note, Lander and Kruglyak considered associations to be "significant" (or more precisely, of genome-wide significance) only if the degree of statistical evidence shown for them would occur by chance across all markers in an entire GWAS only 5% of the time. This significance criterion is essentially identical to the p-value threshold of 5×10^{-8} that we mentioned above.

Adhering to this standard ensures that the vast majority of associations from GWASs are not due to mere chance. Nonetheless, as a sobering reminder, recall that even at this level of stringency, one would expect to see a spurious association at the rate of one per 20 studies conducted. Therefore, a common practice in many GWASs is to require replication of findings in an independent cohort (using a looser Bonferroni correction in the replication sampled, based only on the number of attempted replications) before they are accepted to be of likely biological significance. Independent replication using samples in an independent population, sometimes using different genotyping technology, can provide a robust way to ensure that an association signal has broad biological significance.

Poor Data Quality

Today, SNP genotyping chips are among the most accurate biological assay technologies available, with accuracy and reproducibility well above 99% in many cases. However, keep in mind that GWASs typically analyze more than 500,000 SNPs in more than 1000 individuals, for a total of more than 500 million genotypes; even a 0.1% error rate (a lower-bound estimate for even the best platforms) therefore yields 500,000 genotype errors. Furthermore, errors in SNP array data are often not randomly distributed across SNPs or across samples, and even slight biases in genotyping error rates can manifest as highly significant but biologically inauthentic genetic associations when these biases are associated with case-control status.

A simple example occurs in case-control studies in which there exists some systematic difference in the genotyping of cases and controls, as may occur when shared or pregenotyped control sets are used. The use of different DNA extraction protocols and/or genotyping platforms for case and control samples can result in differences in genotype calling in the two sets of samples. Also, genotyping samples are often processed in batches, and if errors occur in one or more batches of samples that include particularly high proportions of either cases or controls, this situation can result in unintended associations. Although incorrect calls can be bad news, spurious associations can arise even if no incorrect calls are made! Most modern genotype-calling algorithms will abstain from calling a genotype in cases where the call seems ambiguous. But if "no calls" do not occur equally frequently among the various genotypes for a given SNP, this would have an effect on allele frequencies that might again be susceptible to differences between cases and controls.

Several steps are commonly taken to avoid data quality problems in a GWAS. The first is to make sure to use a high-quality genotyping chip! Another good idea is to randomly genotype cases and controls together so as to avoid systematic differences between case and control genotyping. Once the data have been collected, use of proper quality-control measures, such as those mentioned in the section "Genotype Calling," should also help to limit the effect of bad SNP calls (i.e., enforcing a minimum minor allele frequency, avoiding SNPs with high no-call rates, removing SNPs with significant departures from Hardy–Weinberg equilibrium). Another useful diagnostic used to check for biases due to differential no-calling is to test for the association of no-call status with phenotype; SNPs that show up as significant are potential culprits for otherwise unexplained statistical associations.

Finally, Manhattan plots can sometimes be used to help diagnose genotyping errors. A Manhattan plot is a two-dimensional (2D) graph in which the horizontal axis corresponds to genomic position, the vertical axis represents the strength of an association signal, and each point in the graph represents the association strength of a single SNP. Manhattan plots are useful for providing a high-level summary of a GWAS in that they allow quick visualization of the regions with very significant associations. They are also useful as quality-control diagnostics because LD near association peaks in a Manhattan plot will typically cause SNPs near those peaks to show moderate association with the phenotype, and the strength of other nearby associations will taper off with increasing distance from the main association signal (see Fig. 7). The absence of this characteristic "trail" of associations near an association peak is often an indicator that the association is a false positive due to genotyping error.

Population Stratification

The term population stratification refers to situations in which differences in the genetic characteristics of case and control groups unrelated to the phenotype of interest lead to a significant association signal. For example, the case and control

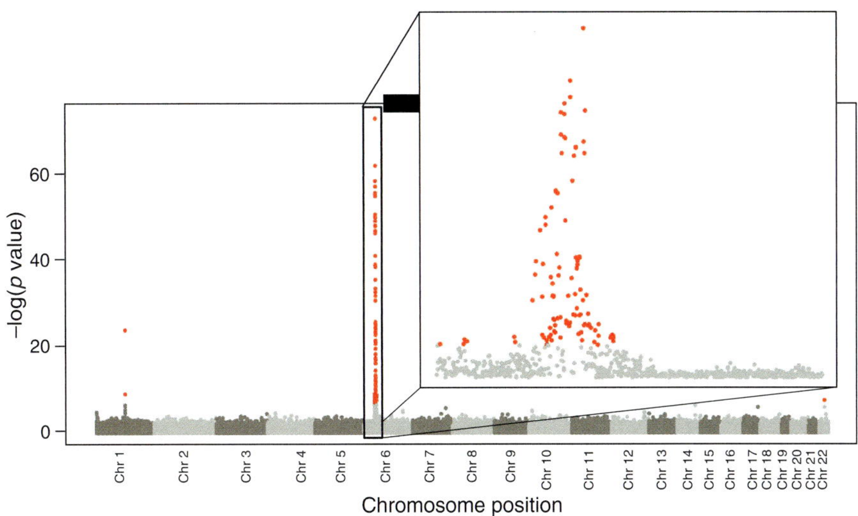

FIGURE 7. Manhattan plot. Each point in the diagram represents an SNP tested in a GWAS, where the horizontal position of the point indicates the location of the SNP in the genome, and the vertical position is the negative logarithm (base 10) of the SNP's association p value. Highly associated regions appear as spikes in the diagram. (Red) SNPs meeting the criteria for genome-wide significance (typically, $p < 5 \times 10^{-8}$). As shown in the *inset*, these spikes tend to consist of multiple SNPs in an LD block that are all correlated with the disease phenotype owing to their high correlation with each other.

groups may differ in their composition with respect to gender or ethnic backgroud. Mathematically speaking, an association due to stratification represents a statistically significant genetic difference between cases and controls. From the perspective of a GWAS practitioner, however, such differences are usually not interesting because they are usually not directly relevant to the phenotype being tested.

A classic example of population stratification can be seen in the association of SNPs in the *LCT* gene (involved in lactase persistence) with height in European American populations (Campbell et al. 2005). Here, the association is driven, not by any actual effect of *LCT* variations on height, but rather by the simple fact that both *LCT* genotype and height vary widely across different European populations. Ancestry-based confounding can be particularly troublesome for interpreting genetic associations with phenotypes such as height, because individuals living in different countries are exposed to different environmental factors (e.g., diet), and these environmental differences are responsible for the phenotypic variation. In this example, confounding made it difficult to determine whether individuals from northern Europe are taller than individuals from southern Europe because of their differing environments or because of genetic differences at the *LCT* gene.

Population stratification is often considered the issue most likely to confound statistical analysis in a GWAS. This presumption derives from the fact that biologically uninteresting associations due to population stratification are often more subtle and difficult to detect than those arising from other factors and has led to the use of terms such as "cryptic population structure."

A standard approach for detecting population stratification is to build what is commonly known as a **quantile–quantile** (QQ) plot. QQ plots are based on the property that in a standard association test, in situations in which no true associations exist, the *p* values observed in practice will be uniformly distributed between 0 and 1. (This property is the reason that 5% of the time, one would expect to see a *p* value of 0.05 or lower by chance.) In most genetic association studies, it is usually safe to assume that of the hundreds of thousands or millions of SNPs being tested, the vast majority do not have true associations with phenotype. Hence, most SNPs should have *p* values that are, for the most part, roughly uniformly distributed between 0 and 1.

QQ plots provide a simple visual way to compare the distribution of observed *p* values seen in a GWAS (across all SNPs) with a theoretically uniform *p*-value distribution. They work by pairing the sorted values of $-\log_{10}(p)$ observed in a GWAS against the sorted values of $-\log_{10}(p)$ for a uniform distribution. Under the null hypothesis, this plot should roughly resemble a 45° line on a 2D graph. When population stratification is present, however, more SNPs have substantially lower *p* values than expected by chance, and this shows up as a significant deviation from the 45° line (see Fig. 8). This deviation is also sometimes characterized by a quantity known as the **genomic inflation factor** (λ) for a study, which is the ratio of observed median χ^2 test statistic from a GWAS to the expected median χ^2 test

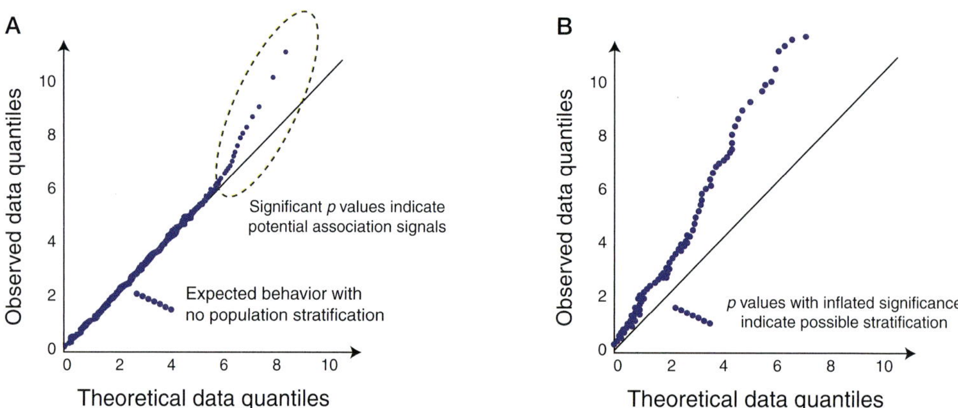

FIGURE 8. Quantile–quantile (QQ) plot. (*A*,*B*) Illustrative QQ plots for a GWAS in the absence (*A*) and presence (*B*) of population stratification, respectively. The *x*-axes of each plot show the expected distribution of $-\log_{10}(p)$ values according to the null hypothesis, and the *y*-axes show the observed $-\log_{10}(p)$ values. Because most SNPs measured in a GWAS would not be expected to be associated with the phenotype being studied, one would expect to observe a good fit to the null model for vast majority of SNPs. (*A*) This result is shown in the *left* plot, where most SNPs are shown to adhere closely to the 45° line in the QQ plot. (*B*) In contrast, in the *right* plot, the observed $-\log_{10}(p)$ values are larger than would be expected according to the null model, suggesting a potential role for population stratification.

statistic under the null hypothesis. In situations in which population stratification is absent, $\lambda \approx 1$.

Correcting for population structure when it is present can be done in several ways (Cardon and Palmer 2003; Price et al. 2010). For example, some statistical tests for association include a way to disregard additional information whose correlation with the phenotype should be factored out of any genotype–phenotype association; this approach is known as **covariate analysis**. Popular ways of determining appropriate covariates to control for population structure include using self-identified or computationally identified ethnicity (through the use of genetic clustering algorithms [Alexander et al. 2009] or principal components analysis [Price et al. 2006]). Another strategy is to perform a matched analysis in which each case is paired with a fixed number of control individuals of similar ethnic background, age, and gender. Alternatively, one can perform a stratified analysis in which results from separate association analyses performed within "strata" of individuals of similar ethnic background, age, and gender are combined together in a single meta-analysis. Other methods have also been developed for controlling population structure, such as the genomic control approach in which the test statistics used in a standard association test are rescaled to remove the effects of population stratification bias (Devlin and Roeder 1999). Fundamentally, each of these strategies attempts to remove uninteresting differences between cases and controls from the returned associations.

Causal Genetic Factor

After the other possible explanations for an apparent genotype–phenotype association have been ruled out, one can more seriously consider the possibility that the association is of biological interest (i.e., that the gene in question is directly involved in generating the phenotype). As always, there are caveats to interpretations of causality. For instance, SNPs lying near one another in the genome tend to be highly correlated with each other because of LD. So, when one does observe a strong genotype–phenotype association in a GWAS, it is often unclear whether the associated SNP is actually a causal factor in determining disease risk or whether it just happens to lie nearby the true causal SNP (which is often untyped). The latter possibility is often the most likely, given the very small proportion of SNPs tested on most genotyping panels. Furthermore, many modern genotyping panels are designed specifically to provide good coverage of the genome via LD rather than to include SNPs that are likely to have major functional roles.

Even though causality cannot be guaranteed for an identified association, the region near an association signal is typically a good place to start looking for the true causal genetic factor. Fine-mapping studies that rely on genotype imputation can provide more precise localization of association signals. Cross-referencing an SNP against the abundance of functional annotation data available through efforts such as the Enyclopedia of DNA Elements (ENCODE) project (Birney et al. 2007) can provide additional clues as to the potential biological significance of an observed association single. Computational servers for performing automated SNP annotation, such as SIFT (sorting intolerant from tolerant) (Ng and Henikoff 2001) and PolyPhen (Ramensky et al. 2002; Adzhubei et al. 2010), can also offer valuable insights.

Ultimately, however, the final test of the biological significance of an association is to perform direct experimental studies (often in animal models to evaluate the functional roles of the genes near an association signal). Because this last stage is extremely labor-intensive, computational strategies for prioritizing association hits can be crucial for reducing the number of false trails explored (Karchin 2009; Cantor et al. 2010).

CONCLUDING THOUGHTS

In the interest of keeping things simple, there are topics that we do not address in this chapter. For instance, we have omitted a number of topics that are usually covered in reviews of the practical aspects of GWASs (e.g., multistage designs, Bayesian techniques, the transmission-disequilibrium test, and gene-based tests).

In recent years, scientists have begun to look beyond GWASs in the search for additional power to detect statistically significant genotype–phenotype associations. These efforts have led to the development of sophisticated haplotype analysis techniques for studying multi-SNP associations, aggregation-based tests for finding rare variants, gene–gene and gene–environment interaction studies for moving beyond the traditional one genotype–one phenotype models, and methods for handling the increasing deluge of whole-genome and exome sequencing data.

As prices drop, SNP genotyping chips may be replaced by whole-genome sequencing as the data collection method of choice. As sample sizes increase, the current focus on developing sophisticated genotyping methods may be supplanted by the need for techniques capable of analyzing the huge amount of data generated by those methods. As more is discovered about the genetics of health, genetics-based predictive models may become an integral part of a personalized approach to healthcare. It is difficult to predict what the future will bring, but it should be clear from this chapter how genome-wide association studies have contributed to our knowledge and how they will likely continue to have a role in bringing new insights into the biology of human genetics.

REFERENCES

Adzhubei IA, Schmidt S, Peshkin L, Ramensky VE, Gerasimova A, Bork P, Kondrashov AS, Sunyaev SR. 2010. A method and server for predicting damaging missense mutations. *Nat Methods* **7:** 248–249.

Alexander DH, Novembre J, Lange K. 2009. Fast model-based estimation of ancestry in unrelated individuals. *Genome Res* **19:** 1655–1664.

Altshuler D, Durbin RM, Abecasis GR, Bentley DR, Chakravarti A, Clark AG, Collins FS, De La Vega FM, Donnelly P, Egholm M, et al. 2010. A map of human genome variation from population-scale sequencing. *Nature* **467:** 1061–1073.

Balding DJ. 2006. A tutorial on statistical methods for population association studies. *Nat Rev Genet* **7:** 781–791.

Birney E, Stamatoyannopoulos JA, Dutta A, Guigo R, Gingeras TR, Margulies EH, Weng Z, Snyder M, Dermitzakis ET, Thurman RE, et al. 2007. Identification and analysis of functional elements in 1 human genome by the ENCODE pilot project. *Nature* **447:** 799–816.

Browning SR, Browning BL. 2011. Haplotype phasing: Existing methods and new developments. *Nat Rev Genet* **12:** 703–714.

Burton PR, Clayton DG, Cardon LR, Craddock N, Deloukas P, Duncanson A, Kwiatkowski DP, McCarthy MI, Ouwehand WH, Samani NJ, et al. 2007. Genome-wide association study of 14,000 cases of seven common diseases and 3,000 shared controls. *Nature* **447:** 661–678.

Campbell CD, Ogburn EL, Lunetta KL, Lyon HN, Freedman ML, Groop LC, Altshuler D, Ardlie KG, Hirschhorn JN. 2005. Demonstrating stratification in a European American population. *Nat Genet* **37:** 868–872.

Cantor RM, Lange K, Sinsheimer JS. 2010. Prioritizing GWAS results: A review of statistical methods and recommendations for their application. *Am J Hum Genet* **86:** 6–22.

Cardon LR, Palmer LJ. 2003. Population stratification and spurious allelic association. *Lancet* **361:** 598–604.

Carlson CS, Eberle MA, Kruglyak L, Nickerson DA. 2004. Mapping complex disease loci in whole-genome association studies. *Nature* **429:** 446–452.

Cichon S, Craddock N, Daly M, Faraone SV, Gejman PV, Kelsoe J, Lehner T, Levinson DF, Moran A, Sklar P, et al. 2009. Genomewide association studies: History, rationale, and prospects for psychiatric disorders. *Am J Psychiatry* **166:** 540–556.

Daly MJ, Rioux JD, Schaffner SF, Hudson TJ, Lander ES. 2001. High-resolution haplotype structure in the human genome. *Nat Genet* **29:** 229–232.

de Bakker PI, Ferreira MA, Jia X, Neale BM, Raychaudhuri S, Voight BF. 2008. Practical aspects of imputation-driven meta-analysis of genome-wide association studies. *Hum Mol Genet* **17:** R122–R128.

Devlin B, Roeder K. 1999. Genomic control for association studies. *Biometrics* **55:** 997–1004.

Dewan A, Liu M, Hartman S, Zhang SS, Liu DT, Zhao C, Tam PO, Chan WM, Lam DS, Snyder M, et al. 2006. HTRA1 promoter polymorphism in wet age-related macular degeneration. *Science* **314:** 989–992.

Donnelly P. 2008. Progress and challenges in genome-wide association studies in humans. *Nature* **456:** 728–731.

Eiberg H, Troelsen J, Nielsen M, Mikkelsen A, Mengel-From J, Kjaer KW, Hansen L. 2008. Blue eye color in humans may be caused by a perfectly associated founder mutation in a regulatory element located within the HERC2 gene inhibiting OCA2 expression. *Hum Genet* **123:** 177–187.

Frazer KA, Ballinger DG, Cox DR, Hinds DA, Stuve LL, Gibbs RA, Belmont JW, Boudreau A, Hardenbol P, Leal SM, et al. 2007. A second generation human haplotype map of over 3.1 million SNPs. *Nature* **449:** 851–861.

Gabriel SB, Schaffner SF, Nguyen H, Moore JM, Roy J, Blumenstiel B, Higgins J, DeFelice M, Lochner A, Faggart M, et al. 2002. The structure of haplotype blocks in the human genome. *Science* **296:** 2225–2229.

Hindorff LA, Sethupathy P, Junkins HA, Ramos EM, Mehta JP, Collins FS, Manolio TA. 2009. Potential etiologic and functional implications of genome-wide association loci for human diseases and traits. *Proc Natl Acad Sci* **106:** 9362–9367.

Hirschhorn JN, Daly MJ. 2005. Genome-wide association studies for common diseases and complex traits. *Nat Rev Genet* **6:** 95–108.

International HapMap Consortium. 2005. A haplotype map of the human genome. *Nature* **437:** 1299–1320.

Karchin R. 2009. Next generation tools for the annotation of human SNPs. *Brief Bioinformatics* **10:** 35–52.

Kayser M, Liu F, Janssens AC, Rivadeneira F, Lao O, van Duijn K, Vermeulen M, Arp P, Jhamai MM, van Ijcken WF, et al. 2008. Three genome-wide association studies and a linkage analysis identify HERC2 as a human iris color gene. *Am J Hum Genet* **82:** 411–423.

Kruglyak L. 2008. The road to genome-wide association studies. *Nat Rev Genet* **9:** 314–318.

Lander E, Kruglyak L. 1995. Genetic dissection of complex traits: Guidelines for interpreting and reporting linkage results. *Nat Genet* **11:** 241–247.

Li Y, Willer C, Sanna S, Abecasis G. 2009. Genotype imputation. *Annu Rev Genomics Hum Genet* **10:** 387–406.

Manly KF, Nettleton D, Hwang JT. 2004. Genomics, prior probability, and statistical tests of multiple hypotheses. *Genome Res* **14:** 997–1001.

Marchini J, Howie B. 2010. Genotype imputation for genome-wide association studies. *Nat Rev Genet* **11:** 499–511.

McCarroll SA, Altshuler DM. 2007. Copy-number variation and association studies of human disease. *Nat Genet* **39:** 37–42.

McCarthy MI, Abecasis GR, Cardon LR, Goldstein DB, Little J, Ioannidis JP, Hirschhorn JN. 2008. Genome-wide association studies for complex traits: Consensus, uncertainty and challenges. *Nat Rev Genet* **9:** 356–369.

Mills RE, Luttig CT, Larkins CE, Beauchamp A, Tsui C, Pittard WS, Devine SE. 2006. An initial map of insertion and deletion (INDEL) variation in the human genome. *Genome Res* **16:** 1182–1190.

Ng PC, Henikoff S. 2001. Predicting deleterious amino acid substitutions. *Genome Res* **11:** 863–874.

Oliphant A, Barker DL, Stuelpnagel JR, Chee MS. 2002. BeadArray technology: Enabling an accurate, cost-effective approach to high-throughput genotyping. *BioTechniques* Suppl: 56–58.

Pearson TA, Manolio TA. 2008. How to interpret a genome-wide association study. *JAMA* **299:** 1335–1344.

Price AL, Patterson NJ, Plenge RM, Weinblatt ME, Shadick NA, Reich D. 2006. Principal components analysis corrects for stratification in genome-wide association studies. *Nat Genet* **38:** 904–909.

Price AL, Zaitlen NA, Reich D, Patterson N. 2010. New approaches to population stratification in genome-wide association studies. *Nat Rev Genet* **11:** 459–463.

Ramensky V, Bork P, Sunyaev S. 2002. Human non-synonymous SNPs: Server and survey. *Nucleic Acids Res* **30:** 3894–3900.

Risch N, Merikangas K. 1996. The future of genetic studies of complex human diseases. *Science* **273:** 1516–1517.

Sherry ST, Ward MH, Kholodov M, Baker J, Phan L, Smigielski EM, Sirotkin K. 2001. dbSNP: The NCBI database of genetic variation. *Nucleic Acids Res* **29**: 308–311.

Spencer CC, Su Z, Donnelly P, Marchini J. 2009. Designing genome-wide association studies: Sample size, power, imputation, and the choice of genotyping chip. *PLoS Genet* **5:** e1000477.

Storey JD, Tibshirani R. 2003. Statistical significance for genomewide studies. *Proc Natl Acad Sci* **100:** 9440–9445.

Sturm RA, Duffy DL, Zhao ZZ, Leite FP, Stark MS, Hayward NK, Martin NG, Montgomery GW. 2008. A single SNP in an evolutionary conserved region within intron 86 of the HERC2 gene determines human blue-brown eye color. *Am J Hum Genet* **82:** 424–431.

Wang WY, Barratt BJ, Clayton DG, Todd JA. 2005. Genome-wide association studies: Theoretical and practical concerns. *Nat Rev Genet* **6:** 109–118.

Weale ME. 2010. Quality control for genome-wide association studies. *Methods Mol Biol* **628:** 341–372.

Xu J, Turner A, Little J, Bleecker ER, Meyers DA. 2002. Positive results in association studies are associated with departure from Hardy–Weinberg equilibrium: Hint for genotyping error? *Hum Genet* **111:** 573–574.

8

Next-Generation Sequencing Technologies

Jesse Rodriguez[1] *and George Asimenos*[2]

[1]Stanford University School of Medicine, Biomedical Informatics Training Program, Stanford, California 94305

[2]Stanford University, Computer Science, Stanford, California 94305

DNA sequencing has become the cornerstone for most biological research. Determining the nucleotide bases in a DNA molecule is essential not only for basic biological experiments but also for extended applications such as forensics, diagnostics, genetic testing, food safety, and others. Originally developed by Fred Sanger in the 1970s, the chain-termination method (commonly referred to as Sanger sequencing) became the predominant sequencing technology (Sanger et al. 1977). Modern automated instruments use capillary electrophoresis to separate DNA fragments by size and fluorescently labeled chain terminators to determine nucleotide content. State-of-the-art Sanger sequencers can produce 0.5 Gb/day in reads of up to 1000 bases long. This extended read length allows easy sequencing of PCR amplicons and facilitates de novo genome sequencing thanks to higher read overlaps. Sanger sequencing was used to sequence the human genome and is still in use because of its broad deployment. However, it is somewhat labor-intensive, and the relatively limited throughput and high cost have shifted the focus away from this classical method and toward newer technologies.

The new generation of sequencing technologies offers very high throughputs and low costs, at the expense of read length. A comparison of the leading next-generation sequencing (NGS) solutions (as of November 2010) is shown in Table 1; each technology is explained in further detail in the following sections. For a more detailed discussion of each technique, you are directed to the review article by Elaine Mardis (Mardis 2008) and the overview by Elaine Mardis and Dick McCombie (Mardis and McCombie 2012).

TABLE 1. Summary of the major next-generation sequencing technologies

Technology	Read length (bp)	Throughput (Gb/day)	Reagent cost[a]
454 FLX	400[b]	1	$16/Mb
Illumina HiSeq2000	100[c]	25	$60/Gb
ABI SOLiD 5500xl	75	30[d]	$33/Gb

[a]Based on list price for reagents; excludes library prepping and initial cost of instrument.
[b]Roche has announced plans to increase read length to 1000 bp.
[c]Illumina has announced plans to increase read length to 150 bp and throughput to 50 Gb/day.
[d]ABI has announced plans to increase throughput to 45 Gb/day.

454 FLX

In 2004, the 454 system was the first next-generation sequencer to be introduced commercially. It allows for DNA molecules to be massively sequenced in parallel after being attached to beads, amplified with emulsion PCR (amplification of the DNA inside water droplets in an oil solution), and deposited into wells. Sequencing is performed using Pyrosequencing, a method that works by synthesizing the complementary strand of the original DNA molecule, which releases pyrophosphates during nucleotide incorporation. This results in a light signal produced by luciferase that is detected by a CCD camera; the intensity of the light generated is proportional to the number of nucleotides incorporated. A common weakness of this process is the imprecise strength of the light signal related to the incorporation of a homopolymer stretch (a run of many identical nucleotides). This often causes an incorrect estimation of the exact length of the homopolymer stretch, leading to increased read error rates in homopolymers in the form of extra inserted or deleted letters (when these reads are aligned back to the reference genome). Although 454 FLX allows for longer reads than competing technologies, reagent costs are substantially higher, limiting this technology to applications that benefit from the increased read lengths, such as de novo sequencing, HLA typing, food safety and biosafety, and metagenomics.

Illumina

In the Illumina sequencing pipeline, DNA molecules from a prepared sequencing library are first amplified into clonal clusters on a flow cell using the Illumina cBot Cluster Generation System. The flow cells are then sequenced in the Illumina sequencing instrument. The sequencing technology (originally developed by Solexa) uses reversible terminator chemistry to sequence DNA molecules by synthesizing them cycle-by-cycle with fluorescently labeled terminators. At the end of each cycle, the dye and terminating groups are cleaved, allowing the process to move to the next cycle. Of the technologies mentioned in Table 1, the Illumina sequencing technology is currently the most widely adopted. Low reagent costs combined with acceptable read lengths and a very high throughput make this technology ideal for most NGS applications.

ABI SOLiD

The ABI SOLiD platform is based on sequencing by ligation, using two-base-encoded probes that are fluorescently labeled. At each cycle, a ligation reaction connects the correct probe to the primer (or to a previously incorporated probe). Ligated probes are detected using four-color imaging. The process continues for a few cycles, reading encoded dinucleotide combinations every five bases. This whole round is performed a total of five times, with the initial primer shifted by one location each time so as to interrogate all the bases. Because there are 16 possible dinucleotide combinations but only four different dyes, each color detected in the output represents four possible dinucleotides in the input molecule (Fig. 1). This is known as color space encoding, with the results of the sequencing reported in colors (0 = blue, 1 = green, 2 = orange, 3 = red) rather than nucleotides. A sequence of colors can be decoded to the respective DNA sequence with knowledge of the primer letter, but a single color mistake in the readout can render the decoded sequence incorrectly from that point on. For that reason, downstream bioinformatics processing (such as alignment to a reference genome) is preferably performed in color space. The use of color space generally requires more complicated algorithms and occasionally confuses researchers who are used to working with nucleotide sequences. Nevertheless, the ABI SOLiD technology offers all the advantages of NGS and is an affordable solution.

Alignment

Regardless of the sequencing method used, two major questions drive all NGS studies. First, many people want to use NGS to answer the question: Which portions of the genome were present in my particular DNA sample? Often, this is the result of a

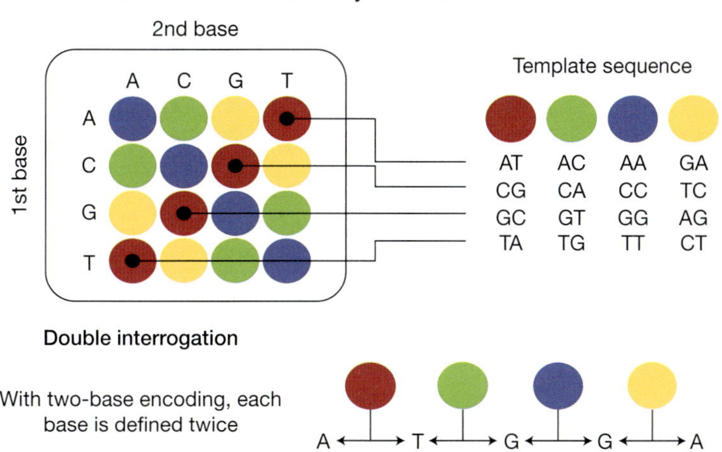

FIGURE 1. The ABI SOLiD color space encoding. (Reproduced, with permission, from Mardis 2008, © Annual Reviews.)

specific experiment such as immunoprecipitation. Others want NGS to answer the question, "What are the contents of my sample's genome at particular locations?" This is useful for performing variation analysis to determine a phenotypic connection to a genotype or for evolutionary tracking, and so on. The sequencing machine outputs only the contents of random bits of the sample's genome, but at the time the machine outputs them, we do not know which bits of the genome were sequenced to produce those reads.

To answer these questions, we must go beyond examining the contents of the reads alone. We can answer both of these questions if we know exactly which read came from which location within the sample's genome. In this way, we could reconstruct the sample's genome at all locations from which reads were produced and would know what parts of the genome were present in our sample. The task to identify where each read originated within the genome is called **alignment** or **mapping**. To do this, we can compare the contents of our reads with the contents of a reference genome (i.e., the consensus genome for a particular species). Mapping is the fundamental task that is required before most downstream analyses with NGS data. Many methods exist today to map NGS reads, but, to understand them, it is helpful to understand a bit about the concepts and the history behind them.

A naive approach to mapping reads would be simply to consider each read independently and ask whether it maps to each location of the genome independently. This would require one to consider (number of reads) * (number of genomic locations). When the size of the genome is 3 billion letters and the number of reads in a sequencing run is on the order of a billion reads, this determination would require 3×10^{18} comparisons. Even if a modern CPU could perform 1 comparison per CPU operation (the fastest CPUs today can perform $\sim 3 \times 10^{9}$ operations/sec), it would take around 30 years to map the reads! The chief problem to mapping short reads is that there are too many reads and too many possible genomic locations to which the reads could map.

In this section, we first discuss some of the motivation for newer short-read-specific methods for aligning reads, then we will cover the approaches of these methods, and finally we will discuss some of the popular methods and the output of those methods.

Can NGS Reads Be BLASTed?

Today, most molecular biologists are familiar with BLAST (basic local alignment search tool) and regularly use it to compare sequences with vast databases of all known genomes relatively quickly. This raises the question: why not just use BLAST to align all the reads to the genome? Originally, this used to be a strategy for traditional Sanger DNA sequencing reads, which produces reads upward of 1000 bp. However, BLAST was originally designed as a way to compare a relatively short sequence (roughly gene-sized) with a genome or any large database of known sequences, but it also works for aligning these long sequencing reads. BLAST can align two DNA sequences of any size and produce alignments with gaps and matches, which makes it incredibly powerful for matching a sequence to a database. This also makes BLAST reasonably appropriate to map the 454 reads, which are longer and contain a relatively high number of insertions or deletions (around homopolymers). For Illumina and SOLiD reads, however, the computational cost of using BLAST is prohibitively large.

APPROACHES TO SHORT-READ MAPPING

Although BLAST is technically capable of aligning the short reads of Illumina and SOLiD, it is impractical to use BLAST on typical short-read data sets because the number of reads is simply too large. BLAST was originally designed to map a small number of longer sequences with a large number of differences rather than a large number of short sequences with a small number of differences. Newer methods have been developed that were specially optimized for the shorter reads in NGS to make them orders of magnitude faster.

In this section, we cover the two major general steps that most methods use in mapping short reads to a reference genome: (1) identifying candidate mapping locations, and (2) scoring and filtering to select the most likely mapping locations for the read.

However, before we get to that, let us consider the properties of NGS reads that make it possible to map them efficiently to very large genomes at all.

Characteristics of Short Reads

When a sequencer outputs a letter, it also outputs a probability that this letter is incorrect. That is, it gives the probability that the letter output by the machine is *not* the same nucleotide of the sample DNA molecule that it is reading. This probability is often encoded with a number called the **quality score**, or **Phred score** (Q) (Ewing et al. 1998a,b), which follows the Phred probability (P) encoding as follows:

$$Q = -10 \log_{10}(P),$$

where P is the base-calling error probability. Higher quality scores mean higher confidence that the base is correct. For example, a Phred score of 20 represents a 99.0% probability that the base is correct, whereas a Phred score of 30 represents 99.9% (i.e., there is a 1 in 100 and 1 in 1000 probability of an incorrect base call, respectively). Typically, Phred scores reach a maximum of 40 because the machines cannot estimate the probability to more than two digits past the decimal. The important thing to note about sequencers and read quality scores is that the bases at the beginning of the read tend to have high-quality scores, which tend to decrease toward the end of

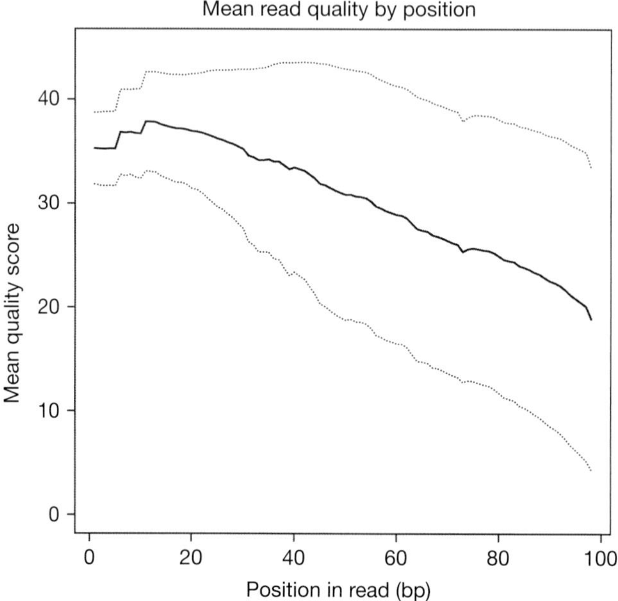

FIGURE 2. Read base quality scores by read position. The solid line represents the mean base quality score for each position within a read. The *upper* and *lower* dotted lines represent the mean plus and minus one standard deviation, respectively.

the read (Fig. 2). This means that when we see the letters of a read, we can usually trust that the letters at the beginning of the read are correct and assume that the letters at the ends of reads may be incorrect.

This property is very important for short-read mapping because it facilitates the first step toward mapping a read: determining a set of possible locations to which the read may map. The following sections describe the steps involved in genome scanning and genomic location determination.

The First Step: Seeding

Conceptually, the first step is to scan the genome to determine a set of genomic locations that are very likely to contain the true mapping location for the read. This first step can be performed with a technique called **seeding**.

Because of the high quality at the start of the reads, we can expect the first ~35 bp of a short read to match nearly perfectly with the genome, allowing for the relatively rare sequencer-induced error or biological single-nucleotide polymorphism (SNP) in the sample. The first step to determining the origin of the read is to extract the first "*s*" (~35 bp) bases of a read (usually called the read's **seed**) and find locations in the genome that are nearly identical to the seed. If we assume that there are no insertions or deletions between the two sequences (e.g., the read and the genome) such that they are the same length, it turns out that there are several very efficient ways of computationally determining whether the sequences have at most "k" differences from one another (when k is small). This fact allows us to quickly identify many regions in the genome where the seed matches with up to a handful of mismatches. These locations are sometimes known as **seed matches** or **candidate mapping locations**. Candidate mapping locations will allow us to dramatically narrow down the set of possible locations in the genome to which the read can map. This quick narrowing down of possible locations through seeding is what makes it possible to map NGS reads, and the ability to use seeding is possible because of the high-quality beginning region of NGS reads.

The Second Step: Scoring and Filtering

We know that each molecule in the sequencing machine originated from exactly one location in the genome, so we would like to assign the read to the single "best" genomic location among our candidate's mapping locations. To do this, we must be able to score each candidate location to compare them against one another to determine which is best. Again, assuming there are no insertions or deletions, we can extend the seed alignment by walking down the read and comparing all of the read's letters against the genome. High-scoring locations will be ones where most of the letters of the read match the letters of the genome. The actual mathematical function to score a given candidate varies widely among methods, but the simplest formula is to count the number of locations that are mismatched. At this point, we need only pick the highest-scoring location and assign the read to that location.

Unfortunately, it is very common that there are ties between multiple candidate locations. In this situation, this read is considered to be **repetitive**. This can occur because genomes, especially large ones such as those of mammals, contain many sequences (longer than reads) that are very similar to one another for various reasons, including transposons, centromeric tandem duplications, gene duplications, or noncoding RNA (ncRNA) targets. In this situation, there are many options that differ among methods. Some will report all the best locations, some, or one location randomly (but will report that the read came from a repetitive region). When the scores are identical, we have no true way of knowing which of the best-scoring locations the read came from, so we have to live with the inherent ambiguity of the result.

Some methods go beyond reporting the top-scoring location and report the relative likelihood of the read originating from one particular candidate location when compared with all other candidate locations. A simple approach to do this is to compute the total score of the read by summing the scores of all candidate locations for each read. Then, to normalize the score of each candidate location, we can divide the score of the location by the total score of the read. After normalization, all the scores for a given read will sum to 1. Under most circumstances, when a read comes from a unique portion of the genome (such as an exon), it will map **uniquely** to the genome. That is, one candidate location will have a very good score (a low number of differences), and any other candidate locations will have a small score. With the normalization scheme described above, the normalized score of the best candidate location for a uniquely mapped read will have a score that is close to 1. If a read comes from a repetitive part of the genome such as a transposable element, it will usually map repetitively to the genome. In this case, it will have many candidate locations with a low score, such that the normalized score of the best candidate location will be close to 0.

Mapping Quality/Posterior Probability

Some methods use a probabilistic framework to interpret scores of the alignments. Instead of the simple normalized score for each location, they want to convey the **posterior probability** that the read mapped to the location. That is, given that the read maps *somewhere* in the genome, what is the probability the read mapped to a specific location? To calculate this, we can use the raw score (i.e., the nonnormalized score) to compute a probability of the read mapping to the specific location (i.e., the probability that it maps to the genome *at all* and that it maps to a specific location) of each candidate location. Then we can normalize all the probabilities by determining the sum of the total probability of all candidate locations. Just like with the normalized score, uniquely mapping reads will have a best location with posterior probability close to 1.0, and repetitively mapping reads will have a best location with posterior probability close to 0. Note that the sum of the posterior probabilities of all the candidate locations will sum to 1.0 for a given read (Fig. 3).

With this posterior probability, we can capture more nuanced situations beyond just unique and repetitive. For example, suppose a read has one candidate location with posterior probability 0.5 and 50 candidate locations each with posterior probability 0.01. If we only looked at the "best" location, we would be ignoring the fact that there is a decent chance that this read mapped to a repetitive region of the genome that had 50 copies in the genome.

On a technical note, some methods encode the posterior probability of a candidate location using the Phred-encoding scheme from above. When represented in

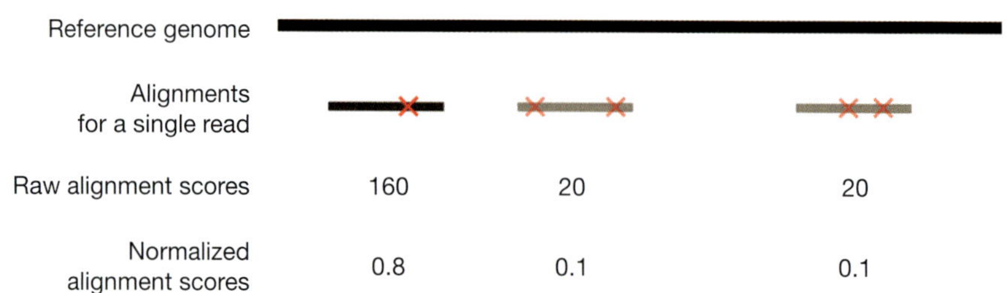

FIGURE 3. Normalizing alignment quality scores. A read is shown with three possible alignments. The red × corresponds to a mismatch in the read with respect to the reference, and the shade reflects the alignment quality score.

this way, it is often referred to as the **mapping quality** of the particular candidate location. Just like with the quality scores of individual read letters, high mapping quality scores indicate high confidence that the candidate location is the correct mapping (they have a low probability of being incorrect).

Using Quality Scores in Mapping

Although all the popular methods use the technique described above, some methods are specialized for higher-quality alignments by using the quality scores within the reads to help break ties between candidate locations that have the same number of mismatches. Because the quality score of a read letter communicates how likely the letter is to be correct, we would prefer not to mismatch letters that have high confidence. For example, suppose a read has two candidate locations—the first has a mismatch at the beginning of the read at a letter with a high-quality score, and the second has a mismatch at the end of the read with a low-quality score. We would prefer the latter one because it is more likely that the read actually originated from this location, but the machine introduced the error because the machine was less confident about the mismatched letter.

Indel Alignment

So far, we have discussed the simple case of ungapped alignment where we do not allow any insertions or deletions in the read, with respect to the genome. A problem arises when using an ungapped method if the read has an **indel** [mutation(s) that results in the insertion or deletion of nucleotides resulting in a net change in the total number of nucleotides] because it causes a shift in all the letters of the read, potentially causing our ungapped method to fail. See below for an illustration of why this is the case.

Sample genome sequence:

GATTGATTACACCATTGCACCACACGGATACATCAGTTTAC

A read from the sample:

CCATTGCACCACACGGA

Reference genome sequence has an extra **T**:

GATTGATTACACCATTG**T**CACCACACGGATACATCAGTTTAC

Ungapped alignment:

GATTGATTACACCATTG**T**CACCACACGGATACATCAGTTTACCCATTGCACCACACGGA

Gapped alignment:

GATTGATTACACCATTG**T**CACCACACGGATACATCAGTTTACCCATTG-CACCACACGGA

If a sample contains an indel with respect to the reference genome, the reads produced that contain that read are very unlikely to be mapped, and, if they are, they may map to the incorrect location. This usually results in sharp drops in read coverage in the vicinity of the indel because, even if the sample contained reads from this region, they cannot be reliably mapped to it.

However, it turns out that ungapped alignment is sufficient for aligning most reads because the vast majority of reads map to the genome without any gaps necessary. This is because (1) indels are fairly rare in biology (most of the organisms that have been sequenced have few indels when comparing one individual's genome to another), and (2) sequencing machines have an extremely low rate of producing erroneous insertions or deletions in a read. However, if one is interested in identifying indels in a sample with respect to a reference (see Variation section) or one is performing an analysis that requires a very smooth distribution of reads without drops in coverage, then mapping with indels is necessary.

Unfortunately, it is much more computationally intensive to identify indels. For this reason, not all methods have indel-mode enabled by default because they compete for which method can map the fastest. Often, methods will have a limit of mapping with one indel per read in an effort to increase the speed of indel alignment. In this scenario, when there are two indels very close to each other (less than one read's length apart), the method will probably be unable to map reads produced from this region of the genome, causing the same characteristic drop in coverage.

Paired-End Alignment

A **paired-end alignment** is one in which a standard single-read DNA library is modified, to facilitate reading both the forward and reverse template strand during one paired-end read. Alignment for paired-end or mate-pair libraries is very similar to that of single-end or fragment libraries. Paired libraries have the advantage that the pairing information can be used to increase the number of reads mapped and detect structural variation in the sample. The main approach is to align both **partners** of a pair independently in the fashion described above because the majority of reads map uniquely to the genome even without pair information. In the event that one read maps uniquely to the genome but the other pair maps repetitively, one can use the unique alignment as an **anchor** that decides which mapping of the repetitive partner to use (Fig. 4). To do this, one must first identify how far from the anchor the repetitive partner is expected to map based on what is known about the insertion size used in the paired library. Therefore, one will first calculate the average **paired distance** (the genomic distance between mappings of partners of a pair) among all pairs where both partners map uniquely on the same chromosome and choose the mapping for the repetitive partner that is most in accordance with the paired distance. Furthermore, when one partner maps uniquely and the other partner does not map at all, the unmapped partner may not map because of high variation that causes more

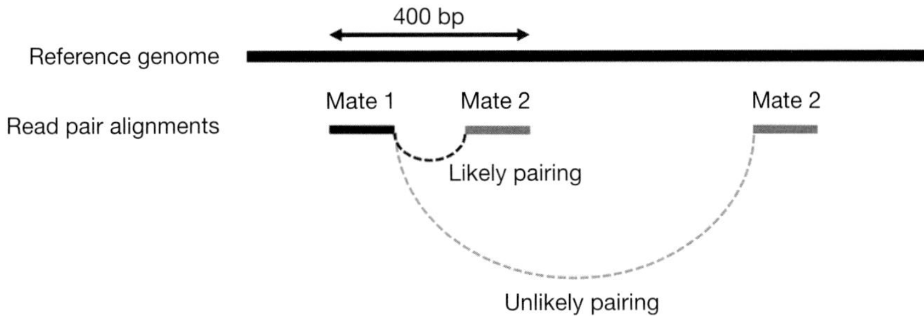

FIGURE 4. Resolving ambiguous paired-end alignments. The alignments for two reads (Mate 1 and Mate 2) that form a read pair are shown here. Mate 1 is shown to have a confident alignment to the reference, but Mate 2 ambiguously maps to two locations equally well. Because the DNA fragments used in this sequencing protocol were ~400 bp, the left alignment for Mate 2 forms a more likely pairing than the right alignment.

mismatches than allowed for the aligner. To get around this, one can perform a more sensitive alignment (e.g., BLAST) to the genomic region near the unique partner to attempt to align the unmapped read. Using the anchored partner to reduce the region of the genome to which to align the unmapped partner is essential because it is slow to align reads to the entire genome with a BLAST-like method and it would produce many false-positive alignments, which gives paired sequencing runs a distinct advantage over standard single-ended libraries.

Mapping Output

Having discussed the techniques used in mapping reads, we can now discuss what to expect from the output of a mapping method. From an abstract point of view, mapping needs to communicate (1) the location(s) from which the read originates (where it maps), (2) the confidence that we have in our prediction, and (3) the details of the alignment (i.e., where are there gaps and/or mismatches?). Although there are many file formats to encapsulate mapping results (and each method tends to output in its own specific format), the most widely used format is the SAM and BAM file formats. The SAM format is a text file that you can read without any special tools. BAM, on the other hand, is a binary file format that is designed to be small and computer readable, so you will need a special program to view the contents of a BAM file. The content of the file formats is the same, however, and one can readily convert between one format and the other. Let us take a look at the basic information you will find in a SAM file. Each entry represents a candidate location as we discussed above, so you might see several entries indicating several locations for a particular read.

- ***Read name*** (or pair name if the reads are paired)
- ***Flag.*** Special number that encodes some details about the alignment
- ***Chromosome.*** The name of the chromosome/contig/sequence to which the read was mapped

- ***Position.*** Leftmost position in the chromosome to which the read overlaps
- ***Mapping quality.*** Phred-encoded score indicating the confidence of this candidate location
- ***CIGAR.*** Details of the alignment: Where are all the mismatches and indels?
- ***Mate chromosome*** (if paired)
- ***Mate position*** (if paired)
- ***Inferred insert size*** (if paired)
- ***Sequence of the read*** (copied from the original input file)
- ***Quality string of the read*** (copied from the original input file)
- ***Various columns.*** Indicating number of mismatches, whether a read is repetitive, and so on

SAM files have a great deal of flexibility because of the last columns allowing for custom flags and properties so that each program that outputs SAM files can store its unique set of properties about each alignment. The most popular tool for working with these files is **SAMtools**, which provides various programs for manipulating, filtering, converting, and visualizing (in a text-based command-line terminal) SAM alignment files. It can also be used to perform some tertiary analyses such as calling variants and the consensus sequence of a sample.

Practical Considerations

The many commonly used short alignment programs include the Burroughs–Wheeler aligner (BWA), Bowtie, ELANDv2, and SOAP (Short Oligonucleotide Analysis Package) (Table 2). Each has its own particular set of capabilities, parameters, and options, but they are similar in many ways. All represent the latest generation of short-read aligners that are capable of sensitively aligning reads that are smaller than ~100 bp with small indels (note—with the exception of Bowtie) and multiple mismatches with exceptional computational efficiency. For longer reads that are several hundred base pairs in length, like 454 reads, a different class of

TABLE 2. Alignment programs and references

Popular aligners for short reads (Illumina/SOLiD)		Popular aligners for long reads (454)	
BWA	Li and Durbin 2009	BLAT	Kent 2002
Bowtie	Langmead et al. 2009	BLAST	Altschul et al. 1990; Ning et al. 2001
ELANDv2	MJ Bauer, AJ Cox, DJ Evers (unpubl.)	SSAHA2	Ning et al. 2001
SOAP	Li et al. 2009	BWA-SW	Li and Durbin 2010
MOSAIK	Lee et al. 2013; arXiv:1309.1149		

aligners must be used. Although BLAST is capable of aligning these reads, it is far from the fastest method. The most common longer read aligners are BLAT (BLAST-like alignment tool) and SSAHA2 (sequence search and alignment by hashing algorithm), which align reads much more slowly than short-read aligners owing to the length of the reads. Newer methods such as BWA-SW (Burrows–Wheeler aligner, Smith–Waterman alignment) and MOSAIK have been designed to be more computationally efficient at aligning longer reads and may significantly reduce the amount of resources needed to align longer reads.

Despite the varying approaches and read sizes between these aligners, they all have similar parameters to control the sensitivity, specificity, and the running time of the aligner (Li and Homer 2010). These common parameters control the number of allowed mismatches, limit the number and length of allowed indels, and provide various heuristics such as built-in read trimming. When many scientists purchase a sequencing run, they often want to extract the most information from the data, so they will choose parameters to increase the sensitivity of these programs, ultimately increasing the number of reads mapped. Setting alignment parameters to be more permissive of mismatches and indels can be necessary to map reads that occur in regions of extremely high biological variation (e.g., those regions having many SNPs within 50 bp of each other), but it also introduces many more spurious mappings of lower-quality reads that contain errors due to sequencing, not biological variation. Depending on the application, these polluting reads can throw off downstream analyses, so it is prudent to consider the effect on the specificity of the mapper when tuning it to achieve higher sensitivity. Furthermore, because permissive settings significantly increase the computational time of these programs, if one is in a hurry or has limited computational resources, using more standard or restrictive settings is advisable.

RNA-seq

RNA-seq is an emerging technology that is a versatile tool for analyzing many different properties of the transcriptome, the set of all transcripts of an organism including mRNAs and small RNAs (Wang et al. 2009). The protocol for RNA-seq is to isolate RNA, convert it to cDNA, and sequence the cDNA library with NGS. The method has many applications for studying the transcriptome ranging from examining the catalog of transcribed sequences of a species to performing a comparative analysis of the transcriptome between individuals or across conditions. The major uses for RNA-seq today are novel transcript discovery; transcript quantification; transcript structure analysis such as determining transcripts' beginning, end, and splice locations; and allele-specific expression analysis.

RNA-seq has several advantages over microarrays and qPCR and can be used for novel applications for which neither can be used. First, because RNA-seq does not require genomic primers or probes like PCR and microarrays, RNA-seq can be used for species that do not have a reference genome available. Second, it has a higher dynamic range for transcript quantification than microarrays because it produces an absolute count of read sequences rather than fold change over background, which allows for the detection of rare transcripts expressed at very low levels. Third, it can achieve single-base-pair resolution when analyzing transcript structure, compared with dozens or hundreds of base pairs for microarrays. Finally, if longer-read technologies are used, complex multiexon structures can be elucidated with RNA-seq.

With these diverse applications of RNA-seq come various computational challenges that we will now discuss.

Approaches to Identifying Transcript Structure

With a Reference Genome

If a reference genome exists for the organism of interest, the general approach to identifying the genomic boundaries and splice sites of an organism begins with mapping the reads to either a putative reference transcriptome or the reference genome. We define the **reference transcriptome** as the sequences of the set of annotated and predicted exons along with all annotated and predicted splice junctions. Using the reference transcriptome has the advantage that one can use a standard read mapper because all transcripts will be present as spliced, contiguous sequences. One downside to this approach is that it is limited to known and putative transcribed sequences and splice sites, making it more difficult to capture novel transcriptional structures beyond what has been observed or predicted. The other downside is that one must translate the coordinates of the reference transcriptome to those of the reference genome to perform some analyses. As an alternative, one can map the reads directly to the reference genome, which allows for unbiased discovery of novel splicing events, as well as simplifying analyses based on genomic coordinates. This approach, however, requires that the RNA-seq program perform **split-read alignment** (Fig. 5).

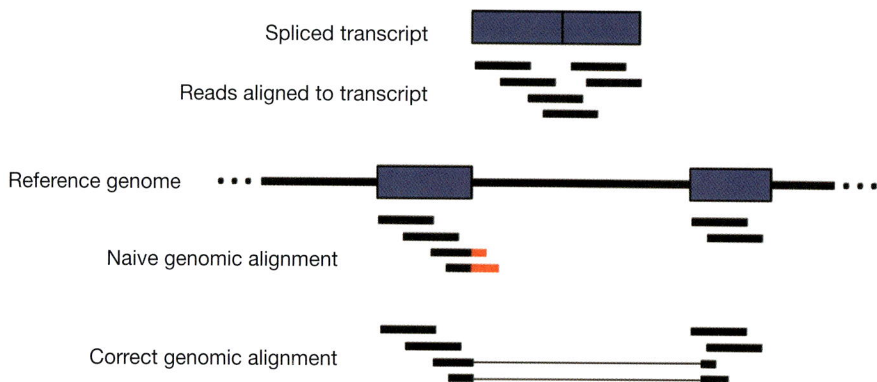

FIGURE 5. Spliced read alignment. The red color indicates the portion of the read that mismatches with the reference genome at that position. A thin line between two parts of a read indicates a long gap in the read's alignment.

Reads that span splice sites must be able to be split into two or more parts and be mapped to separate exons, potentially a great distance away from each other, which makes them computationally more difficult to align than normal reads. RNA-seq programs that attempt to detect novel splice sites on the fly (such as TopHat and QPALMA) do so by first mapping reads to the genome, identifying putative exons by identifying regions enriched with many mapped reads, identifying predicted splice sites (with or without the knowledge of common splice signatures), and, finally, mapping all unmapped reads from the first genomic alignment to the predicted splice junctions. It has been suggested that this approach, although capable of detecting completely novel splice junctions, can miss many known splice sites (Trapnell et al. 2009). TopHat is a hybrid method that is designed to address this problem: It can identify both known splice sites as well as previously unknown sites.

Without a Reference Genome

When a reference genome is not available, the computational task to characterize the transcriptome requires **de novo assembly** of the reads, rather than mapping them to a reference. Assembly programs like Velvet and ABySS aim to construct contiguous sequences (or **contigs**) by identifying reads that **overlap** one another by matching sequences at the ends of the reads as shown below.

Reads overlapping each other:

```
TCTCCTTTGTCGTTATGGAA
          TTGTCGTTATGGAACCGGCGTATC
                    GTTATGGAACCGGCGTATCAATCGG
                             GGAACCGGCGTATCAATCGGCAACAG
                                       GAACCGGCGTATCAATCGGCAACAGT
                                                 ATCGGCAACAGTGGCCAGATAGGGA
```

Consensus contig:

TCTCCTTTGTCGTTATGGAACCGGCGTATCAATCGGCAACAGTGGCCAGATAGGGA

These methods are analogous to the techniques used to construct reference genome sequences like the human genome but are specialized for short reads. In the case of transcriptome sequencing, these assemblers will output contigs that correspond to the spliced transcript sequences, which can be used in downstream analysis to discover genes or splicing characteristics. A caveat to using these methods is that they were originally designed to assemble genomic sequences from reads that are evenly distributed along the entire genome. With RNA-seq, the coverage of transcripts varies over many orders of magnitude, which may cause these programs to be less accurate for assembling transcriptomes, and they may not capture the entire diversity of the splicing repertoire of a sample. On a practical note, it is advantageous to use longer reads such as those from 454 to perform de novo assembly because shorter reads produce more ambiguously overlapping regions. One can exclusively use these longer reads at the expense of lower coverage or one can use a combination of short and longer reads to aid the assembly process.

Transcript Quantification

To use RNA-seq for transcript quantification, one can first map reads to the transcriptome and simply count the number of reads that overlap each transcript. The number of reads that map to each transcript is connected to its size, however, because one long transcript contributes more sequencable fragments to the sequencing library than one short transcript. Therefore, to compare the expression levels of two transcripts in the same experiment, you must divide this count by the length of each transcript yielding expression levels in units of reads-per-base pair. To make the numbers a bit more manageable and round, the length of the transcripts can be recorded in kilobases and we can divide the whole quantity by 1 million, yielding the commonly used unit of the **RPKM**, or reads per kilobase per million reads sequenced. Using RPKM values for each transcript, one can rank the expression levels between genes in the same experiment. However, because there is a great deal of variance between experiments due to RNA quality and library prep conditions, it is unwise to directly compare RPKM values across two RNA-seq experiments: The values must be normalized. Expression value normalization in RNA-seq is extremely important for intergene and intersample comparisons and is a problem very similar to the normalization required for microarrays. One can use one of the common normalization techniques that were originally used in microarray analysis to normalize the RPKM values of RNA-seq. See the microarray chapter (Chapter 6) for more details on these techniques.

Another notable caveat to calculating RPKM values is the problem of repetitive reads: How should they be counted in the RPKM value? Because a read may map to multiple locations when it actually originates from exactly one location, we do not know which transcript's read count to add the read to. Instead of choosing one transcript and adding 1 to its sum, we can distribute a fraction of the value to all transcripts to which the read maps proportional to the number of other reads that map to the transcript.

Table 3 presents a collection of popular software for mapping reads from RNA-seq as well as quantifying transcripts (modified from Pepke et al. 2009). TopHat and Cufflinks (which relies on TopHat) are tools for mapping and detecting novel transcript isoforms for organisms with an assembled reference sequence. QPALMA and ERANGE (which relies on QPALMA) are among the first methods developed for

TABLE 3. Popular software for mapping RNA-seq reads and quantifying transcripts

Program	Primary category	Transcript discovery	Need genomic assembly	Associated read mapper	Splice junctions	Quantitation
ERANGE	Existing and novel gene quantitation	Yes	Yes	BLAT, Bowtie, ELAND	From existing models (novel with BLAT)	RPKM from gene annotations and novel transfrags
QPALMA	Spliced read mapper	Yes	Yes	Integrated	Predicted from transfrags	No
TopHat	Spliced read mapper	Yes	Yes	Bowtie	Predicted from transfrags and optionally from existing models	RPKM from supplied annotations
Cufflinks	Transcript isoform discovery and quantification	Yes	Yes	Bowtie and TopHat	See TopHat	RPKM, FPKM, differential expression
Velvet/ Oases	Short-read assembler	Yes	No	NA	Assembled	Fold coverage
Trans-ABySS	Short-read assembler	Yes	No	NA	Assembled	Read coverage

Modified from Pepke et al. 2009.
(NA) Not applicable.

RNA-seq analysis and are not as computationally efficient as TopHat and Cufflinks. Velvet and ABySS are popular tools for performing genomic assembly of short reads and can also be used for assembling transcriptomes of organisms without a reference genome by using them in conjunction with the Oases and Trans-ABySS packages, respectively.

Advantages of RNA-seq

RNA-seq has proven to be a useful tool for studying many aspects of the transcriptome and will eventually replace microarrays entirely as sequencing technologies improve and prices decrease. A valuable aspect of RNA-seq data is that it can be used and reused for many purposes. We may be initially interested in identifying splice variants, but we can also use the data to perform differential splice variant quantification at the same time. Alternatively, we can identify sequence variations such as SNPs and indels in transcripts, which can then be used to measure allele-specific gene expression. Finally, of course, these analyses can be performed even for species without a reference genome, which makes it ideal for scientists studying nonmodel organisms. As a relative newcomer for transcriptome analysis, RNA-seq will continue to grow as sequencing technologies improve and new methods are developed to analyze this unique and versatile data.

CHIP-seq

Chromatin immunoprecipitation, or ChIP, is a popular technique for isolating DNA from genomic regions of interest that are bound to some protein molecule. A DNA-binding protein (DBP) of interest is cross-linked to its DNA binding partner in cells (typically performed using formaldehyde fixation because it is heat-reversible). The cells are then lysed, followed by fragmentation of the DNA into short pieces. The protein and its short bound DNA fragments are isolated by immunoprecipitation, and the cross-linking is reversed in order to separate the DNA from the protein. At this point, the DNA fragments can be used either to quantify or identify the presence of some region of interest.

Common DBPs of interest used with ChIP include transcription factors (TFs), RNA polymerase, and histones; however, ChIP can be used to extract the genomic region bound by any protein that binds DNA. This extraction can facilitate identifying regulatory relationships of transcription factors, identifying the transcription of noncoding RNA, or examining which regions of the genome are being transcriptionally repressed by histones with a specific posttranslational modification.

When one is interested in quantifying how much binding activity occurs at a particular location of interest, the gold standard technique is **quantitative PCR** (qPCR). By quantifying the number of fragments in the library that contain a location of interest, one can infer how many DNA molecules had a protein bound at that location in the original sample. The primary shortcoming of qPCR is the throughput. One must produce primers and perform qPCR separately for every region of interest, which can take a great deal of time and effort. A newer, higher-throughput method is **ChIP-on-chip** (or ChIP-chip), which allows the multiplexing of thousands of regions of interest simultaneously in one experiment by using a microarray to quantify the abundance of each region of interest. The primary drawback to ChIP-chip is that the data produced are much noisier, providing only rough approximations of the relative abundance of regions rather than absolute abundance like qPCR.

ChIP-seq, as an alternative to ChIP followed by qPCR and ChIP-chip, uses DNA sequencing followed by alignment to the genome to quantify the fragments. One need only count the number of reads that map to a particular location of interest. ChIP-seq is both high throughput and provides absolute quantification. It has the further advantage that it allows one to depart from the paradigm of querying regions of interest: The whole genome is used to identify the location bound to DNA. This allows one to look for a variety of signals. For instance, TFs tend to bind tightly to very small regions, RNA polymerase binds loosely to a wider region (because it actively walks down the DNA strand), and histones bind to vast portions of the genome. Furthermore, because the bound-region boundaries can be identified precisely, this facilitates motif discovery to identify the common DNA sequences bound by a protein, allowing one to make predictions about other binding sites in the genome that were not actively bound in one particular experiment.

The raw data produced by ChIP-seq are not as readily interpretable as are the data from qPCR or ChIP-chip. qPCR generally measures the abundance of one amplicon per gene, and ChIP-chip measures the abundance of a small number of distinct probes for each gene. With ChIP-seq, we are given an alignment of reads across the whole genome where each read contributes, so one must use specialized software to interpret the alignment in order to analyze the data. Here, we discuss how this software works and what caveats to consider when using it.

ChIP-seq Algorithms

Step 1: Smoothed Signal Creation

The input to a ChIP-seq algorithm is the alignment of reads originating from fragments that overlap a region bound by a protein. The algorithm outputs the center and/or the boundaries of bound regions. Intuitively, one would simply need to calculate the coverage over the genome and look for local peaks of high coverage, indicating enrichment in the sample. Unfortunately, there are two issues with the data that prevent a simplistic approach like this. The first problem is that the raw coverage produces a very peaky signal, which can lead to many false peaks that do not correspond to biologically enriched regions. Therefore, all ChIP-seq algorithms perform a **smoothing** operation over the raw coverage to produce a smoothed signal (Fig. 6). One approach to smoothing is to create a **sliding window** of fixed size around

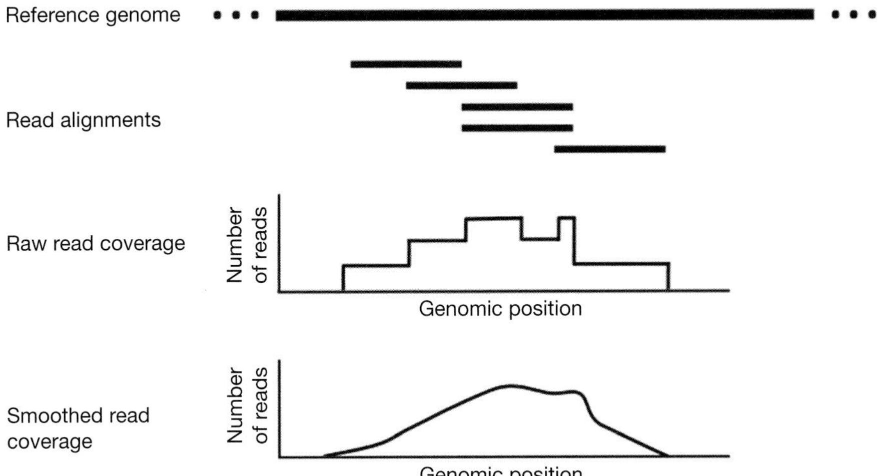

FIGURE 6. From read alignments to smoothed coverage plot. A smoothing function was applied to the raw read coverage plot to produce the smoothed coverage plot.

each position of the genome and set the value of the signal at that position to be the average coverage of all positions in the window. The size of the window can be chosen depending on the desired amount of smoothness and/or the assumed width of the bound region (e.g., smaller for TFs, larger for histones).

Another approach is to spread out the coverage contribution of each read directly using an operation known as **kernel smoothing**. When computing the raw coverage signal, we add 1 to the value of the signal at each location for which a read overlaps, and this is performed for all reads. If the length of the read is N, then each read contributes a total of N to the signal, spread out uniformly across the N positions covered by the read. More formally, we say that the read contributes a density of N uniformly over N positions. Instead, one can distribute the density using a normal distribution centered at the center of the read. One can adjust the width of the distribution again, increasing it for a smoother signal aimed at detecting larger regions like histone regions and decreasing it for a less smooth signal suitable for detecting smaller regions like TFs.

Because most popular methods are aimed at identifying TF-binding signals, we first consider the problem of **peak finding**, which aims at identifying the center of a small binding site, and then we cover **region finding** for broader binding signals.

Step 2: Signal Shifting

Because of the way that ChIP-seq reads are produced, the data require another transformation before they can be used to identify coverage peaks. The problem is that the reads are produced by sequencing the ends of fragments of DNA rather than the entire fragment. Thus, when a fragment overlaps a region, its ends tend to "stick out" past the boundaries of the bound region (Fig. 7A). This has the effect of producing a coverage signal that has two separate peaks: reads that come from the left side of the fragments and reads that come from the right side of the fragments (Fig. 7B). This is problematic because neither peak is centered on the actual binding site. If one is attempting to count the number of DNA molecules bound to a binding site, each peak contains only half of the reads that are actually bound, giving an underestimate of the number of molecules bound, even if the location were correct. However, it turns out that because reads are always sequenced 5′ to 3′, the "left" reads (those with the lower genomic coordinate) map to the forward strand and the "right" reads (those with the higher genomic coordinate) map to the reverse strand (see Fig. 7).

Therefore, we can merge the two separate peaks on the forward and reverse strands into one large peak in between them. We **shift** all the reads in the forward strand composing the "left" peak to the right by (insert_size/2), and shift the reverse reads composing the "right" peak to the left by (insert_size/2). This has the effect of creating one large peak with the correct center on the binding site (Fig. 7C).

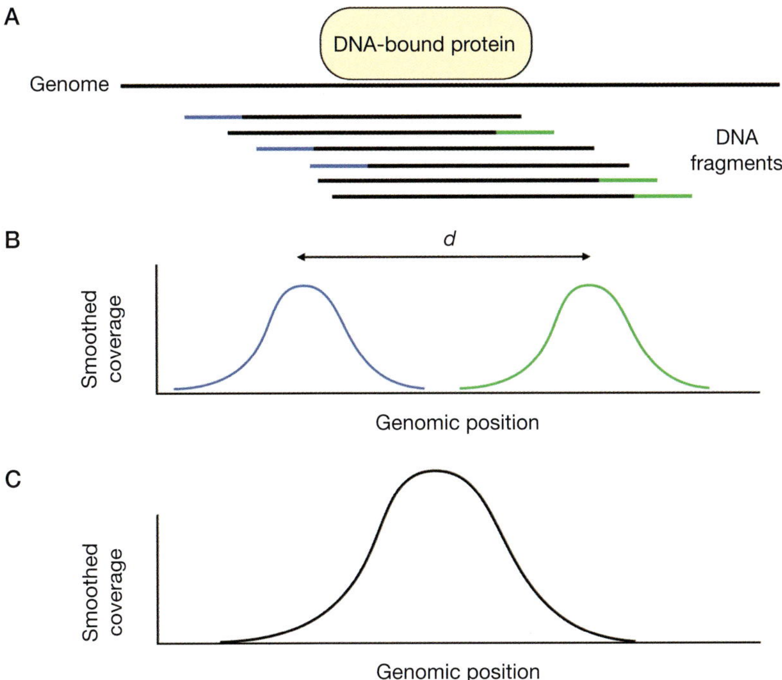

FIGURE 7. Peak shift from a bound protein. (*A*) DNA long fragments are pulled down from a DNA-bound protein, and the ends are sequenced with short reads. The strand of the read is indicated by the color (blue for forward, green for reverse). (*B*) Coverage plots for reads from each strand are calculated separately. (*C*) The separate strand-specific coverage plots are shifted by $d/2$ and summed.

We typically apply this shifting operation to all reads in the data set, regardless of the presence of a local peak. Because nonspecifically sequenced reads do not have a coverage pattern that resembles a binding site, shifting the reads does not have a boosting effect on the coverage—it simply moves the location of the noise signal. Therefore, shifting increases the signal-to-noise ratio by boosting the signal of the true peaks from binding sites while keeping the signal magnitude of the noise constant on average.

Many methods can automatically detect the correct insert size to perform this shift, or they can be set manually if the user knows, with high accuracy, the mean insert size in the sequencing library.

Step 3: Identify Regions of Enrichment

After the signal has been smoothed and shifted, the next step is to identify a set of candidate locations that are enriched for a high number of reads. Because random (nonchip) sequencing can produce regions with many reads simply because of chance, we must identify **significantly enriched** regions that are high coverage with respect to what we expect by a random distribution of the reads. One approach for this step is to measure, at each location, the height of the signal divided by the average signal if one distributed the reads evenly across the genome. This measurement is known as the **fold enrichment** because it indicates the amount of enrichment that the region has when compared with the average signal height. Then, we can specify a threshold in advance above which we consider a location to be a potential binding site.

Step 4: Filtering

At this stage, candidates must be narrowed down to confident peaks. When a background sample is present, some programs allow peaks to be filtered based on the ratio of the experimental signal-to-background signal.[1] The intuition behind this approach is that some parts of the genome may have many reads even in the background, so they should be discarded from the final output.

Some of the candidate locations may be due to artifacts in the sequencing process itself, so they must be filtered out. One common artifact is **one-strand peaks**, which are peaks that contain reads from only one strand. In low coverage, one-strand peaks are expected to occur by random chance fairly frequently, but at the high levels of coverage that the candidate locations have, this is extremely unlikely to occur because of the natural sampling process in sequencing. Another common artifact is **identical-read stacks** caused by PCR amplification bias, where the same amplicon is sequenced with a frequency that is vastly disproportionate when compared with other DNA fragments, causing many reads with the exact same contents to map to the same location with enough frequency to be detected by the enrichment-detection step (Fig. 8). Many methods specifically remove candidate regions that contain either reads from only one strand or overrepresented read sequences from amplification bias.

Step 5: Ranking by Significance

Because ChIP-seq experiments often yield thousands of peaks, the use of a method to rank the peaks is necessary so that the users can quickly identify the "top" peaks in their experiments. Significance determination depends on the particular method and whether a background example is included in the experiment. In the one-sample experiment that does not have a background, one can use a mathematical model as an approximation for a background sample. The most common approach is to calculate a p value for each peak under the assumption of a Poisson distribution. This p value represents the probability that a peak contains at least N reads, assuming that the reads from the ChIP-seq experiment were randomly distributed across the genome according to a Poisson distribution. In the two-sample experiment, we

---

[1] As an alternative to using the background to filter in a postprocess fashion, some methods perform a subtraction of the background from the foreground to reduce the signal used for peak calling.

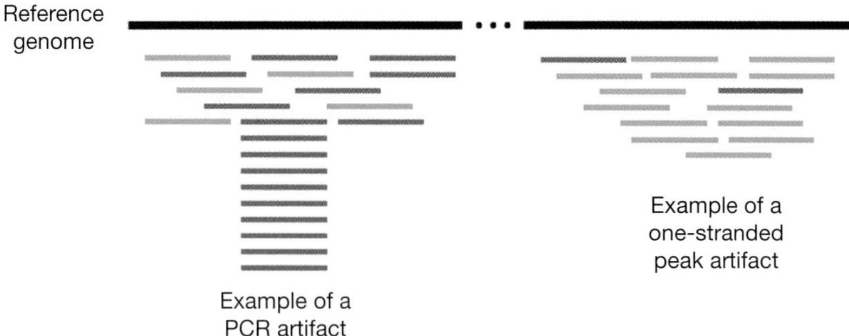

FIGURE 8. Examples of two common types of ChiP-seq artifacts.

can also calculate a **false discovery rate** (FDR) in addition to the *p* value. To do this, we can identify peaks and calculate *p* values in the background sample in the same way we do it for the experiment to identify a set of peaks that "look" significant, but that we know are false positives. Then, we can calculate the FDR for a peak in the experiment based on the number of false positives that occur with a *p* value that is at least as significant as the *p* value for the experimental peak. The list of peaks can then be filtered by a minimum FDR.

Practical Considerations

Now that we have covered how ChIP-seq algorithms work, we can discuss the practical aspects of choosing and running one. There are major factors to consider in choosing a program for ChIP-seq analysis. The first is **performance**, or the sensitivity and specificity of the method. It turns out that for TF peak-finding applications, many of the popular methods perform very similarly, despite having different algorithmic approaches to the different stages of the basic ChIP-seq algorithm. Although the number of peaks output by the methods can vary significantly, when considering the higher-ranking peaks, the methods perform very similarly.

The second factor is the ability of the program to use a control sample to calculate an FDR in order to provide an accurate confidence estimate and help tune parameters to use when running a program. Although almost all methods are capable of accurately estimating relative confidence values (e.g., *p* values) that yield an accurate ranking of peaks, this ranking does not provide an answer to the question, "How many of the reported peaks are estimated to be false positives?" The FDR provides exactly that. It greatly helps in deciding what parameters/thresholds to use when running a ChIP-seq program, as well. With FDR, one can adjust the parameters and observe how many of the called peaks are estimated to be false, in addition to how many peaks are called. Without the FDR, users can only use the number of peaks as a very rough gauge for the accuracy of the called peaks. Note, however, that FDR calculations require a two-sample experiment with a ChIP and a control sample.

The third factor is the **usability** of the program. Most of the programs are written as command-line utilities designed to run on Linux, which may make running the program difficult for many biologists without extensive computer expertise. A few of the more recently developed programs provide graphical user interfaces that are much more user friendly, allowing users to run the programs more quickly and easily.

With these factors in mind, we recommend three ChIP-seq programs that perform very well and are able to calculate an FDR: **MACS** (Zhang et al. 2008), **QuEST** (Valouev et al. 2008), and **CisGenome** (Ji et al. 2008). MACS is probably the most popular tool in use today, and it was among the earliest to be developed. It uses a sliding-window count with a Poisson distribution as the null model to calculate *p* values. One drawback is that it only filters PCR-duplicate peaks and does not filter single-stranded peaks. By default, MACS outputs a large number of peaks, which can be useful for users who are interested in using even low-ranking peaks. QuEST is based on kernel smoothing that also uses a Poisson distribution and can filter both PCR duplicates and single-stranded peaks. QuEST tends to be a bit more conservative than MACS but still outputs a relatively large number of peaks. One drawback to QuEST is that it is more conservative in its requirements for calculating an FDR, needing a fairly large number of control reads to provide an accurate FDR estimation. CisGenome is a sliding-window method like MACS that uses the significant binomial distribution as a null model and can filter both kinds of artifacts. It outputs significantly fewer peaks than the other two methods, by default, which can be advantageous for applications requiring high specificity. Despite having slightly lower performance than the other two methods, the major advantage of CisGenome is that it provides a graphical user interface, making it a great choice for many biologists. For a more in-depth review of ChIP-seq peak detection algorithms, you are directed to Wilbanks and Facciotti (2010).

Figure 9 lists the features of many ChIP-seq software packages in addition to those described here.

Advantages of ChIP-seq

ChIP-seq is a powerful tool for identifying protein-binding locations that leverages NGS. Despite ChIP-seq being less noisy than ChIP-chip, it is very important to use control samples when performing ChIP-seq experiments. The use of controls is critical to estimate the FDR accurately and to filter out peaks with high background signal; note that multiple replicates can be used if maximum peak-calling accuracy is required. Although many of the peak-calling algorithms fare comparably in accuracy, their additional features such as FDR calculations, filtering, and user interface are useful in deciding which one to use. It should be noted that the task of identifying broader regions bound by histones can be performed with the peak finders discussed here, but there are many tools that are more specialized for the tasks that follow a similar approach to peak finding.

Program	Reference	Version	Graphical user interface?	Window-based scan	Tag clustering	Gaussian kernal density estimator	Strand-specific scoring	Peak height or fold enrichment (FE)	Background subtraction	Compensates for genomic duplications or deletions	False discovery rate	Compare to normalized control data (FE)	Compare to statistical model fitted with control data	Statistical model or test
				Generating density profiles			Peak assignment		Adjustments with control data		Significance relative control data			
CisGenome	1	1.1	X*	X				X	X		X		X	Conditional binomial model
Minimal ChipSeq Peak Finder	2	2.0.1			X			X				X		
E-RANGE	3	3.1			X			X				X	X	Chromsome scale Poisson distribution
MACS	4	1.3.5		X				X			X		X	Local Poisson distribution
QuEST	5	2.3				X		X			X**		X	Chromsome scale Poisson distribution
HPeak	6	1.1		X				X					X	Hidden Markov model
Sole-Search	7	1	X	X				X		X			X	One sample *t*-test
PeakSeq	8	1.01			X			X					X	Conditional binomial model
SISSRS	9	0.4		X			X					X		
spp package (wtd & mtc)	10	1.7		X			X		X	X′	X			

X* = windows-only graphical user interface or cross-platform command line interface

X** = optional if sufficient data is available to split control data

X′ = method excludes putative duplicated regions; no treatment of deletions

FIGURE 9. Features of ChIP-seq software packages. References: (1) Ji et al. 2008; (2) Johnson et al. 2007; (3) Mortazavi et al. 2008; (4) Zhang et al. 2008; (5) Valouer et al. 2008; (6) Qin and Shen 2009; (7) Blahnik et al. 2010; (8) Rozowsky et al. 2009; (9) Jothi et al. 2008; (10) Kharchenko et al. 2008. (Reproduced from Wilbanks and Facciotti 2010.)

VARIATION DETECTION

NGS technologies are suited for genome resequencing projects that aim to identify **genomic variants**. Genetic variation is responsible for some of the differences observed between different individuals of the same species, for genetic disorders and susceptibility to certain diseases, and for a lot of other observed traits. Knowledge of genetic variation is important for human ancestry studies, cancer genomics, genome-wide association studies, and other medical research.

Genomic variants can be classified based on their scale.

- At the nucleotide level, there are SNPs consisting of a difference in a single nucleotide, as well as small insertions and deletions, and changes in the lengths of short tandem repeats and microsatellites. A multitude of assays and methods have already been developed to query a limited set of predefined locations in the genome (an experiment known as SNP genotyping). These include hybridization-based solutions such as the Agilent SurePrint and the Affymetrix GeneChip microarrays, primer-extension-based solutions like the Illumina Infinium and the Sequenom iPLEX, and Sanger-based methods like the ABI SNPlex. These methods are suitable for low-cost validation of known SNP locations but cannot scale to discovery of variants in the whole human genome, something for which NGS technologies are quite appropriate.
- On a larger scale, there is structural variation, including **copy-number variants** (CNVs). Copy-number variants are pieces of DNA that differ in their number of copies; these pieces may have been deleted in one or both chromosomes (for diploid genomes) or copied in a tandem fashion. Other structural variants include mobile element insertions, inversions, translocations, and segmental duplications. Copy-number variants can be detected by using **comparative genomic hybridization** (CGH), but this method does not detect translocations, inversions, or other copy-number-preserving types of variation. Structural variation can be identified using NGS technologies, especially when using sequencing protocols that produce **pairs of reads**.
- On the largest scale, whole chromosomal arms may have been fused, deleted, or replicated. Such chromosomal abnormalities are usually detected by karyotyping and are not so much the focus of NGS experiments.

Detecting Nucleotide-Level Variation

Sequencing Considerations

At the nucleotide level, variation detection is performed by aligning the reads to the reference genome and detecting locations where the sequence of the reads differs from the reference. This detection requires a minimum number of reads in order to be accurate, and in general, the higher the sequencing coverage of a sample, the

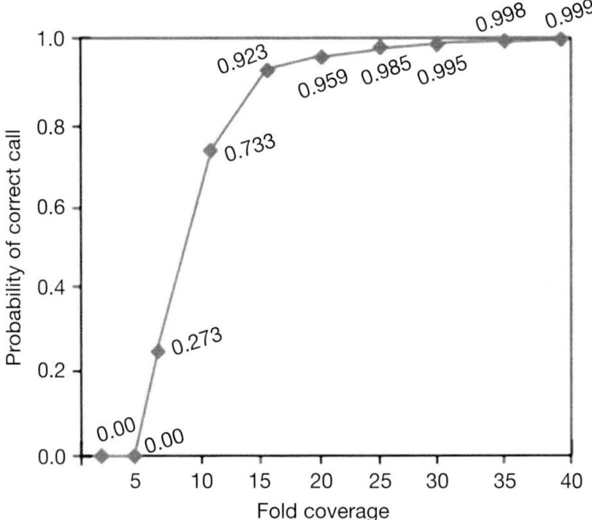

FIGURE 10. Probability of correct SNP call per fold coverage. (Redrawn, with permission, from Illumina 2010, © 2014 Illumina, Inc. All rights reserved.)

more accurate the variation detection will be. Therefore, coverage is usually the most important consideration when performing sequencing for detecting variants. For diploid genomes, a certain variant may be homozygous (the two alleles contain the same sequence) or heterozygous (the two alleles contain different sequences). Because reads are randomly originating from both alleles, heterozygous positions require higher coverage than homozygous because more reads are needed to ensure that both alleles have been sufficiently covered. Figure 10, provided by Illumina, illustrates the probability of correctly detecting a heterozygous variant given a certain fold coverage (number of reads overlapping the variant).

As can be seen, Illumina recommends sequencing a sample at 40× coverage in order to correctly detect variants with 99.9% probability. If sequencing a whole genome at that level of coverage cannot be afforded, researchers can focus on the subset of the genome that they believe is most important and perform a sequencing experiment on that subset alone. This is done by creating a sequencing library that enriches for the regions of interest, using target enrichment techniques such as Agilent's SureSelect and Nimblegen's SeqCap EZ/Sequence Capture Arrays. In particular, these products also offer whole human exome enrichment, allowing customers to sequence all coding regions of the human genome in high coverage.

Methods

As mentioned above, the methodology for detecting variants at the nucleotide level includes aligning the reads to the reference genome and scanning through the alignments for columns that show differences. All companies that develop sequencing technologies have also developed software for detecting variation, which can be used to analyze the output from their instruments. These include Illumina's CASAVA and ABI's BioScope, but open source alternatives are also available. The most popular solution is SAMtools, which includes an aligner, a SNP caller, and a text viewer to browse the results. It is capable of detecting SNPs as well as small insertions and deletions, and reports various statistics that can be of interest, including strand bias (when an allele is not equally distributed between forward and reverse reads) and map quality bias.

Although detecting new variants is the most popular workflow, researchers often find the need to go back to a set of samples and ask whether they contain a certain variant of interest. Because analysis tools tend to report only locations with variation, most people assume that if their variant is not listed, then it is not present in their sample. However, sometimes a variant is not reported because of inadequate sequencing coverage in that region and not because the location contained reads that were equal to the reference genome. Users can go back to the original alignment to see if there was enough coverage for their location of interest, and certain solutions (such as the variation report produced by Complete Genomics) include a distinction between low coverage ("no-call") and sufficient coverage but without variation ("ref").

For cancer samples, detecting variation is more challenging because tumors may amplify or delete certain regions, quickly losing the assumed 50%–50% balance between the two alleles. For cancer genome variation, specialized techniques are still under development, a popular example of which is **SNVMix** (Goya et al. 2010).

Detecting Larger-Scale Variants

Detecting CNVs is usually performed by comparing the coverage of a region with the average level of the chromosome. If a certain region is not covered, and assuming there is no library or sequencing bias that could have led to underrepresentation of that region, then the region may have been fully deleted in the original genome. For diploid genomes, if the coverage of a region is roughly one-half of the average coverage, the region may have been deleted in a heterozygous way. Similarly, if the coverage is a multiple of the average coverage, the region may have been duplicated in the original genome. In practice, these calculations are not as straightforward because natural biases in sequencing technologies (due to GC content and other factors) lead to nonuniform coverage even in the absence of copy-number variation. Moreover, when sequencing is not performed at the whole-genome level but target enrichment techniques are used instead, then the level of coverage can fluctuate so much across targets that CNV detection becomes less accurate. Nevertheless, several methods exist for detecting CNVs, the most popular being **CNV-seq** (Xie and Tammi 2009) and **CNVer** (Medvedev et al. 2010). Complete Genomics also includes a CNV report, calculating ploidy for every window of 2000 bases, and a summary report with gene and repeat annotations.

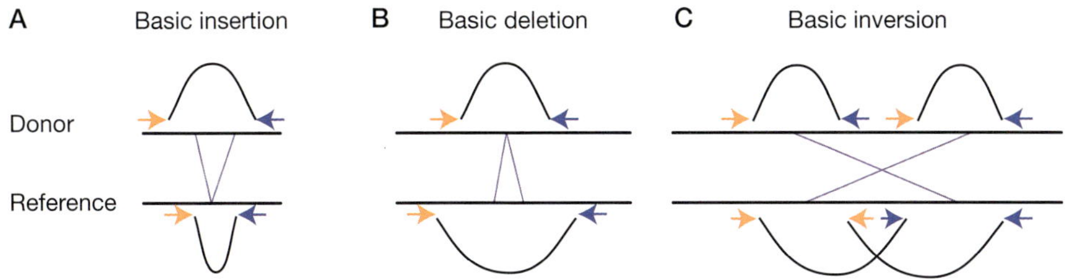

FIGURE 11. An example of structural events. (Redrawn, with permission, from Medvedev et al. 2009, © Macmillan.)

The detection of structural variation relies on sequencing of paired libraries, where reads are produced in pairs (Fig. 11). The main principle is the following: when the two read mates of a pair align to distant locations in the reference genome, this is an indication of potential structural variation, because the two locations are adjacent in the sample but distant in the reference genome (Fig. 11B,C). This is generalized to look for any distance that is not following the distribution of the paired library. When performing sequencing for structural variation detection, the most important consideration is therefore the paired library distance distribution (also known as the insert size distribution). Under the assumption that the insert size in the sequencing library follows a Gaussian distribution, most algorithms require distances of aligned reads that are two or three standard deviations away from the mean insert size in order to accurately detect anomalous structural events. For this reason, it is beneficial to create "tight" libraries whose insert size does not vary much.

Methods and algorithms for detecting structural variants are still being developed, and, thus far, there is no consensus solution adopted by the community. Algorithms include **PEMer** (Korbel et al. 2009), **VariationHunter** (Hormozdiari et al. 2010), **Pindel** (Ye et al. 2009), **MoDIL** (Lee et al. 2009), **BreakDancer** (Chen et al. 2009), and **SVDetect** (Zeitouni et al. 2010). Complete Genomics also includes a *junction* report, listing pairs of nonadjacent locations in the reference genome that appear adjacent in the sequenced sample, and includes assembly of any novel sequence spanning the junction.

SEQUENCING AS A SERVICE

The technologies we have described are deployed in the form of sequencing instruments sold by the respective companies. Although many customers directly acquire such sequencing machines to use for in-house sequencing, others prefer to buy sequencing as a service. Sequencing facilities provide NGS services at relatively low costs and are often an attractive alternative for those who do not want to invest in a sequencing instrument. Moreover, new companies have emerged with proprietary technology that is offered exclusively as a service. Both **Knome** and

Complete Genomics offer whole human genome resequencing and analysis. Complete Genomics aims to decrease the cost of resequencing a human genome down to four digits and delivers assembled and analyzed data, including a list of differences in the resequenced genome.

Existing NGS technologies are constantly being improved, and novel ones keep emerging. The Archon X Prize for Genomics offers $10 million to "the first team that can build a device and use it to sequence 100 human genomes within 10 days or less, with an accuracy of no more than one error in every 100,000 bases sequenced, with sequences accurately covering at least 98% of the genome, and at a recurring cost of no more than $10,000 (US) per genome" (http://www.xprize.org/prize-development/life-sciences). The winner remains to be seen, but the noteworthy competition between sequencing vendors is benefiting end users, who are continuously presented with both better and more cost-effective sequencing solutions.

THE NEXT GENERATION OF SEQUENCING

The sequencing instruments and strategies presented in this chapter have transformed both the questions and experimentation in many areas of research: genetic, genomic, and biomedical studies. The capacity of the original platforms described here is very high, but so is their cost of operation. Since their introduction, much has changed in the sequencing world. In particular, some of the major companies have introduced a new generation of sequencers, some of which are smaller and more efficient (and often faster). These may be lower throughput, but they are also lower-cost technologies. Illumina has introduced HiSeq 2500 and MiSeq; Life Technologies has brought us Ion Torrent PGM and Ion Torrent Proton (somewhat similar in concept to 454). And a new company, Pacific Biosciences, has produced an instrument, PacBio RS II, that uses nanotechnology to detect and collect data. Finally, Roche has reported it will shut down its 454 production. Because sequencing technology is so cutting edge, it is quickly evolving. Therefore, some of the information in this chapter is historical, but the concepts that we have considered set the frame of reference for future innovations in the field.

REFERENCES

Altschul SF, Gish W, Miller W, Myers EW, Lipman DJ. 1990. Basic local alignment search tool. *J Mol Biol* **215:** 403–410.

Blahnik KR, Dou L, O'Geen H, McPhillips T, Xu X, Cao AR, Iyengar S, Nicolet CM, Ludäscher B, Korf I, Farnham PJ. 2010. Sole-Search: An integrated analysis program for peak detection and functional annotation using ChIP-seq data. *Nucleic Acids Res* **38:** e13.

Chen K, Wallis JW, McLellan MD, Larson DE, Kalicki JM, Pohl CS, McGrath SD, Wendl MC, Zhang Q, Locke DP, et al. 2009. BreakDancer: An algorithm for high-resolution mapping of genomic structural variation. *Nat Methods* **6:** 677–681.

Ewing B, Hillier L, Wendl MC, Green P. 1998a. Base-calling of automated sequencer traces using phred. I. Accuracy assessment. *Genome Res* **8:** 175–185.

Ewing B, Hillier L, Wendl MC, Green P. 1998b. Base-calling of automated sequencer traces using phred. II. Error probabilities. *Genome Res* **8:** 186–194.

Goya R, Sun MG, Morin RD, Leung G, Ha G, Wiegand KC, Senz J, Crisan A, Marra MA, Hirst M, et al. 2010. SNVMix: Predicting single nucleotide variants from next-generation sequencing of tumors. *Bioinformatics* **26:** 730–736.

Hormozdiari F, Jajirasouliha I, Dao P, Hach F, Yorukoglu D, Alkan C, Eichler EE, Sahinalmp SC. 2010. Next-generation VariationHunter: Combinatorial algorithms for transposon insertion discovery. *Bioinformatics* **26:** i350–i357.

Illumina 2010. Calling sequencing SNPs. Technical note: Systems and software. http://res.illumina.com/documents/products/technotes/technote_snp_caller_sequencing.pdf. Illumina, San Diego.

Ji H, Jiang H, Ma W, Johnson DS, Myers RM, Wong WH. 2008. An integrated software system for analyzing ChIP-chip and ChIP-seq data. *Nat Biotechnol* **26:** 1293–1300.

Johnson D, Martazavi A, Myers R, Wold B. 2007. Genome-wide mapping of in vivo protein–DNA interactions. *Science* **316:** 1497–1502.

Jothi R, Cuddapah S, Barski A, Cui K, Zhao K. 2008. Genome-wide identification of in vivo protein–DNA binding sites from ChIP-seq data. *Nucleic Acids Res* **36:** 5221–5231.

Kent WJ. 2002. BLAT—The Blast-like alignment tool. *Genome Res* **12:** 656–664.

Kharchenko PV, Tolstorukov MY, Park PJ. 2008. Design and analysis of ChIP-seq experiments for DNA-binding proteins. *Nat Biotechnol* **26:** 1351–1359.

Korbel JO, Abyzov A, Mu XJ, Carriero N, Cayting P, Zhang Z, Snyder M, Gerstein MB. 2009. PEMer: A computational framework with simulation-based error models for inferring genomic structural variants from massive paired-end sequencing data. *Genome Biol* **10:** R23.

Langmead B, Trapnell C, Pop M, Salzberg SL. 2009. Ultrafast and memory-efficient alignment of short DNA sequences to the human genome. *Genome Biol* **10:** R25.

Lee W-P, Stromberg M, Ward A, Stewart C, Garrison E, Marth GT. 2013. MOSAIK: A hash-based algorithm for accurate next-generation sequencing read mapping. arXiv:1309.1149.

Li H, Durbin R. 2009. Fast and accurate short read alignment with Burrows-Wheeler transform. *Bioinformatics* **25:** 1754–1760.

Li H, Durbin R. 2010. Fast and accurate long-read alignment with Burrows-Wheeler transform. *Bioinformatics* **26:** 589–595.

Li H, Homer N. 2010. A survey of sequence alignment algorithms for next-generation sequencing. *Brief Bioinform* **11:** 473–483.

Li R, Yu C, Li Y, Lam T-W, Yiu S-M, Kristiansen K, Wang J. 2009. SOAP2: An improved ultrafast tool for short read alignment. *Bioinformatics* **25:** 1966–1967.

Mardis ER. 2008. Next-generation DNA sequencing methods. *Annu Rev Genomics Hum Genet* **9:** 387–402.

Mardis E, McCombie WR. 2012. DNA sequencing. In *Molecular cloning: A laboratory manual*, 4th ed. (Green MR, Sambrook J), Vol. 2, pp. 735–763. Cold Spring Harbor Laboratory Press, Cold Spring Harbor, NY.

Medvedev P, Stanciu M, Brudno M. 2009. Computational methods for discovering structural variation with next-generation sequencing. *Nat Methods* (suppl.) **6:** S13–S20.

Medvedev P, Fiumel M, Dzambal M, Smith T, Brudno M. 2010. Detecting copy number variation with mated short reads. *Genome Res* **20:** 1613–1622.

Mortazavi A, Williams BA, McCue K, Schaeffer L, Wold B. 2008. Mapping and quantifying mammalian transcriptomes by RNA-seq. *Nat Methods* **5**: 621–628.

Ning Z, Cox AJ, Mullikin JC. 2001. SSAHA: A fast search method for large DNA databases. *Genome Res* **11**: 1725–1729.

Pepke S, Wold B, Mortazavi A. 2009. Computation for ChIP-seq and RNA-seq studies. *Nat Methods* **6:** S22–S32.

Qin S, Shen J. 2009. HPeak: A HMM-based algorithm for defining read-enriched regions form massive parallel sequencing data. www.sph.umich.edu/csg/qin/HPeak.

Rozowsky J, Euskirchen G, Auerbach RK, Zhang ZD, Gibson T, Bjornson R, Carriero N, Snyder M, Gerstein MB. 2009. PeakSeq enables systematic scoring of ChIP-seq experiments relative to controls. *Nat Biotechnol* **27:** 66–75.

Sanger F, Necklen S, Coulson AR. 1977. DNA sequencing with chain-terminating inhibitors. *Proc Natl Acad Sci* **74:** 5463–5467.

Trapnell C, Pachter L, Salzberg SL. 2009. TopHat: Discovering splice junctions with RNA-seq. *Bioinformatics* **25:** 1105–1111.

Valouev A, Johnson DS, Sundquist A, Medina C, Anton E, Batzoglou S, Myers RM, Sidow A, 2008. Genome-wide analysis of transcription factor binding sites based on ChIP-seq data. *Nat Methods* **5:** 829–834.

Wang Z, Gerstein M, Snyder M. 2009. RNA-seq: A revolutionary tool for transcriptomics. *Nat Rev Genet* **10:** 57–63.

Wilbanks EG, Facciotti MT. 2010. Evaluation of algorithm performance in ChIP-seq peak detection. *PLoS ONE* **5:** e11471.

Xie C, Tammi MT. 2009. CNV-seq, a new method to detect CPY number variation using high-throughput sequencing. *BMC Bioinformatics* **10:** 80.

Ye K, Schultz MH, Long Q, Apweiler R, Ning Z. 2009. MoDIL: Detecting small indels from clone-end sequencing with mixtures of distributions. *Nat Methods* **6:** 473–474.

Zeitouni B, Boeva V, Janoueix-Lerosey I, Loeillet S, Legnoix-Ne P, Nicolas A, Delattre O, Barillot E. 2010. SVDetect: A tool to identify genomic structural variations from paired-end and mate-pair sequencing data. *Bioinformatics* **26:** 1895–1896.

Zhang Y, Liu T, Meyer CA, Eeckhoute J, Johnson DS, Bernstein BE, Nusbaum C, Myers RM, Brown M, Li W, Liu XS. 2008. Model-based analysis of ChIP-seq (MACS). *Genome Biol* **9:** R137.

WWW RESOURCES

http://bioinformatics.bc.edu/marthlab/wiki/index.php/Software MOSAIK

http://res.illumina.com/documents/products/technotes/technote_snp_caller_sequencing.pdf Illumina 2010. Calling sequencing SNPs. Technical note: Systems and software. Illumina, San Diego

http://www.iscb.org/uploaded/css/58/17109.pdf Bauer MJ, Cox AJ, Evers DJ. 2010. ELANDv2—Fast gapped read mapping for Illumina reads. Poster at 18th Annual International Conference on Intelligent Systems for Molecular Biology, sponsored by the International Society for Computational Biology, Boston, MA. Poster no. J04

http://www.xprize.org/prize-development/life-sciences XPRIZE

9

Proteomics

Amit Kaushal and Tiffany J. Chen

Stanford University School of Medicine, Biomedical Informatics Training Program, Stanford, California 94305

The central dogma of molecular biology states that, in general, information flows from DNA to RNA to protein via the processes of transcription and translation. The resulting set of proteins is tremendously diverse in its chemical properties, structures, and functions. The field of proteomics deals with identifying, quantifying, and understanding the role of these proteins in their various forms in different tissues under various conditions.

With so many high-throughput (HT) genomic techniques available to study DNA and RNA, one may wonder whether study of the proteome is even worthwhile. There are several reasons for studying proteins and the proteome directly.

- Temporally, an increase in RNA abundance may imply a future increase in protein abundance, whereas direct observation of protein states and abundance is informative regarding the present state of the organism.
- Proteins may be separated in time and space from the RNA that made them. A protein may be translated and stored for subsequent use or a cell may produce a protein for secretion into an extracellular compartment, such as the blood. In these cases, understanding protein function requires looking at the proteins themselves.
- Protein levels are affected not just by their rate of synthesis but also their rate of destruction. Even for an actively translated protein, the RNA level might not accurately reflect the amount of protein present.
- Protein functions and activities are often regulated by posttranslational modifications (PTMs), such as proteolysis (often signaled by ubiquitination tagging), phosphorylation, glycosylation, and hundreds of others. These modifications are not encoded in the DNA or RNA templates of the proteins but can dramatically alter a protein's functional characteristics and be indicative of pathologic states.

Experimental approaches to perform proteome-scale analyses face the challenge that proteins are quite varied and dynamic at a physical level. They are composed of 20 different amino acids, each with different physical properties. Proteins may be long or short; acidic or basic; and hydrophobic, hydrophilic, or both. They have primary, secondary, tertiary, and in some cases, quaternary structures. They may undergo PTMs at many different points. Finally, protein concentrations can span several orders of magnitude in a given subcellular compartment or biological sample. In this chapter, we discuss two experimental technologies that can gather and quantify data regarding many proteins at once: mass spectrometry (MS) and flow cytometry (which includes fluorescence-activated cell sorting, or FACS).

Although MS has a variety of uses, in the context of proteomics, it is typically used in a research setting for identification of peptides and proteins in a complex mixture and, more recently, quantification of these proteins or their constituent peptides. It is useful for cataloging the proteins present in various tissues, understanding differential abundance of proteins and PTMs between two disease states, and investigating systems-level protein phenomena. Methods for quantification are evolving, both at the assay and computational levels. Many methods are now being developed to enable profiling of protein modifications using MS.

Flow cytometry has been around since the 1960s and 1970s. With the use of fluorescently or metallically (in the case of mass cytometry) labeled cells, beads, or organisms, flow cytometry is able to accurately detect and count cells based on the abundance of the labeled elements, often proteins. As of this writing, tens of proteins can be assayed simultaneously via flow cytometry, and recent advances have allowed for the labeling of not only membrane proteins but also intracellular proteins or their modifications. Although prior knowledge of the labeled protein of interest is required, this method is a fast, inexpensive, and accurate means of cell counting based on protein properties. It is reliable enough to be used in many medical fields for cell profiling, diagnosis, and prognosis.

Just as with other HT assays, computational tools and techniques are key to taking the large amounts of data generated and turning them into meaningful biological insights. We view each of these experimental approaches with the eye of the informatician, highlighting the computational challenges and solutions at different steps of the experimental process.

MASS SPECTROMETRY

Although mass spectrometers have been used to analyze the chemical composition of countless compounds for almost 100 years, their mainstream use as a powerful tool for probing proteome content and quantifying protein abundance has emerged only in the last couple of decades. The mass spectrometer is ideally designed to analyze a large quantity of a single molecular entity of low molecular weight for chemical

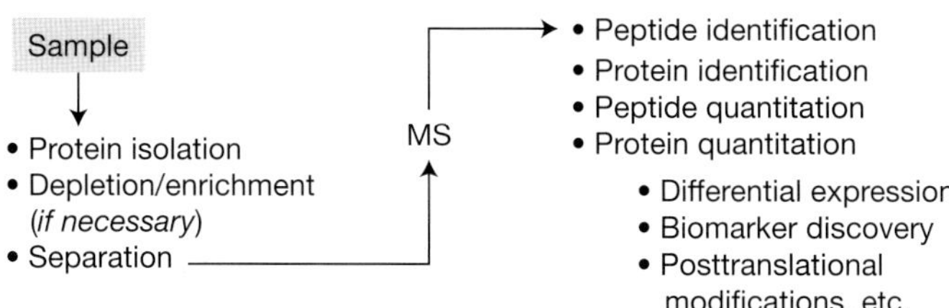

FIGURE 1. Overview of MS proteomics workflow.

composition. This would make it, at first glance, a poor choice for quantitative analysis of a complex heterogeneous biological sample such as blood or urine, with protein concentrations spanning several orders of magnitude. However, sample preparation and separation approaches have allowed scientists to take advantage of the mass spectrometer's tremendous analytic capabilities at the proteome level. In this section, we step through the various stages of MS proteomics to better understand the computational challenges at each step. Figure 1 provides an overview of these stages and can be used as a reference to help guide the reader through this section.

Consider an example problem for which we might use a proteomics approach. Following a kidney transplant, some patients experience rejection of the donated organ. The gold-standard approach to diagnosing kidney rejection is a biopsy—a procedure involving inserting a long needle into the sedated patient to extract a sample of kidney for further analysis. This procedure is associated with significant discomfort to the patient and has its own complications.

As a budding proteomics practitioner, one may think, "There might be a better way! Maybe the protein content of a patient's urine can provide markers for kidney rejection." A simple experiment is devised, whereby urine samples are collected from several patients in the posttransplant weeks. Subsequently, 20 patients are selected who were shown to not have undergone rejection and 20 who did, and MS proteomics is used to determine the proteins present in the samples and their relative abundance (a typical goal of MS proteomics experiments). The samples are sent off for mass spectrometry analysis. What happens once the samples are analyzed and how does one use the data received? Let us follow the sample through its journey, identifying computational challenges at each step.

Protein Purification, Separation, and MS Analysis

Although a thorough description of sample preparation methods is beyond the scope of this discussion, we will cover a few key points. To prepare a sample for analysis, the protein content of the sample is first isolated. In certain complex mixtures, such as plasma, proteins can span more than 10 orders of magnitude in concentration. In these mixtures, a depletion or enrichment step is often considered to avoid the analysis being swamped by a few very highly abundant proteins.

Because mass spectrometers have an easier time ionizing smaller molecules, the next step is usually to digest proteins into peptides using an enzyme with well-understood digestive properties, often trypsin. Some groups do study the intact protein, which is referred to as the "top-down" approach (versus the "bottom-up" approach that we have described). However, it is not as common, and we will address only the bottom-up approach. Although bottom-up proteomics does overcome a technical hurdle of MS, the trade-off is that the downstream analysis of results will be on individual peptides, not intact proteins. Tryptic digestion typically results in peptides between five and 50 amino acids in length, with most peptides being shorter than 30 amino acids.

Once the proteins are digested into small peptides, each peptide must somehow be fed to the mass spectrometer with as few confounding peptides at a time as possible. To accomplish this, the samples are usually run through a separation apparatus before feeding them into the mass spectrometer. Initially, two-dimensional (2D) gels were used for this purpose, but liquid chromatography (LC) is now more commonly used to separate the peptide mixture.

To obtain the highest-quality mass spectrometry results, the sample would ideally be separated as finely as possible, and each minuscule amount of separated sample would be run separately through the mass spectrometer. However, the greater the separation of the initial mixture, the more MS time is required to analyze the whole sample, which means a smaller number of samples being run overall. It is important to realize that a trade-off exists between the quality of separation of each sample and overall sample throughput. This is a key decision in experimental design, which is influenced by factors such as the number of samples needed to produce a meaningful biological result and the degree of separation needed to generate data of sufficient quality. It is often worthwhile to generate pilot data with samples to better simulate these trade-offs.

The sample is now ready for MS analysis. It is separated and then fed through the mass spectrometer bit by bit, where it passes through the ionization device coupled to the mass spectrometer. This device subjects the molecules to an ionizing force to turn them into charged ions. Once ionized, these ions pass through a region known as the mass analyzer, in which a force of known magnitude induces the ions to move to a detector. By measuring how the motion of the ion is affected by the force, a precise mass-to-charge ratio (m/z) can be calculated for the ion.

MS analysis for a particular protein will result in a dominant m/z value for each parent peptide that is ionized and detected in sufficient quantity as well as the fragment information for each peptide. From this accurate mass and charge information from the parent peptide and fragments, we can deduce the set of amino acids that make up each peptide. We do not know the sequence of those peptides.

MS can further select and trap a peptide and break the peptide into many fragments with the fragments passing through the mass analyzer again and all of the m/z of the fragments being detected, a technique called tandem mass spectrometry, or MS/MS. The MS/MS spectrum for peptide identification is plotted with m/z of

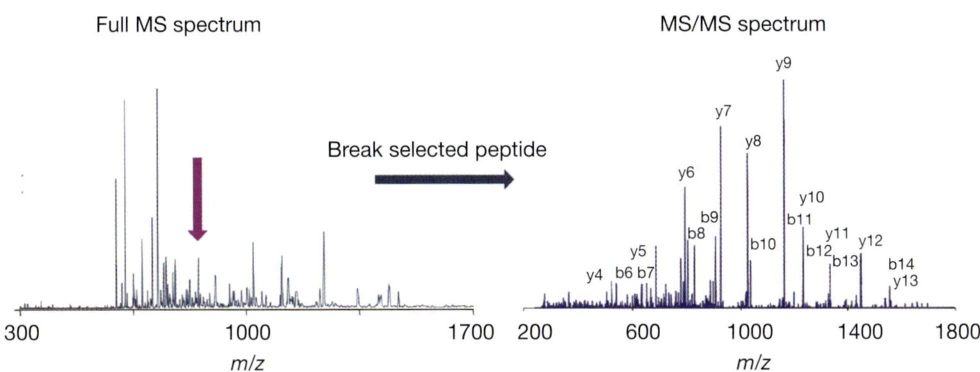

FIGURE 2. Example MS and MS/MS spectra. (Adapted, with permission, from Xie et al. © 2011 American Society for Biochemistry and Molecular Biology.)

the fragments on the *x*-axis, and a measure of abundance on the *y*-axis (Fig. 2). In a single LC-MS analysis, thousands of MS/MS spectra corresponding to individual peptides will be generated. This is the typical data from an initial MS experiment.

Peptide Identification

Our first goal is to use MS spectra to identify which peptides were detected in our mixture. The approaches to this problem break down into two main groups. The first and most commonly used group attempts to map observed proteins to those in some known reference database, whereas the second approach attempts to reconstruct sequence de novo, using the data alone. Although the de novo approach is more computationally challenging, it is important to remember that there are several limitations to the database-driven approach, the most important being: If a protein does not already exist in the reference database, its peptides cannot be identified in the sample.

Database-Driven Approaches

The most common approach to peptide identification dates back to the SEQUEST algorithm published by John Yates and colleagues in 1994 (Eng et al. 1994). The general principle is to compare the spectra of the current peptide against a reference database. This reference database contains thousands of proteins specific to the organism of study. Each of these proteins is "digested" in silico to predict peptides that might be observed. For each of these peptides, SEQUEST then computes its mass and a predicted spectrum. The algorithm then compares each observed spectrum with its database of peptides with similar mass and predicted spectrum. The statistic used for comparing spectra is cross-correlation. If the observed spectrum matches a predicted spectrum in the database, then the spectra are considered "identified." See Box 1, The SEQUEST Algorithm, for more details.

Several newer approaches to this problem exist, such as Mascot, OMSAA, and X! Tandem. These algorithms differ in the model used for searching predicted spectra, whether they are free or commercially available, and web-accessible versus standalone. It is important to realize that in each of these approaches, if the protein is not in the database, its peptides cannot be identified. This is worth repeating: Database-driven peptide identification, although efficient, limits the experimenter to detecting only proteins in the reference database. Peptides from novel proteins will not find a matching spectrum. Additionally, if a protein is in the reference database but the observed peptide has a posttranslational modification in the sample that results in a mass change, the peptide will not be detected unless that modification is explicitly searched for.

One work-around to this problem is to try to broaden the reference database. In the emerging field of proteogenomics, rather than use a database of known proteins as the reference database, a reference protein database is derived from the genome of

BOX 1. The SEQUEST Algorithm

Problem

How can we take the hundreds of thousands of spectra generated by modern MS proteomics experiments and identify which proteins are present?

Approach

Reference database:

1. Create a database with all known protein sequences that might be found in the sample.
2. Computationally "trypsin-digest" these proteins to generate possible peptide fragments that might be found in the sample.
3. For each predicted peptide in the database, predict the mass and spectra of the peptides. These are the possible matches.

Sample Analysis:

For each spectrum observed by MS: Calculate the similarity to the entire database of peptides using the cross-correlation score, which can measure the degree of similarity between two waveforms. Alternatively, rather than comparing the peptides to the entire database, we may restrict comparisons to peptides in the reference database that are close to that of our current spectrum. The observed spectrum will be identified as the peptide from the reference database that has the predicted spectrum with the highest cross-correlation with the observed spectrum.

the organism. All six reading frames of the organism's genome are computationally "translated" to produce possible proteins. This expanded set of theoretically possible proteins forms the new reference database. Proteogenomics has been used to enable discovery of new genes or correct annotations of existing ones.

Estimating False Positives: The Target-Decoy Approach

Spectra do not match exactly. Instead, they are given a score to show their "best match" in the data set. However, the "best-match" peptide might not always be the correct peptide, as in the following cases:

1. The observed spectrum may be generated by a peptide that is not in the protein database, or it might not have been generated by a protein species at all. In this case, the correct answer should be "no match," but the peptide might still be assigned an identity to a peptide with similar mass in the reference database. This is a false-positive identification.
2. The observed peptide was generated by a peptide in the database, but the incorrect peptide might have been assigned.

For a given data set, it is useful to estimate what percentage of peptide assignments might be incorrect at given parameter settings for each search algorithm. The current solution to this problem is known as the target-decoy approach. Instead of searching each observed peptide against the usual target database, we add several fictitious ("decoy") proteins to the database as well.

In the ideal case, the observed peptides would never be assigned a decoy peptide match (because these come from fictitious proteins). By observing how many peptides are actually assigned matches to decoys, we have an estimate of the rate of false-positive identification in our actual data.

Of course, the method used to create the decoy proteins will impact the number of decoy identifications. The decoy proteins should be similar to the target database in terms of number of proteins, protein length, and amino acid distribution. Although there are several ways to create a good decoy database, one of the most commonly used methods is to take each protein in the target database and form a corresponding fictitious protein by reversing its sequence. This database will have the same number of decoy as target proteins, and the decoys will have the same length and amino acid distribution as the real proteins but will have spectra unrelated to the original database because the digestion sites have been scrambled.

Subsequent Lookups: The Accurate Mass and Time Tag (AMT) Approach

Performing tandem MS on each and every sample can be time- and resource-consuming, limiting the number of samples that can be analyzed in a given amount of machine time. One way to get around this is the accurate mass and time tag (AMT) approach, where the setup uses high-performance liquid chromatography (HPLC) as the separation mechanism. The general principle behind this approach is that after the first round of MS, some information is known regarding the peptide. We know the time at which the peptide eluted from the HPLC machine, a property that is quite reproducible between samples. We also have the spectrum and, using MS, we have an accurate measure of the mass of the peptide. Using these data, MS/MS sequencing can be performed on that peptide just once. We can then store the mapping feature in a database, such as "the spectrum with mass *x* and elution time *t* has sequence *y*." If we encounter a peptide with mass *x* and elution time *t* in subsequent experiments, we can identify it as sequence *y without doing tandem MS.* This is similar to the computational concept of a hash or lookup table: We store the peptide sequence under the two-part lookup key (mass, elution time) and can use this key for subsequent lookups.

This is a good approach to consider if one needs to run several samples of similar origin and the facility supports this method. The first step is to build the lookup table. This is often performed by taking an aliquot from several samples, mixing them together, and doing a single run for high-resolution HPLC-MS/MS. This enables a cataloging of elution times and sequences for the peptides in the sample. Then, each subsequent sample can be run in a simple HPLC-MS setup, and the peptide identification can be assigned through the lookup table. This method relies on the elution time and mass being very consistent from sample to sample.

De Novo Sequencing

Databases themselves are not perfect. They are evolving entities, often incomplete, and often wrong. Incorrect protein sequences, mappings, or annotations are all common problems and can adversely affect peptide identification searches. Additionally, when searching peptide spectra against a database of known proteins, we will only find species that have been seen before. For these reasons, several groups are working in the area of de novo searching, which we now consider.

De novo peptide identification attempts to identify peptide sequences without the aid of existing sequence databases. Indeed, although searching known databases can help us find known proteins, we are unable to discover novel proteins, new splice forms of known proteins, or even posttranslational modifications of already identified peptides. In a de novo search, rather than limiting our search space to that of known proteins, we use the known masses of the building blocks of a peptide—the 20 amino acids and 200 or so posttranslational modifications—and attempt to match using this approach. Because the search space of a completely de novo approach is far too large to be completed in any reasonable amount of time, these approaches often use heuristics to help prune the search space (Table 1). As of this writing, a de novo search is not widely used for the analysis of complex mixtures.

TABLE 1. Comparison of database-driven and de novo approaches to protein identification

	Types of approaches	
	Database driven	De novo
Use case	Known genome, not searching for novel proteins	Unknown genome, or organism has incomplete genome, possibly searching for novel protein or specific modification. Low-complexity sample
Example tools	SEQUEST, MASCOT, X!Tandem, OMSSA	PEAKS
Approach	Generate theoretical spectra from proteins in reference database, match	Combinatorially generate sequences of amino acids and modifications that could produce the observed spectrum

Peptide Modifications

The issue of peptide modifications also complicates peptide identification. Because a mass spectrometer is a high-resolution mass detector, anything that causes the peptide to have a mass different from the mass of a peptide in the database (or causes an amino acid to have a different mass from those predicted in the de novo search algorithm), such as a PTM, will cause the peptide's spectra to change and therefore cause the peptide to not be identified.

Computational algorithms to search for specific modifications have been developed. The general approach has been to rely on precompiled databases of modifications with known mass. The reference database essentially expands to include all possible modifications at all possible positions. This is computationally expensive, thus making the modification problem a challenging one in proteomics today. Searching for a single modification of interest is easier, but searching for a peptide and all of its modifications across an entire proteome is not yet entirely feasible.

Protein Identification

At this stage, we have identified a set of peptides present in each of our samples. Our next task is to map these peptide sequences to known proteins. On the surface, this task should be straightforward: The peptides all came from proteins in the database; thus, a simple reverse mapping should do the trick. Although this is largely true, there are two issues that come up in the protein identification phase: The first is minimizing false-positive protein identifications, and the second is the case of mapping a peptide sequence to multiple protein sequences.

False-Positive Proteins

The number of peptides detected for each protein can vary widely across a data set. Most proteins will be detected with a few peptides, but some proteins will be detected with tens or even hundreds of peptides. It has been shown that when only a single peptide is detected from a protein, it is much more likely that that peptide (and, therefore, that protein) was not actually observed, but rather is a false-positive peptide call. A conservative but widely accepted heuristic in the field is to require that all proteins be observed by two or more peptides before calling a protein present to avoid false positives. This approach is commonly used even though more recent works have shown that such a conservative criterion is not always necessary and that there are cases in which a protein may be reliably identified by a single peptide.

One Peptide, Many Proteins

A simple question that we might want to answer is, "How many proteins were detected in a given sample of transplant patients?" This task is complicated by the fact that one peptide can map to multiple proteins. Phenomena such as alternative splicing and posttranslational cleavage can result in very similar proteins that each have a unique entry in the reference database. Although it is possible for a short peptide to map to two unrelated proteins, in practice, the majority of multiple protein mappings result from the case of highly similar proteins with each receiving a separate entry in the reference database.

If we were to include every such multiple mapping when counting proteins in our data set, we would be saying more regarding the state of the database than the state of our sample. It is usually desirable to collapse multiple similar proteins into one representative protein for the purposes of peptide mapping. Postprocessing software tools will usually handle this task by collapsing similar proteins into a single protein group. Some approaches generate a set of best-fit nonredundant proteins for a given data set, whereas others do not take into account the peptides observed in a given data set but, rather, predetermine protein groups using a BLAST approach.

Peptide Quantitation

Now that we have identified the peptides and proteins present in each sample, our next question is, "How abundant are each of the peptides?" In our search to find a marker for kidney rejection, this will surely be useful information. There are two main MS approaches used when quantitation of peptide abundance is desired: label-free and labeled. In each case, the quantitative estimates for a given peptide only serve to compare the abundance of that specific peptide across samples.

Label-Free Quantitation

In label-free proteomics, estimates of peptide abundance from sample to sample are extracted directly from the MS data. Two measures of abundance are commonly used. The first approach is to use the peak height (or area under the curve) from the peptide's MS spectrum. The idea here is that the more of a peptide is present, the more ions will be generated and analyzed and the greater the resulting peak height.

A second, simpler approach uses spectral counts. In this approach, peptide abundance is measured simply by the number of spectra detected. Although this approach may seem naïve, in fact, it has been shown to have good correlation with peptide abundance.

Regardless of the method used, a peptide's abundance is only useful to compare the abundance of that peptide in other samples. We cannot definitively compare the abundance of one peptide against another because different peptides have different performances in the mass spectrometer. If peptide A has a higher intensity than peptide B, it is not possible to tell whether that is because peptide A is truly more abundant or peptide A is a less-abundant peptide that happens to perform particularly well in the mass spectrometer.

Labeled Quantitation

The second approach, in contrast, involves stable isotope labeling for pairwise comparison or using one labeled sample as the reference standard. The reason for this is because the light and heavy labeled species have the same chemical properties and will cancel all potential biases introduced in downstream processing and analysis, resulting in more accurate quantification. There are at least two possible reasons to use this approach. First, it can be used to do a direct paired comparison of samples. In our running example, we have 20 rejection (R) patients and 20 nonrejection patients (NR). We could label the 20 NR patients (we discuss how to label below) and simultaneously run R1 and NR1 in our first example, R2 and NR2 in our second, and so forth. Running both samples at the same time means that any machine-specific variation will theoretically be equally applied to both samples, so the samples can be directly compared with one another. In addition, this allows us to get meaningful information from 40 samples using only 20 MS runs.

Another design would be to create a commonly labeled reference sample (REF) and run this with every run (Fig. 3). In our case, we would have 40 MS runs: R1/REF, R2/REF, ..., R20/REF, NR1/REF, NR2/REF, ..., NR20/REF. In this approach, the reference channel serves as an internal control for all of the runs. We want a reference sample that has a similar peptide composition to the samples of interest; thus, the reference sample is often made by simply taking an aliquot of each of the samples to be analyzed.

A successful labeling approach causes a slight mass shift in all labeled peptides without affecting their chemical properties. By keeping the other chemical properties the same, we ensure that a given peptide from both the labeled and the unlabeled samples will elute at the same time from the LC separation mechanism and therefore enter the MS at the same time. Thus, both spectra will appear on the same data set.

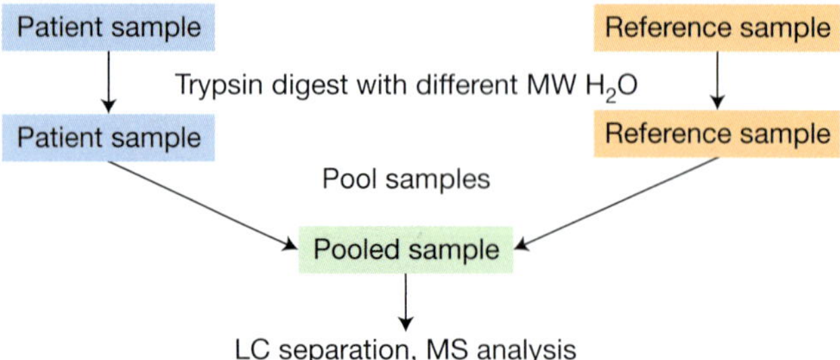

FIGURE 3. Schematic of labeled approach to proteomics.

There are several approaches for this labeling. A sample can be tryptically digested in heavy water (^{18}O), resulting in peptides that are heavier than unlabeled peptides by 4 unified atomic mass units (U) (or 2 U in the case of amino- or carboxy-terminal peptides). In cell culture, an approach known as SILAC (stable isotope labeling by amino acids in cell culture) can be used. In this approach, cells are grown in the presence of heavy arginine, for example, resulting in a 6-AMU shift per arginine on mass spectrometry. Other approaches such as ICAT (isotope coded affinity tags) label all cysteines with a tag, whereas iTRAQ (isobaric tags for relative and absolute quantitation) labels the amino termini and lysine residues.

To obtain a quantitative assessment of peptide abundance in a labeled approach, the ratio (or log-ratio) of the unlabeled to labeled intensities of a peptide is often used. Unfortunately, this approach is doubly affected by the missing data problem described in the label-free approach. If either the labeled or unlabeled peptide is missing, we can no longer take a ratio and will have difficulty interpreting the observed value.

Selected Reaction Monitoring (SRM): Quantitation at the Assay Level

Selected reaction monitoring (SRM), also known as multiple reaction monitoring (MRM), allows for high-resolution quantitative assessment of a subset of proteins within the MS portion of the experiment (Fig. 4). Using a type of mass spectrometry known as a triple quadrupole, the mass spectrometer is set to focus on very specific m/z values. For each peptide of interest, two m/z values are set. First, the m/z value of

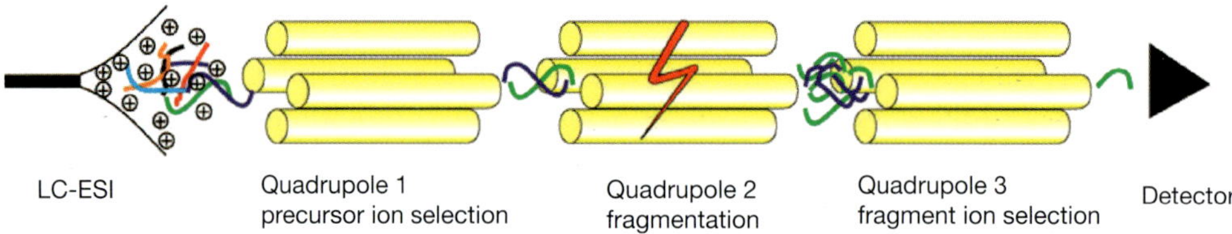

FIGURE 4. Selection reaction monitoring assay. (Reprinted from Lange et al. 2008, *EMBO* and *Nature* Publishing Group.)

the dominant ion of the peptide is set, and these ions are selectively filtered and passed for further analysis. To ensure specificity, these ions are again fragmented, and the *m/z* of the dominant fragment is also specified. Because it is very unlikely to have two species that elute at the same time from LC, have the same mass, and when fragmented again, have a resulting ion that also has the same mass, this dual mass filtering provides a means of very specifically following a particular species through the MS process. By spiking in elements at varying concentrations, a concentration curve can be created, and absolute quantitation of the species can be achieved. At this point, SRM assays are capable of analyzing tens to hundreds of species in a given experiment.

Protein Quantitation

Once we have peptide-level quantitative data from our peptides, we also want protein-level quantitative data. Challenges here include that, as mentioned above, different peptides from the same protein may have different performances in MS, complicating aggregation. In a typical proteomics experiment, we will see approximately half of the proteins detected by only one or two peptides and the rest detected by three or more (even up to 100+) peptides. How can we provide reliable quantitation in the presence of such heterogeneity? Missing data again have a role here. Even when we do have complete data, what is the best way to aggregate? What happens when peptides from the same protein disagree?

Because of these complications, most experimentalists use simple averaging of peptides or simply pick the best peptide (the "proteotypic" peptide) to represent the abundance of a protein, where *best* means the peptide with either the highest observed intensities for that protein or the least missing values. The choice varies from data set to data set, but each approach is cautiously accepted in the field at this time.

After Protein Quantitation

With a matrix of quantitative data, whether peptide level or protein level, we are able to perform the usual statistical tasks with a variety of methods similar to those used in the microarray community, such as normalization, differential expression, and classification. These are discussed extensively elsewhere in this book (see, e.g., Chapter 6). Proteomics has the additional wrinkle of having data at the peptide level. What does it mean when one peptide is differentially expressed, but the rest of the protein is not (e.g., a protein expressed from a spliced variant of a given mRNA)? Coming up with biological explanations for the concordance between peptides within the same protein over a set of samples, or the lack thereof, plays a significant role in the analysis of proteomics data across samples from different conditions.

FLOW CYTOMETRY

Flow cytometry has emerged as a technology that can quantitate the amount of cell-surface proteins as well as intracellular marker levels. Although the majority of scientists mainly quantify cell-surface proteins, a recent scientific benefit is that flow cytometers are able to detect single-cell, intracellular protein modifications. This is performed over thousands or even millions of cells over a very short period of time. As a result, this increase in data is facilitating the emergence of flow cytometry as an HT technology, in which statistical methods previously used for other HT technologies (microarrays, etc.) can now translate to flow cytometry data analysis. Because of this sudden growth in data, however, there are few standards for flow cytometry data analysis aside from manual annotation and analysis. When measuring two or three markers, visualization and manual delineation of cell subsets are simpler for an individual researcher to perform. With newer, large-scale experiments using flow cytometry technologies, however, measuring thousands of cells over tens of parameters results in data that are impossible to visualize in all dimensions at the same time. Trying to find cell subsets in this high-dimensional space is difficult. Informatics methods are a necessary component of flow cytometry data analysis. Furthermore, the cell states generated by manual analysis can vary between experts. This variability also underscores the need for computational methods. Computational analysis is especially useful when trying to think of new methods to find potential cell subsets, or subpopulations in high-dimensional space.

Background of Flow Cytometry

Flow cytometry has a rich history. There exist excellent references that describe the reasons and design behind flow cytometers, including *Practical Flow Cytometry* by Howard Shapiro (2005). As a result, this chapter focuses on looking at the data from the viewpoint of a bioinformatician and how an informatician would view and analyze flow data. Furthermore, because a large number of problems in flow cytometry are currently analyzed using manual analysis, this chapter focuses mainly on computational methods that facilitate more complex workflows, including the elucidation of statistically relevant differences within and across sample types. This section mainly focuses on fluorescence-based flow cytometry, although the majority of these methods can be applied to mass cytometry as well. Mass cytometry, or cyTOF, is a newer technology in which antibodies are labeled with lanthanide series metals instead of fluorescent tags (Bendall et al. 2011). Subsequently, instead of measuring for fluorescence, mass cytometers use a mass spectrometer to detect the level of metal tags as a proxy for protein levels. Something to keep in mind throughout this section is that the field of flow cytometry analysis is continuously evolving. Subsequently, many of the current analytic workflows used are variations on other methods that may be encountered throughout this book, including those in the machine-learning chapter (Chapter 4).

From a broad viewpoint, many investigators use flow cytometry to find subsets of events (where events are often cells) or *states*. What is a state? There are many different types of states. From a cellular standpoint, a group of cells in the same state could be of a specific cell type (B cell, T cell, etc.). A group of cells could also have the same signaling state, where a set of particular intracellular pathways are similarly activated across a set of cells. Other distinguishing factors include those cells showing similar levels of RNA expression for the same gene or cells at the same point of the cell cycle (e.g., quiescent cells in G_0 phase). Flow cytometry can be used to identify or distinguish among various states—not only single cell but aggregate states—in a cell supernatant across various conditions. Finally, flow cytometry can be used to determine organismal states, even at the multicellular level.

Discovery and quantification of these states are currently in use, both in research and in medical laboratories. The ability to detect the presence of different states has had a large impact in the clinical setting, especially hematology. Flow cytometers are used to quickly detect and count specific immunophenotypes based on cell-surface marker detection. To detect the overall presence of proteins, flow cytometers count immunofluorescent beads that have been exposed to cell supernatant. Flow has even been used to separate small organisms (*Caenorhabditis elegans*, etc.) or detect cells with specific intracellular protein modifications.

For the purposes of this chapter, we focus on understanding the output of a flow cytometer and how these data can be interpreted computationally. In Figure 5, this is represented by processes shown along the right part of the figure. Many steps go into preparation of a sample for flow cytometry, however, and are not covered in this chapter. In addition, lysate-based methods (including Luminex assays, etc.) are not covered in this figure, although similar protocols are sometimes used.

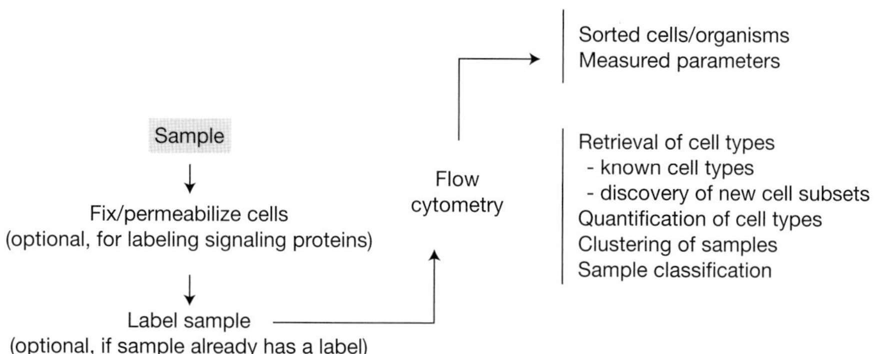

FIGURE 5. Overview of the stages of flow cytometry.

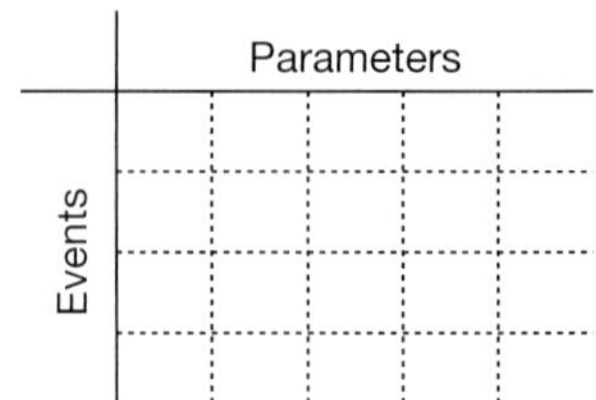

FIGURE 6. General scheme for .fcs file format for display of the output from a flow cytometer.

The .fcs File: Standard Format for Flow Cytometry Data

The output from a flow cytometer is typically supplied as an .fcs file (the general grid for the display is shown in Fig. 6), where the output has the extension .fcs at the end of the file name. To maintain consistency across flow cytometers and for analysis purposes, standardization of .fcs files is set forth by the International Society for Advancement of Cytometry, or ISAC. As of the writing of this chapter, the latest existing version is FCS 3.1, although it is likely that a user reading this chapter will encounter files using version FCS 3.0.

Because the output of a flow cytometer into .fcs format is compressed, opening an .fcs file in a commonly used text editor (e.g., Notepad on Windows, TextEdit on Mac, etc.) results in a file that is practically impossible for human interpretation. Therefore, .fcs readers parse and decompress these files into human-readable form. A list of some of the current available software for .fcs file parsing (free and licensed) is shown in Table 2.

The .fcs file contains many sections of information. For simplicity, we will focus on a few main components. First is the expression values, or experimental data that are extracted from a raw .fcs file. Although the content within a .fcs file may vary significantly between files, the general structure for the plot of the data is as shown in Figure 6.

Each row in Figure 6 displays an *event*, which may be a cell, bead, organism, or other individual detectable item (and is possibly sortable, depending on the machine). Each column represents a *parameter*. A parameter is the measured quantity (e.g., fluorescence intensity, or FI) of a tag for a particular molecule of interest (as examples: a labeled surface marker, the expression of a gene, or the content of DNA). For mass cytometry, a parameter is the measured quantity of a particular metal. Each entry in the table is a numeric value representing the measured FI of a parameter for

TABLE 2. Features of commonly used software for .fcs file parsing

	Software programs			
	Cytobank	**FCS Express**	**FlowCore (using R)**	**FlowJo**
Free version available?	Yes	Reader: free; otherwise, 30-d trial	Yes (open source)	30-d trial
Type of interaction with data?	Web interface	Application interface	Code (through the Bioconductor package)	Application interface

each corresponding event up to the saturation level of the detector. During analysis, there is an assumption that the measurement detected by a particular channel has a direct relationship with the amount of the target to which it is bound. For each event, multiple parameters can be measured. In whole-cell flow cytometry, these can be extracellular or intracellular markers. These tables, once processed using an .fcs reader, can be opened and visually read by an investigator in many types of software, including spreadsheet-based applications. In addition to the software suites for manual analysis of .fcs data, there exist several packages in the programming language R that are available for parsing .fcs files into these tables. These packages can be found using Bioconductor, a repository that contains a number of useful packages for analyzing biological data in R. Publically available FCS readers in other programming languages (particularly, python and matlab) also exist. The plots in this chapter are generated through the use of Cytobank as well as the programming language R.

Data Visualization of Flow Data: A Brief Overview

Before diving into the details of flow cytometry preprocessing and data analysis, let us take a general look at how flow cytometry data are visualized for each individual parameter, because visualization is currently an integral part of manual and some computational flow cytometry analysis. Readers of this chapter who have worked manually with flow data will recognize the plot shown in Figure 7. Flow cytometry data are typically viewed in one or two dimensions. For one dimension (shown in

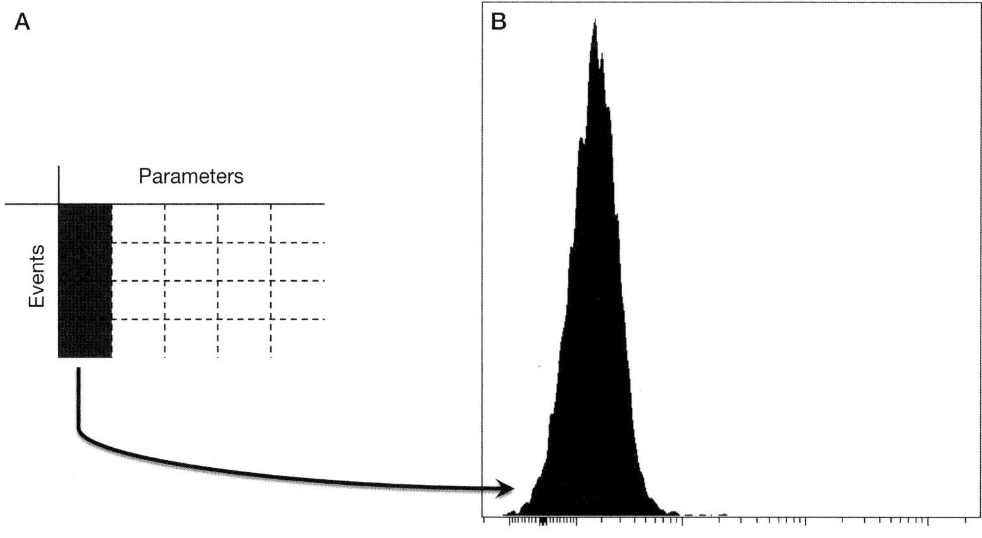

FIGURE 7. Flow cytometry output viewed in one dimension. (*A*) The .fcs file displays the output as a series of events for one parameter. (*B*) The histogram represents the distribution of collected events.

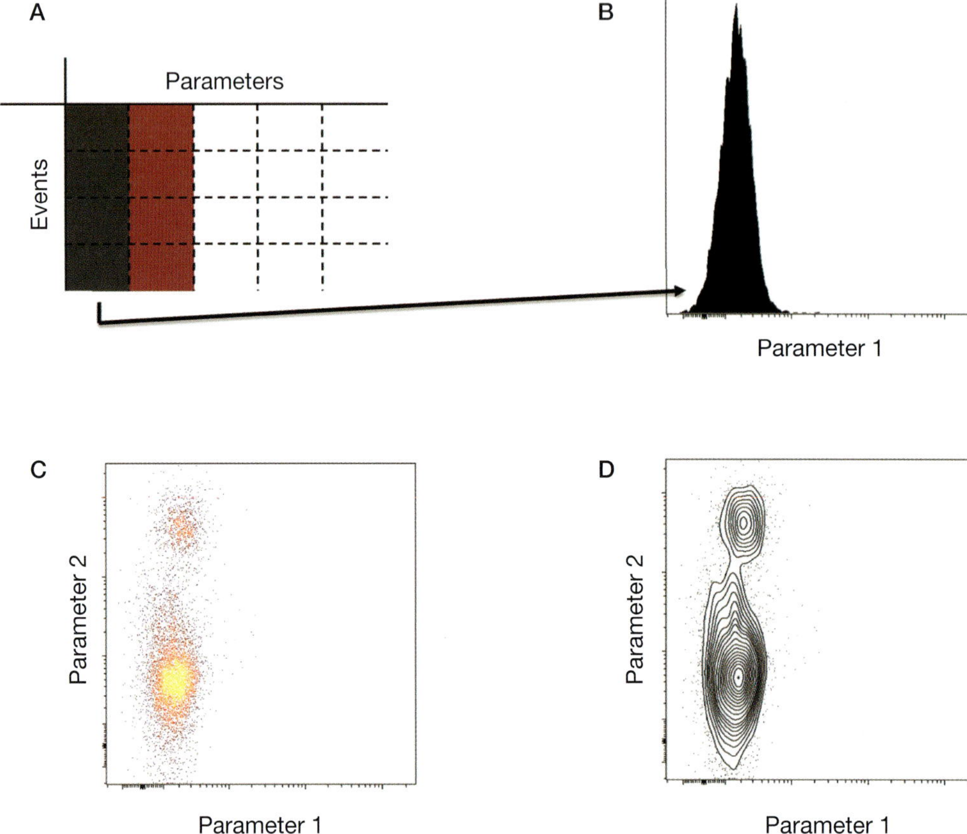

FIGURE 8. Flow cytometry output viewed in two dimensions. (*A*) The .fcs file displays the output as a series of events for two parameters. (*B*) The histogram represents the distribution of collected events. (*C*) 2D density dot plot of the scatter data. (*D*) Contour plot of the scatter data.

Fig. 7A for a single parameter), data can easily be represented as a histogram that represents the distribution of collected events (Fig. 7B).

In two dimensions, it is possible to pinpoint each individual cell within a scatterplot (shown in Fig. 8). In this example, two columns (representing the two parameters) are plotted in the .fcs file (Fig. 8A). The addition of a second parameter facilitates easier detection of subpopulations or states of cells that may have been previously embedded in the histogram; note the difference between Figure 8B and Figure 8C,D. Often, these plots are converted to density-based representations either by area or contours, as can be seen in Figure 8C,D. In these plots, each axis represents one parameter, and each point represents an individual cell.

There are even more ways to visualize three-dimensional (3D) plots, the simplest of which are clouds of points. In these 3D point clouds, it is challenging to define the boundaries of subpopulations in this space visually. Other methods include keeping the 2D plots, but coloring or emphasizing points with respect to their third dimension. Manually, it is difficult to select subpopulations of interest when looking at more than two dimensions. Choosing subpopulations of interest in plots (in one dimension [1D], two dimensions [2D], etc.) is known as *gating*. Traditionally this method is performed by drilling down through a sequence of polygonal shapes drawn by hand to delineate the boundaries of event subpopulations within one or two dimensions. Gating is described in more detail in the section on 1D and 2D analysis.

Traditionally, most flow cytometry in the clinic has focused on the presence of certain immunophenotypes for the diagnosis of hematological malignancies. Often, cells are viewed as having positive or negative populations in one or two dimensions. However, with the increase in the number of markers being measured, from a more general, computational perspective, increasingly more complex biological questions can be asked. These questions can be asked at a high resolution, looking for overall changes between samples. One question, for example, is whether or not the distribution or profile (overall shape) of events changes between samples. Conversely, another goal can be the specific elucidation and identification of individual event subpopulations that are responsible for differentiating between samples. Both approaches are useful and can even include the analysis of traditionally annotated populations found using manual analysis.

We address many methods that can be useful to answering these questions throughout this chapter. For example, when a cancer cell line of interest is exposed to an external stimulus (e.g., a cytokine or a drug), a researcher may want to determine if any signaling changes occur. We investigate this problem as well as related problems throughout the chapter. For those who are interested in looking into this question further, the majority of the data illustrating many steps of flow cytometry analysis comes from phospho-flow experiments in the U937 cell line, and many of the plots were derived online at Cytobank.

Analyzing Flow Cytometry Data: Finding Biologically Relevant Information

Taking a step back, it is important to realize that it is not trivial to move from raw data to graphs and plots that can then be analyzed manually or computationally. Overall, to answer the questions we listed, much of the computational flow analysis is grouped into multiple steps after data collection. Sometimes, some of these steps can be skipped, depending on the question or the data set collected. The steps are as follows:

1. Data preprocessing.
2. Subpopulation finding and feature extraction.
3. Comparing across samples.

There are many ways to perform these steps, ranging from the very simple to the complex, depending on what is needed. Choosing the method is often dependent on the biological problem. In this chapter, we start with simpler analyses and work our way to more complex types of analysis. Let us return to the example in which we have exposed a cancer cell line to a stimulus and want to see if the cancer cells have changed. The only information we have thus far is the table from the .fcs file, the structure of which we have already described. Although we are using a stimulated sample as an example, one could imagine that these methods would work for other problems, such as finding populations of cells with varying cell-surface markers. For the purpose of this chapter, we give many examples using only a few parameters. It is possible, however, to apply many of these methods in higher parameter space. One potential issue that can arise from more-complex, higher-dimensional analyses, however, is the lack of sufficient data in this high-dimensional space to find any biological results that are statistically significant. This difficulty is also known as the *curse of dimensionality*, a problem that is a challenge in flow cytometry data analysis and something to consider when analyzing your own data.

Preprocessing Steps

What are the required preprocessing steps that get us past human-readable data to this human-usable data? Besides converting it into readable form, preprocessing often includes compensation and transformation of the data. *Compensation* is a process that is needed when more fluorescent labels (columns in the .fcs file) are being used at the same time, because there will start to be some spectral overlap between labels. *Spectral resolution* is the ability to distinguish two colors (spectra) from each other when they are detected. There are areas of potential overlap between different spectra that need to be deconvolved. When two colors overlap, the overlaps must be subtracted so that artificial intensities will not be detected. With mass cytometry, compensation is not performed because of the stark contrasts in detectability between rare earth metals. A graph of spectral overlap is shown in Figure 9.

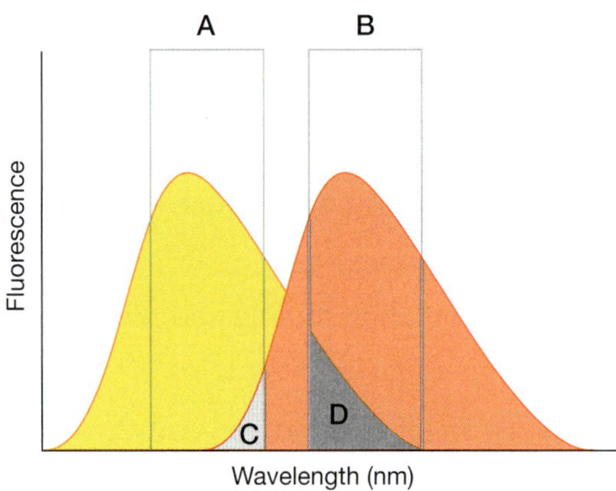

FIGURE 9. Example of spectral overlap. (*A*,*B*) These areas show the filter boundaries for the yellow and red fluorophores being measured. In short, they represent filters that transmit the photons (colors) for detection. Subsequently, *C* and *D* are the parts of the yellow and red spectra that overlap into the other filters and must be subtracted out using compensation.

As a result, in compensation, a series of equations is defined that explains the spillover of fluorescence between neighboring channels. Some flow machines can perform hardware compensation, but usually compensation is performed after data collection. The result is a matrix that defines how to correct for this spillover.

Calculating the compensation matrix is something easily achieved. After running multiple control samples that have been stained with only one of the probes being used in the full experiment (often beads or even cells), automated compensation can be performed by running one of many pieces of code or software, although researchers may wish to adjust compensation manually.

Transformation or, in our case, the scaling of the data on the axes, is also useful for data visualization and human interpretation. In the past, visualization of flow data was kept either on a linear or a log scale. Without going into the details of how the fluorescent signal is amplified for detection inside the flow cytometer, we simply point out that modern transforms scale the data so that they consist of a transition from linear to log-like. These newer transforms need to take into account negative values because modern machines will often perform an instrument-based correction that results in negative values. Several different transforms can be used. In addition to no transform (linear scale), commonly used transforms include the log, logical/biexponential, and hyperbolic arcsine. Other transforms include quadratic, split scale, hyperlog, and the Box-Cox transformation. Depending on the software, a different transform may be used. In Figure 10, data are shown from a stimulated cancer-cell-line experiment on a linear and a hyperbolic arcsine transform (with a cofactor of 150, often a cofactor of 5 is used for mass cytometry). As one can see, using the hyperbolic arcsine tightens the spread on the population. Figure 10A displays data under a linear transform, and Figure 10B displays data under a hyperbolic arcsine. Notice that there are four populations in Figure 10B that are difficult to see in Figure 10A.

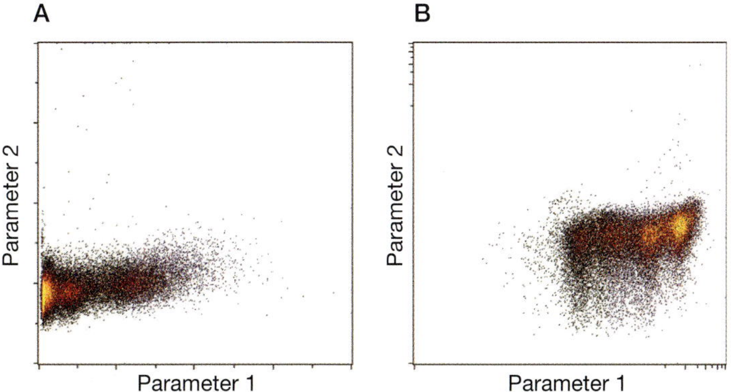

FIGURE 10. Comparison of transformation of data from a stimulated cancer cell line experiment. (*A*) Display of data under a linear transform. (*B*) Display of data under a hyperbolic arcsine.

For stimulated, drug-treated, or general differing sample types (e.g., leukemic vs nonleukemic samples), another preprocessing procedure is to standarize or normalize the treated events with respect to a set of control events that can be a bead or biological sample. Other preprocessing steps include debarcoding of samples. For mass cytometry data, file concatenation is also regularly used.

Subpopulation-Finding and Feature Extraction Methods

One key part of flow cytometry analysis is the extraction of information from raw flow cytometry data in a way that is meaningful and biologically relevant. This information can be at the aggregate level (over many events) or at the subpopulation level. Currently, there exists a vast number of methods to detect and segregate biologically relevant subpopulations. In fact, since 2010, there has been a competition to evaluate current subpopulation-finding methods for flow cytometry data (http://flowcap.flowsite.org).

For the purposes of this chapter, we focus on some general subpopulation-finding methods. First, several subpopulation-finding methods are initially agnostic to the underlying structure of the data (binning methods, etc.). Other methods make assumptions regarding the underlying structure of the data before population discovery. *Gating* is a term that usually refers to the defining of a space where all the cells are believed to be in the same state. Most popular is expert-guided manual gating, in which an expert defines populations of interest. In fact, standard encoding of manually derived gates is performed through the gating-ML standard, which allows different software programs to read or write gates to file. Here, we distinguish subpopulation-finding methods from manual gating either by delineating computation methods as subpopulation finders or automated gating methods.

Regardless of the final populations, feature extraction is necessary to compare across samples. Features may include something as simple as event count or the percent of events in a subpopulation, or other metrics such as the median intensity of a subpopulation. We address this in more detail throughout this section.

1D Methods for Feature Extraction and Subpopulation Finding

Recall the .fcs file format (Fig. 6). Although there can often be a large number of parameters, it is possible to analyze and interpret one parameter at a time. The simplest form of comparing samples is looking at each individually measured parameter. Visualizing each of these dimensions alone reveals information. As a result, visual comparisons between two samples can be made by comparing either two histograms or properties of their distributions (mean, median, etc.). Single-parameter analysis is beneficial because in its aggregate form, it is less subject to noise than higher-dimensional analysis. However, it lacks many of the interparameter relationships that multiparameter flow cytometry allows us to study.

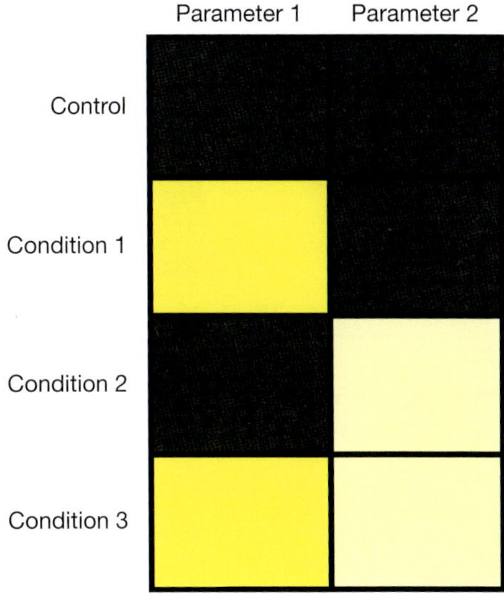

FIGURE 11. Example heatmap is generated in Cytobank. (Bright yellow) High-median intensity; (black) low intensity for the measured fluorophore.

Heatmaps

When visually comparing properties (e.g., median) over many samples, we can extract these features and plot them as a heatmap (Fig. 11). As we can see, one property of one sample parameter is preserved in the heatmap. We can clearly and quickly determine significant changes in a parameter across multiple experiments. The intensity of each element in the heatmap is a direct representation of some property for the current sample. In Figure 11, the example is colored by the median intensity.

With a heatmap, however, we lose some information regarding the shape of the original distributions because each location in a heatmap represents only a summary statistic of the underlying distribution. Although useful, using only the median of a distribution fails to capture a lot of underlying information regarding the data. We lose all of the prior information regarding the histogram shown in Figure 7. Instead, only one value for each histogram is preserved.

1D Binning: Histogram Generation

In addition to summary statistics such as the median, histograms are a common method for representing subpopulations of flow data. Histograms are a series of *bins*, also defined as a calculated range of values. Each event belongs to one bin, lying within a certain range of values. The result is a histogram that describes the data. Figure 12 shows several histograms for one simulated parameter, with varying numbers of bins. The number and bounds of each bin can vary, depending on how they are defined, and the ways in which these bins are constructed can vary. Binning is a method that is used mainly to render the space into discrete areas to make it easier

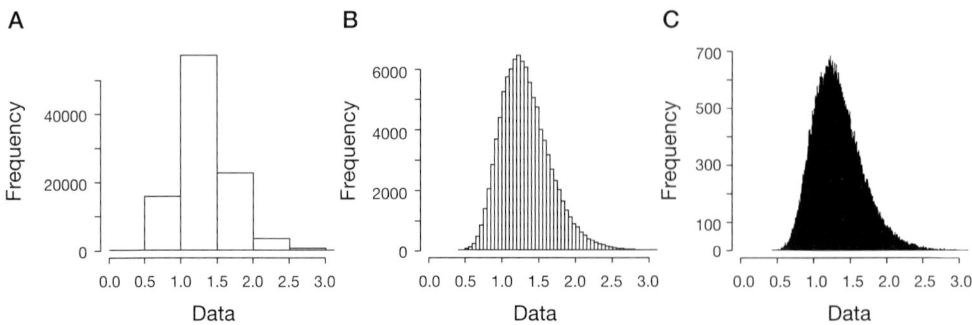

FIGURE 12. Histograms representing a single simulated parameter. (*A–C*) Varying numbers and bounds of bins and how the display of data changes as a consequence.

to compare samples directly with one another. As a result, the cell subpopulations that are derived from binning methods may be difficult to distinguish biologically from cell subpopulations in neighboring bins.

Methods for 1D binning include equal width binning and adaptive binning. Equal width binning is the delineation of *k* distinct bins, where the width of each of the *k* bins is equal. Equal width binning is exactly what the name implies: Data ranges are split into *k* bins (where *k* is an integer, determined by the user) of equal size. Adaptive, also used for probability binning, is a method in which the data are also split into *k* bins, but each bin contains the same number of events per bin. As a result, it retrieves some automated information regarding subpopulations without requiring the complexity of manual gating. Binning methods are especially useful when populations cannot be easily and discretely gated manually. Choosing the number of bins can be difficult, however. If too many bins are chosen, differences in noise may be found when comparing across bins because of the very small number of events per bin.

Once distinct subpopulations of cells have been determined, there are several features that can then be extracted. One common feature is the percent of the total number of cells in an individual bin. This is equivalent to the current number of cells in one bin divided by the total number of cells. Other features include cell count and median fluorescence intensity.

Higher-Dimension Methods for Feature Extraction and Subpopulation Finding

Many of the 2D subpopulation-finding methods considered in this section are also applicable to 1D. They are, however, often used or thought of commonly in two or higher dimensions; therefore, we discuss them here.

Binning can also be extended to higher dimensions. In some ways, binning can be viewed as a way to try to understand an overall picture of the data. Bins are relatively easy to calculate. One of the main benefits of binning is that once bins are

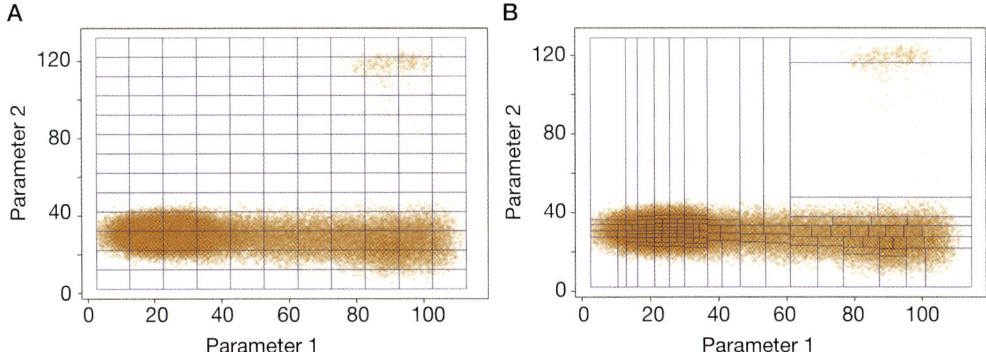

FIGURE 13. Example of binning in two dimensions. (*A*) Equal width binning; (*B*) adaptive binning.

selected, their cell counts or other properties can be compared directly between experiments. In addition, binning can be useful in cases in which it is very difficult to designate a clear difference between subpopulations of interest (which is performed in manual or automated gating). Figure 13 presents an example of binning in two dimensions.

In two dimensions, manual gating is the most widely used method and the current gold standard for finding subpopulations. One of the major benefits of manual gating is that an expert knows and designates which subpopulations are most biologically relevant. Several tools (some of which are listed in Table 1) are very good for performing manual analysis. However, manual gating can be difficult to maintain consistently over experiments and can be difficult to perform over hundreds, if not thousands, of experiments. In addition, once biologically relevant subpopulations have been found (in any type of gating), it is sometimes difficult to match and then compare these subpopulations between experiments.

Cluster Analysis and Mixture Models

For subpopulation finding, several automated gating methods have surfaced. One group of methods involves cluster analysis. Clustering the data is useful when there exist distinct subpopulations. Clustering is a type of unsupervised learning. Unsupervised machine learning is a way of finding the underlying structure of data without prior knowledge of data labels. An example of this is trying to find subpopulations of cells in an overall population without knowing that the population actually consists of a mixture of B cells and T cells.

Cluster analysis is a well-established field because much work has been performed in investigating efficient methods for clustering. One of the main issues with clustering is that many clustering algorithms will find statistically significant, but not necessarily biologically significant, populations. In addition, the same matching problem that we described for manual gating applies to matching these clusters.

There are, however, many benefits to performing clustering on flow data. If subpopulations are relatively distinct and reproducible, clustering is a good technique for finding these subpopulations. Some of these unsupervised techniques find separate distinct groups of cells. Other techniques describe a distribution of these cells.

k-means clustering is a subpopulation-finding method that tries to assign each individual cell in our data to one of *k* clusters. Also described in the imaging chapter (Chapter 5), *k* means is an iterative process. *k* locations are first selected (randomly or based on initial data). These locations are the initial centers of all of the clusters. Next, every cell in the data set is assigned to its closest cluster center. The centers of these cell clusters are then computed, and every cell is assigned again. This process iterates until the cell assignments in these clusters do not change. One of the important factors to remember about *k* means and other clustering techniques is that several clusters are required. This is either preset (*k*) or can be learned. There are methods that can be used for selecting the optimal number of clusters. Many other types of subpopulation finders use clustering after first projecting the data into another space and then clustering in this new space (Zare et al. 2010; Bodenmiller et al. 2012; Shekhar 2014). Other types of clustering try to improve cluster selection by incorporating randomness, weights, or even probabilities. Projecting into a new space helps to reduce the number of dimensions of the data, because many experiments can measure tens of parameters (and thus, tens of dimensions).

Feature extraction from *k*-means clustering includes similar methods to those of binning. Because *k*-means clustering finds distinct subpopulations of cells, cell percent, median intensity, and other similar metrics can similarly be used for *k* means. Recently, methods such as hierarchical clustering have been used in group mass cytometry data before data visualization (Bendall 2011; Qiu 2011).

Mixture modeling via probability distributions (Gaussians, *t* distributions, etc.) is somewhat different from clustering methods. Mixture modeling has been widely used in flow cytometry analysis (Lo et al. 2008; Pyne et al. 2009). One way to think of *k* means is overlaying a structure of clusters on the data. Mixture modeling is similar in that it is trying to overlay a group of distributions on the cell data. These distributions, however, are probability densities that overlap with each other and are not distinct clusters. Actually determining the location of distributions is performed via an iterative approach, also known as the expectation–maximization (EM) algorithm. There are off-the-shelf tools that implement the EM algorithm in order to fit these mixture models to data, so we will not go into the details here. However, the following discussion describes the general idea behind how a mixture model solution is found. For those who do not know about probability distributions, the discussion in Box 2, Probability Distributions, is particularly helpful.

Let *k* be the current number of mixtures that we want to find (similar to the *k*-means clustering problem previously described).

BOX 2. Probability Distributions

What is a probability distribution? A simple example is a normal, or Gaussian distribution. Let us say that we have a dartboard that looks like the one below (Box 2, Fig. 1).

−5	−4	−3	−2	−1	0	1	2	3	4	5

BOX 2, FIGURE 1. Illustration of a dartboard with distribution of areas relative to the target.

If we were to throw a large number of darts at the dartboard, aiming for the center, we would probably not hit the center every time. Instead, we would get a distribution of events. Each dart thrown could land closer or farther from the center. One would expect, however, that by aiming at the center more of the dart hit events would be in the center of the dartboard. We can plot this distribution on a graph, as seen in the figure below (Box 2, Fig. 2).

As we throw more and more darts, the distribution becomes more uniform. If we allow the darts to fall in between values (e.g. −2.5 instead of −2 or −3), then the distribution will fill out even more. As the number of darts approaches infinity, we can plot the probability density function (the eventual distribution as the number of darts reaches infinity), shown below (Box 2, Fig. 3).

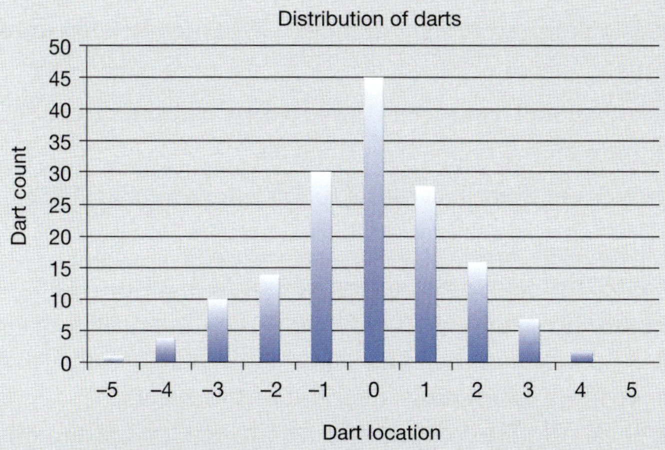

BOX 2, FIGURE 2. Distribution of dart hits by location.

(Continued.)

This shape looks somewhat similar to many of the flow cytometry data histograms we displayed before. As a result, many methods for finding subpopulations in flow data fit different types of these distributions. Some example distributions include gaussian, *t* distribution, skew *t*, and wavelet.

A probability distribution can be represented by just a few numbers. For example in the case of a gaussian distribution, it can be represented by two numbers, because it is symmetric. This is equivalent to an equal chance of throwing a dart and having it go either left or right of the center on the dartboard. These two numbers consist of the mean (average) of the distribution, as well as the standard deviation of the distribution. These are commonly represented by two symbols, the symbol μ (mu) for the average and σ (sigma) for the standard deviation.

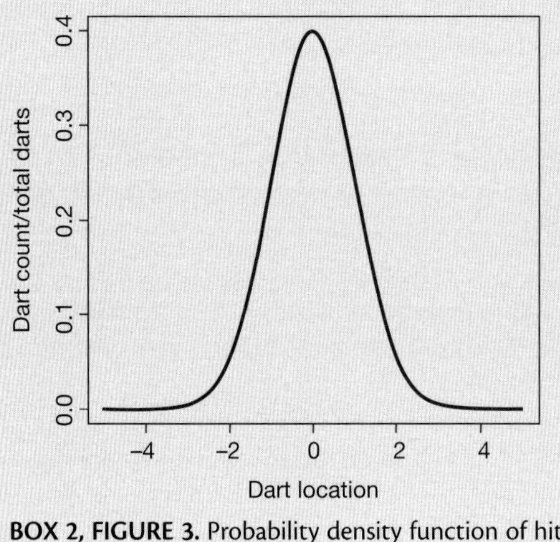

BOX 2, FIGURE 3. Probability density function of hits.

1. Start with *k* distributions (e.g., Gaussians, each with some guessed average and standard deviation [μ, σ]).
2. For every data point, store the "strength" of its membership in each of these *k* distributions.
3. Reestimate the model parameters (all of the [μ, σ] values from Step 1) based on the values from Step 2 to better fit the data.
4. Repeat Steps 2 and 3 until all of the (μ, σ) values remain stable.

Although described somewhat differently from *k* means, using the EM algorithm to find a mixture model that fits the data is similar, such that we are trying to overlay a set of distributions that best capture the data. Instead of assigning events to clusters, we are fitting continuous distributions to events.

As a result, there are added options for feature extraction from mixture models. In addition to the direct assignment of events to individual mixtures, the properties of these mixture models can become features for comparison. These features, for example, include the mean (μ) and standard deviation (σ) of Gaussian distributions (see Box 2, Probability Distributions).

Mixture models can be applied in more than two dimensions, however, the stability of the models can become less reliable as we increase the number of parameters. For markers that are very distinct and have less-complex populations, mixture models work very well. One of the hardest parts of automating flow cytometry analysis is that many auto-population-finding techniques find populations that (1) are not always reproducible and (2) may or may not be biologically relevant. As a result, comparing across these hard-to-define populations is very difficult. Manual gating often wins as the method of choice over automated methods, because (1) the gates are universally applicable, and (2) the populations derived from gating are easy to interpret, because they are matched by eye by the manual gater.

As a result, the newest wave of computational methods take two different approaches. The first approach is to try to find oddly shaped populations the same way an experimentalist would manually gate their data. As a result, the idea is that these populations will be more reproducible and can be directly compared. Usually, this involves either clustering or using a density estimate to find two populations, then merging these clusters into bigger, and often biologically interpretable, clusters (Walther et al. 2009; Aghaeepour et al. 2011). The second approach is to find many reproducible populations that do not necessarily have a lot of pre-understood biological meaning, but for which changes can be tracked across samples.

Nonparametric Population-Finding Methods

Besides mixture modeling, which requires a distribution (also known as parametric approaches), there are nonparametric, or distribution-free methods that do not require fitting a probability distribution to the data. Not only does this include the list of binning methods we described above, but also more complex density-based methods. Some of these methods are very similar to those in image processing segmentation. Instead of thresholds in images, however, thresholds define new gates on the data. A list of existing and newer tools can be found in the online Supplemental Materials. There are benefits and difficulties associated with nonparametric approaches. One of the benefits is that many biologically relevant populations have been discovered using these density-based nonparametric methods. They are able to capture strange shapes that probability distributions cannot. However, comparing across samples can still be difficult because, for some methods, the number of subpopulations may sometimes vary. In addition, many of these clustering and density-based methods are computationally inefficient and require either downsampling of the data or much more computing power than the average personal computer. Finally, many tuning parameters are needed for some of these methods, which may be complicated if they need to be manually tuned.

Finally, there are supervised methods for extracting information from data. These methods incorporate previous knowledge regarding the data into population finding and analysis. These are mainly tools that are commercially available to users who have prior knowledge of the structure of their data.

Feature Extraction in Two and Higher Dimensions

There are several properties that can be used to describe flow cytometry data in multidimensional space, many of which encompass properties that can be measured in one dimension as well. Many of the more straightforward metrics include the cell count per population and the percent of total cells within a population, as well as the median intensity of a population. These metrics are useful both for manual as well as automated analysis and describe the data well. For parametric methods, properties such as mean and standard deviation are also potential features that can be used to describe the data (μ, σ). Featurization of event populations can vary but is often selected based on interpretability. The main utility of feature extraction allows for universal features along which to compare samples. Here, one of the hardest barriers to overcome is that many populations of interest are not easily compared across samples. Subsequently, many methods now first combine flow cytometry data across samples and then perform subpopulation finding, thus matching populations across samples. In these cases, however, it is important to make sure that sample variation due to experimental and machine variation is reduced and that biological variation is captured.

Comparing Across Samples

Within statistics, there exist several methods that facilitate the comparison of samples using the features described in the previous section. Comparing across samples can be categorized using a few analytical questions. Here, we group them into the following problems: (1) sample classification, (2) quantitative difference, and (3) informative event problems. This section aims to explain what the three problems are so that you can move forward with thinking about your data. Scientifically, one may try to answer one, two, or even all three in the same set of experiments. Answering these three questions is still a growing field; thus, the examples given in this section are references to prior work but are not necessarily the standard methods for performing analysis.

The Sample Classification Problem

The sample classification problem is one that is based on an initial need by the investigator to distinguish between different classes of experimental samples. There are two potential approaches: the unsupervised approach and the supervised approach. The unsupervised approach is often used when the investigator does not know how many different classes of samples exist (e.g., within a group of samples, there are several potential unknown sample types). In this case, unsupervised methods such as k-means clustering and mixture modeling described above can be implemented on a sample level after features have been extracted from flow data, finding inherent classes by clustering samples based on their features. The supervised approach is used when the investigator has some prior knowledge regarding the sample classes (e.g., patients who have a disease and those who do not) and wants to be able to predict whether or not a sample belongs in a known class. An example of this is to find a classifier that differentiates between patients with or without a particular hematological malignancy. In particular, the primary goal of an investigator could be to determine a proper classification between two cancer types. Once features have been extracted from cell subpopulations, existing machine learning classifiers can be used and evaluated for the classification problem (see Chapter 4). These include classifiers that have previously been developed for other fields including microarray analysis. One in particular that has been useful is prediction across microarrays (PAM) (Tibshirani et al. 2002; Chen 2012); however, many other classification models have been developed and used (Bashashati 2009).

The Quantitative Difference Problem

This problem focuses on quantitatively determining the differences between various samples using flow cytometry data. Typically, this is performed by directly comparing features across samples using a generalized metric. These quantitative comparisons can either be applied directly to binned/gated intensity values or on more complex feature sets. The goal of approaching flow data in this way is to create a quantitative description between samples, indicating similarity or dissimilarity. These distances can be further used in order to separate samples into classes (via clustering, etc.), but is not the initial goal. True distance metrics can be used to compare these features, including Euclidean, quadratic form, and the earth mover's distance. However, other statistical tests have also been used, including the Kolmogorov–Smirnov test as well as versions of the χ^2 test. Research in this area in the field of flow cytometry has been growing during the last 15 years.

The Informative Event Problem

Although it is important to determine overall classes or similarity between samples, a researcher often seeks to determine an understanding of which cells are responsible for these class differences or dissimilarity between samples. The approach here is to narrow down the features that are robustly predictive of classes or features that change significantly across dissimilar samples. In particular, this group of problems has major implications for understanding the mechanisms of disease because aberrant cell populations are a major problem in disease. At this point in time, the event responsibility problem is a growing field with respect to clinical applications. The discovery of these cell events that differentiate between disease types varies in complexity, from a percentile-based threshold (Kotecha et al. 2008) to a few (Aghaeepour et al. 2011) to 10s of features necessary to distinguish between diseases (Steinbrich-Zöellner et al. 2008). When narrowed down to informative events, these features can be used for survival analysis and other statistical analyses currently used for traditional patient prognosis.

Exploratory Analysis

Finally, exploratory analysis of flow cytometry data is beginning to grow, primarily because of the ability to measure tens of parameters simultaneously. As a result, panel design for flow cytometry assays is no longer restricted to only the direct proteins that are potentially involved in a phenotypical outcome, but other, less understood markers as well. Because of this, a new era of large-scale data analysis and visualization techniques is beginning to emerge in application to flow and mass cytometry data including newer methods such as a minimum spanning-tree-based visualization tool for mass and flow cytometry data entitled SPADE (Qiu et al. 2011).

FUTURE DIRECTIONS

The field of proteomics is advancing quickly. New technologies that are being developed are facilitating a better understanding of the proteome. With these advances, however, several challenges arise, many of which need to be addressed in the near future.

Data Variability

A problem that plagues many high-throughput technologies—data variability due to instrument or operator differences—is also a problem in proteomics. Instruments at different institutions can generate different results while analyzing the same sample under the same protocol. Day-to-day variability in instrument sensitivity is a frequently observed phenomenon; in fact, instruments can even vary over the course of a single run. This variability presents a notable challenge for informaticians, who often need to compare data from different techniques or to aggregate data for meta-analyses. Finding technical controls and other standards of comparison for experiments is an important part of the future of proteomics. Many researchers currently try to standardize based on internal biological controls; however, technical replicates, cell-based controls, and bead-based controls are becoming more important for normalization of data, for example, within mass cytometry (Finck et al. 2013).

Structured Annotation for Data Sharing

Currently, files generated from mass spectrometers and flow cytometers are on the order of the size of megabytes to gigabytes of data per run. There exist many databases that can store these data for collaboration and distribution. The barrier for collaboration and advanced analysis is not the storage and processing of these data, but the need for standards for annotation across platforms. In flow cytometry, standardizing antigen names, as well as file and other experimental naming conventions, would greatly enhance the ability to share and analyze data sets providing the foundation for cross sample and replicate analysis across many experiments and collaborations.

Clinical Applications

We are now beginning to have access to proteomics data of the coverage (hundreds to thousands of proteins), scale (tens to hundreds of samples), and throughput (a few days to weeks per experiment) to bring MS proteomics to the study of clinical questions. The cost of a false positive in clinical applications is high; as mass spectrometry moves to clinical use, care must be taken to ensure that results are properly validated and computationally derived conclusions are sufficiently conservative.

Regardless of these difficulties, we can already ask advanced questions using proteomics data. We can identify and quantify thousands of proteins in a given sample. We can investigate proteins at a very specific level, such as alternative splicing, post-translational modifications, and protein regulation. We can reconstruct functional pathways, even at single-cell resolution. Proteomics will continue to rely on the advancement of computational algorithms and techniques for turning raw proteomics data into biological knowledge.

ACKNOWLEDGMENTS

We thank Steve Briggs, Weijun Qian, and Mario Roederer for review and feedback on earlier versions of this chapter.

REFERENCES

Aghaeepour N, Nikolic R, Hoos HH, Brinkman RR. 2011. Rapid cell population identification in flow cytometry data. *Cytometry A* **79:** 6–13.

Aghaeepour N, Chattopadhyay PK, Ganesan A, O'Neill K, Zare H, Jalali A, Hoos HH, Roederer M, Brinkman RR. 2012. Early immunologic correlates of HIV protection can be identified from computational analysis of complex multivariate T-cell flow cytometry assays. *Bioinformatics* **28:** 1009–1016.

Bashashati A, Brinkman R. 2009. A survey of flow cytometry data analysis methods. *Adv Bioinform* **2009:** 1–19.

Bendall SC, Simonds EF, Qiu P, Amir E-aD, Krutzik PO, Finck R, Bruggner RV, Melamed R, Trejo A, Ornatsky OI, et al. 2011. Single-cell mass cytometry of differential immune and drug responses across a human hematopoietic continuum. *Science* **332:** 687–696.

Bodenmiller B, Zunder ER, Finck R, Chen TJ, Savig ES, Bruggner RV, Simonds EF, Bendall SC, Sachs K, Krutzik PO, et al. 2012. Multiplexed mass cytometry profiling of cellular states perturbed by small-molecule regulators. *Nat Biotechnol* **30:** 858–867.

Chen TJ. 2012. "De novo reconstruction of cell cycle and chemotherapeutic mechanisms in cancer." PhD thesis, Stanford University, Stanford, California.

Eng JK, McCormack AL, Yates JR. 1994. An approach to correlate tandem mass spectral data of peptides with amino acid sequences in a protein database. *J Am Soc Mass Spectrom* **5:** 976–989.

Finck R, Simonds EF, Jager A, Krishnaswami S, Sachs K, Fantl W, Pe'er D, Nolan GP, Bendall SC. 2013. Normalization of mass cytometry data with bead standards. *Cytometry A* **83:** 483–494.

Kotecha N, Krutzik PO, Irish JM. 2010. Web-based analysis and publication of flow cytometry experiments. *Curr Protoc Cytom* **10:** 17.

Kotecha N, Flores NJ, Irish JM, Simonds EF, Sakai DS, Archambeault S, Diaz-Flores E, Coram M, Shannon KM, Nolan GP, Loh ML. 2008. Single-cell profiling identifies aberrant STAT5 activation in myeloid malignancies with specific clinical and biologic correlates. *Cancer Cell* **14:** 335–343.

Kotecha N, Krutzik PO, Irish JM. 2010. Web-based analysis and publication of flow cytometry experiments. *Curr Protoc Cytom* **53:** 10.17.1–10.17.24.

Lo K, Brinkman RR, Gottardo R. 2008. Automated gating of flow cytometry data via robust model-based clustering. *Cytometry A* **73:** 321–332.

Pyne S, Hu X, Wang K, Rossin E, Lin TI, Maier LM, Baecher-Allan C, McLachlan GJ, Tamayo P, Hafler DA, et al. 2009. Automated high-dimensional flow cytometric data analysis. *Proc Natl Acad Sci* **106:** 8519–8524.

Qiu P, Simonds EF, Bendall AC, Gibbs KD Jr, Bruggner RV, Linderman MD, Sachs K, Nolan GP, Plevritis SK. 2011. Extracting a cellular hierarchy from high-dimensional cytometry data with SPADE. *Nat Biotechnol* **29:** 886–891.

Shapiro H. 2005. *Paractical Flow Cytometry*, 4th ed. Wiley, New York.

Shekhar K, Brodin P, Davis MM, Shakraborty AK. 2014. Automatic classification of cellular expression by nonlinear stochastic embedding (ACCENSE). *Proc Natl Acad Sci* **111:** 202–207.

Steinbrich-Zöllner M, Grün JR, Kaiser T, Biesen R, Raba K, Wu P, Thiel A, Sieper J, Burmester GR, Radbruch A, Grützkau A. 2008. From transcriptome to cytome: Integrating cytometric profiling, multivariate cluster, and prediction analyses for a phenotypical classification of inflammatory diseases. *Cytometry A* **73:** 333–340.

Tibshirani R, Hastie T, Balasubramanian N, Chu G. 2002. Diagnosis of multiple cancer types by shrunken centroids of gene expression. *Proc Natl Acad Sci* **99:** 6567–6572.

Walther G, Zimmerman N, Moore W, Parks D, Meehan S, Belitskaya I, Pan J, Herzenberg L. 2009. Automatic clustering of flow cytometry data with density-based merging. *Adv Bioinfor* **2009:** 7.

Zare H, Shooshtari P, Gupta A, Brinkman RR. 2010. Data reduction for spectral clustering to analyze high throughput flow cytometry data. *BMC Bioinformatics* **11:** 403.

FURTHER READING

Achuthanandam R, Quinn J, Capocasale RJ, Bugelski PJ, Hrebien L, Kam M. 2008. Sequential univariate gating approach to study the effects of erythropoietin in murine bone marrow. *Cytometry A* **73:** 702–714.

Aebersold R, Mann M. 2003. Mass spectrometry-based proteomics. *Nature* **422:** 198–207.

Baggerly KA. 2001. Probability binning and testing agreement between multivariate immunofluorescence histograms: Extending the chi-squared test. *Cytometry* **45:** 141–150.

Bagwell C. (2005). Hyperlog: A flexible log-like transform for negative, zero, and positive valued data. *Cytometry A* **42:** 34–42.

Bashashati A, Brinkman RR. 2009. A survey of flow cytometry data analysis methods. *Adv Bioinformatics* **2009:** 584603.

Bernas T, Asem EK, Robinson JP, Rajwa B. 2008. Quadratic form: A robust metric for quantitative comparison of flow cytometric histograms. *Cytometry A* **73:** 715–726.

Boddy L, Wilkins MF, Morris CW. 2001. Pattern recognition in flow cytometry. *Cytometry* **44:** 195–209.

Boedigheimer MJ, Ferbas J. 2008. Mixture modeling approach to flow cytometry data. *Cytometry A* **73:** 421–429.

Costa ES, Arroyo ME, Pedreira CE, García-Marcos MA, Tabernero MD, Almeida J, Orfao A. 2006. A new automated flow cytometry data analysis approach for the diagnostic screening of neoplastic B-cell disorders in peripheral blood samples with absolute lymphocytosis. *Leukemia* **20:** 1221–1230.

Craig R, Cortens JP, Beavis RC. 2005. The use of proteotypic peptide libraries for protein identification. *Rapid Commun Mass Spectrom* **19:** 1844–1850.

Davey HM. 2010. Prospects for the automation of analysis and interpretation of flow cytometric data. *Cytometry A* **77:** 3–5.

Dvorak JA, Banks SM. 1989. Modified Box-Cox transform for modulating the dynamic range of flow cytometry data. *Cytometry* **10:** 811–813.

Elias JE, Gygi SP. 2007. Target-decoy search strategy for increased confidence in large-scale protein identifications by mass spectrometry. *Nat Methods* **4:** 207–214.

Geer LY, Markey SP, Kowalak JA, Wagner L, Xu M, Maynard DM, Yang X, Shi W, Bryant SH. 2004. Open mass spectrometry search algorithm. *J Proteome Res* **3:** 958–964.

Higdon R, Kolker E. 2007. A predictive model for identifying proteins by a single peptide match. *Bioinformatics* **23:** 277–280.

Jain AK, Duin RPW, Mao JC. 2000. Statistical pattern recognition: A review. *Proc IEEE Int Conf Data Min* **22:** 4–37.

Keller A, Nesvizhskii AI, Kolker E, Aebersold R. 2002. Empirical statistical model to estimate the accuracy of peptide identifications made by MS/MS and database search. *Anal Chem* **74:** 5383–5392.

Kuhn E, Wu J, Karl J, Liao H, Zolg W, Guild B. 2004. Quantification of C-reactive protein in the serum of patients with rheumatoid arthritis using multiple reaction monitoring mass spectrometry and ^{13}C-labeled peptide standards. *Proteomics* **4:** 1175–1186.

Lange V, Picotti P, Domon B, Aebersold R. 2008. Selected reaction monitoring for quantitative proteomics: A tutorial. *Mol Syst Biol* **4:** 222.

Lo K, Hahne F, Brinkman RR, Gottardo R. 2009. flowClust: A Bioconductor package for automated gating of flow cytometry data. *BMC Bioinformatics* **10:** 145.

Martínez A, Aymerich M, Castillo M, Colomer D, Bellosillo B, Campo E, Villamor N. 2003. Routine use of immunophenotype by flow cytometry in tissues with suspected hematological malignancies. *Cytometry B Clin Cytom* **56:** 8–15.

Naumann U, Wand MP. 2009. Automation in high-content flow cytometry screening. *Cytometry A* **75:** 789–797.

Nesvizhskii AI, Keller A, Kolker E, Aebersold R. 2003. A statistical model for identifying proteins by tandem mass spectrometry. *Anal Chem* **75:** 4646–4658.

Novo D, Wood J. 2008. Flow cytometry histograms: Transformations, resolution, and display. *Cytometry A* **73:** 685–692.

Parks DR, Roederer M, Moore WA. 2006. A new "logicle" display method avoids deceptive effects of logarithmic scaling for low signals and compensated data. *Cytometry A* **69:** 541–551.

Perkins DN, Pappin DJ, Creasy DM, Cottrell JS. 1999. Probability-based protein identification by searching sequence databases using mass spectrometry data. *Electrophoresis* **20:** 3551–3567.

Qian WJ, Liu T, Petyuk VA, Gritsenko MA, Petritis BO, Polpitiya AD, Kaushal A, Xiao W, Finnerty CC, Jeschke MG, et al. 2009. Inflammation and the host response to injury large scale collaborative research program. Large-scale multiplexed quantitative discovery proteomics enabled by the use of an ^{18}O-labeled "universal" reference sample. *J Proteome Res* **8:** 290–299.

Qian WJ, Petritis BO, Kaushal A, Finnerty CC, Jeschke MG, Monroe ME, Moore RJ, Schepmoes AA, Xiao W, Moldawer LL, et al. 2010. Inflammation and the host response to injury large scale collaborative research program. Plasma proteome response to severe burn injury revealed by ^{18}O-labeled "universal" reference-based quantitative proteomics. *J Proteome Res* **9:** 4779–4789.

Roederer M, Hardy RR. 2001. Frequency difference gating: A multivariate method for identifying subsets that differ between samples. *Cytometry* **45:** 56–64.

Roederer M, Treister A, Moore W, Herzenberg LA. 2001. Probability binning comparison: a metric for quantitating univariate distribution differences. *Cytometry* **45:** 37–46.

Rogers WT, Moser AR, Holyst HA, Bantly A, Mohler ER III, Scangas G, Moore JS. 2008. Cytometric fingerprinting: Quantitative characterization of multivariate distributions. *Cytometry A* **73:** 430–441.

Shen Y, Tolić N, Hixson KK, Purvine SO, Anderson GA, Smith RD. 2008. De novo sequencing of unique sequence tags for discovery of post-translational modifications of proteins. *Anal Chem* **80:** 7742–7754.

Sigdel TK, Kaushal A, Gritsenko M, Norbeck AD, Qian WJ, Xiao W, Camp DG II, Smith RD, Sarwal MM. 2010. Shotgun proteomics identifies proteins specific for acute renal transplant rejection. *Proteomics Clin Appl* **4:** 32–47.

Smith RD, Anderson GA, Lipton MS, Pasa-Tolic L, Shen Y, Conrads TP, Veenstra TD, Udseth HR. 2002. An accurate mass tag strategy for quantitative and high-throughput proteome measurements. *Proteomics* **2:** 513–523.

Xie F, Liu T, Qian WJ, Petyuk VA, Smith RD. 2011. Liquid chromatography-mass spectrometry-based quantitative proteomics. *J Biol Chem* **286:** 25443–25449.

Zeng Q, Wand M, Young AJ, Rawn J, Milford EL, Mentzer SJ, Greenes RA. 2002. Matching of flow-cytometry histograms using information theory in feature space. *Proc AMIA Symp* **2002:** 929–933.

Zhang J, Xin L, Shan B, Chen W, Xie M, Yuen D, Zhang W, Zhang Z, Lajoie GA, Ma B. 2012. PEAKS DB: De novo sequencing assisted database search for sensitive and accurate peptide identification. *Mol Cell Proteomics* **11:** M111.010587.

WWW RESOURCE

http://flowcap.flowsite.org FlowCAP–Flow Cytometry: Critical Assessment of Population Identification Methods

10

Knowledge Base–Driven Pathway Analysis

Purvesh Khatri

Stanford University School of Medicine, Stanford, California 94305

The advent of high-throughput sequencing and profiling techniques (e.g., microarray, proteomics, metabolomics, RNA-seq, etc.) has transformed biological research by allowing for the quantification of many transcripts in a cell in a single experiment. A typical RNA-seq experiment allows for unbiased profiling of tens of thousands of genes. Irrespective of the technology used, analysis of high-throughput data typically yields a list of differentially expressed genes or proteins. Such a list can be extremely useful in narrowing down the search for the genes responsible for a given phenomenon or phenotype. However, it often fails to provide any mechanistic insights into the underlying biology. In other words, the advent of high-throughput profiling technologies presented a new challenge: that of extracting biological meaning from a list of differentially expressed genes and proteins.

One approach to extract underlying biology is to analyze data at a pathway level. Pathway-level analysis of high-throughput molecular measurements is very appealing for two reasons. First, grouping thousands of genes, proteins, and other biological molecules by the pathways in which they are involved reduces the dimensionality considerably, to dozens or hundreds of groups depending on the specific strategy used. Second, identification of pathways that are different between two phenotypes may have more explanatory power than a set of different genes or proteins (Glazko and Emmert-Streib 2009).

WHAT IS A PATHWAY?

According to the *Merriam-Webster Dictionary*, a pathway is "the sequence of reactions by which one substance is converted into another" (see http://www.merriam-webster.com/dictionary/pathway). This generally accepted definition assumes an order ("the sequence") in which reactions should happen, as well as an implicit direction of the conversion.

The term "pathway" has been used in broader contexts in the literature to represent a particular biological phenomenon (or semi-arbitrary slice of some phenomenon). There are various ways of representing or quantifying that phenomenon, such as (but not limited to) gene annotation (gene ontology [GO]), biologically relevant gene sets, or protein–protein interaction networks. However, none of these describes the order and the direction of the processes and interactions. On the other hand, a number of knowledge bases (e.g., KEGG, BioCarta, Reactome, and Pathway Interaction Database [PID]) describe pathways with a specific order of reactions among gene products, along with the inputs and outputs of each reaction, the type of reaction, and the location of the reaction.

In this chapter, we use the term "pathway" to include all of these: GO terms, gene sets, interaction networks, and ordered reactions. Typically, this type of pathway analysis is referred to as "knowledge base–driven pathway analysis," because it is formed based on information that is already familiar to the researcher, rather than specific experimental results. When high-throughput molecular measurements (i.e., experimental data) are used to define pathways, it is referred to as "data-driven pathway analysis," which is the focus of the next chapter.

Before discussing the current knowledge base–driven pathway analysis approaches, we should note that virtually all of the approaches discussed are equally applicable to a variety of high-throughput technologies, including protein microarrays, metabolomics, etc. In this chapter, however, we focus on analyses of gene expression as a particularly important and widely studied category.

It is beyond the scope of this chapter to compare the large number of analytic methods covered by such a broad application of the term pathway. Therefore, instead of discussing every available analysis approach, we group them into three different generations or types based on their function (overrepresentation analysis, functional class scoring, and pathway topology (PT)–based approaches), ending with knowledge base–driven analyses. We discuss the advantages and disadvantages of each generation and then compare some of the existing tools from each generation. Finally, we describe the challenges for knowledge base–driven pathway analysis.

FIRST GENERATION: OVERREPRESENTATION ANALYSIS APPROACHES

The immediate need for analyzing microarray gene expression data and the emergence of GO gave rise to overrepresentation analysis (**ORA**) that uses one or more variations of the following strategy (Fig. 1, upper panel) (Castillo-Davis and Hartl 2002; Khatri et al. 2002; Berriz et al. 2003; Doniger et al. 2003; Draghici et al. 2003; Beissbarth and Speed 2004; Boyle et al. 2004; Martin et al. 2004). First, for each gene in the input list, its pathway annotations are retrieved. Then, for each

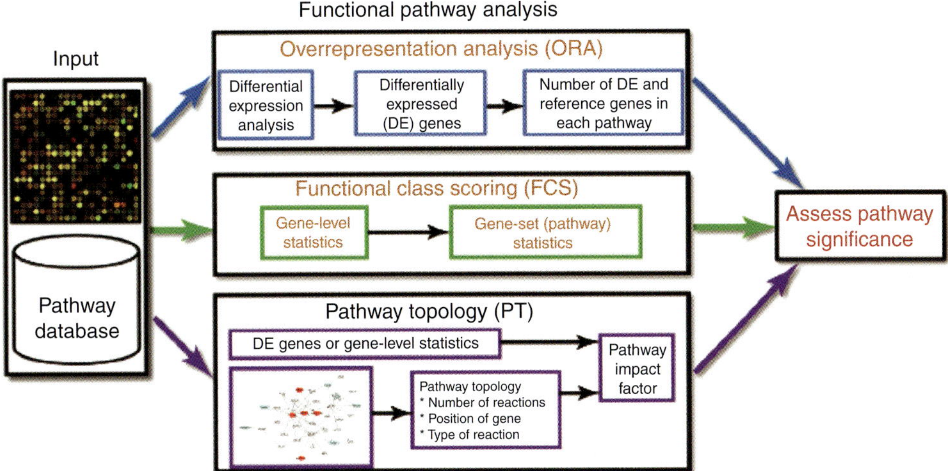

FIGURE 1. Overview of existing pathway analysis methods using gene-expression data as an example. Note that this overview is equally applicable to molecular measurements using proteomics and any other high-throughput technologies. The data generated by an experiment using a high-throughput technology (e.g., microarray, proteomics, metabolomics), along with functional annotations (pathway database) of the corresponding genome, are input to virtually all pathway analysis methods. Whereas ORA methods require that the input is a list of differentially expressed genes, FCS methods use the entire data matrix as input. In addition to functional annotations of a genome, PT-based methods use the number and type of interactions among gene products, which may or may not be a part of a pathway database. The result of every pathway analysis method is a list of significant pathways in the condition under study. DE, differentially expressed. (Redrawn from Khatri et al. 2012.)

pathway, input genes that are known to be in the pathway are counted. This process is repeated for an appropriate background list of genes (e.g., all genes measured on a microarray). Next, pathways are identified that are overrepresented or underrepresented in the list of differentially expressed genes compared with the reference list of genes generated by computing statistical significance for each pathway (e.g., using Fisher's exact test, hypergeometric distribution, or binomial distribution) (Khatri and Draghici 2005) (see Chapter 1). For instance, the p value of having x genes or fewer in pathway P can be calculated using Equation 1 as follows:

$$p = \sum_{i=0}^{x} \frac{\binom{M}{i}\binom{N-M}{K-i}}{\binom{N}{K}}, \tag{1}$$

where N is the number of genes on a microarray, M is the number of genes from the microarray in a pathway P, K is the number of differentially expressed genes, and x is the number of differentially expressed genes in pathway P. This corresponds to a p value for an underrepresented pathway, which can be calculated as

$$p = 1 - \sum_{i=0}^{x} \frac{\binom{M}{i}\binom{N-M}{K-i}}{\binom{N}{K}}. \tag{2}$$

Intuitively, the p value represents how surprising it is to observe so many genes from a particular pathway (i.e., how likely are the experimental results reproducible?). That is, if in an experiment we observe more input genes than expected by chance from a particular pathway, we are more inclined to believe that a pathway is significantly affected by a particular experiment, although this does not indicate the magnitude of the differences observed (discussed in the following section).

Limitations

Despite the availability of a large number of tools and their widespread use, ORA has a number of limitations. First, the statistics used by ORA (e.g., Fisher's exact test, hypergeometric distribution, binomial distribution, χ^2 distribution, etc.) are independent of the magnitude of the measured changes (see Equations 1 and 2). By discarding this information, ORA treats each gene equally, i.e., a gene that changes 10-fold in expression is counted the same as a gene that changes twofold. However, genes are expressed to different extents in any given condition. Data providing information about the extent of regulation (e.g., fold changes, significance of a change, etc.) can be useful in assigning different weights to input genes as well as to the pathways in which they are involved, which in turn can provide more information than simple ORA approaches. To do so, one would need to calculate odds ratios, which measure the size of an effect (Chapter 1). In the microarray example given here, an odds ratio would give us a fold change in gene expression in addition to the discussed p value, which is a measure of reproducibility independent of magnitude.

Second, ORA typically uses only the most significant genes and discards others. For instance, the input list of genes is usually obtained from a microarray experiment using an arbitrary threshold (e.g., genes with fold change ≥ 2 and/or p values ≤ 0.05). When using an arbitrary threshold, marginally less significant genes (e.g., fold change = 1.999 or p value = 0.051) are missed, resulting in information loss. Pavlidis et al. (2004) showed that the use of a predetermined threshold results in inconsistent results.

Third, ORA assumes that each pathway is independent of other pathways, which is not biologically realistic. For example, the cell-cycle pathway as described in the Kyoto Encyclopedia of Genes and Genomes (KEGG; see http://www.genome.jp/kegg/pathway/hsa/hsa04110.html) is strongly correlated with the mitogen-activated protein kinase (MAPK) signaling pathway. None of the existing ORA-based methods account for this dependence between molecular functions in GO and signaling pathways in KEGG.

Fourth, by treating each gene equally, ORA assumes that each gene is independent of the other genes, which again is not biologically realistic. Biology is a manifestation of interactions among gene products that constitutes different pathways to achieve a common biological objective. One goal of gene-expression analysis might be to gain insight into how the interactions among gene products manifest as changes in gene expression. A strategy that assumes the genes are independent is significantly limited in its ability to provide insight into interactions between gene products and the underlying biology.

SECOND GENERATION: FUNCTIONAL CLASS SCORING APPROACHES

The ad hoc division of expression data into significant and nonsignificant genes using threshold, the decoupling of expression magnitude from functional analysis, and the assumption of independence among genes motivated the development of functional class scoring (FCS)-based approaches. FCS-based approaches aim to analyze genes from a given pathway in the context of the entire list of genes while taking into consideration the correlation among them. The hypothesis used by FCS approaches is that although large changes in individual genes can have significant effects on pathways, weaker but coordinated changes in sets of functionally related genes can also have significant effects. It may be useful to consider the example of "enzyme *A* yields a product to enzyme *B*, which yields a final product *C*," i.e., a fourfold knockdown of *B* yields the same amount of *C* as, all else being equal, a twofold knockdown of *A* coupled to a twofold knockdown of *B*.

With few exceptions, FCS-based analysis approaches follow these steps (Fig. 1, middle panel) (Ackermann and Strimmer 2009): (1) Compute a gene-level statistic (local statistic), (2) compute a statistic for a pathway (global statistic), and (3) assess the significance of the pathway-level statistic. Let us say we are interested in computing those pathways that are up-regulated in response to cells being treated with a given compound. First, a gene-specific statistic is computed for each gene on an array comparing expression in the untreated samples and the samples treated with a compound of interest. Then, in Steps 2 and 3 for each pathway, the gene-level statistics are aggregated and the significance is evaluated. So, for the MAP3K pathway, information from *ERK*, *TRAF2*, *Mek*, *BRAF*, *TGFB*, and other genes involved in that pathway of interest is integrated and evaluated.

A large number of methods that apply these steps in different combinations have been proposed. For instance, the gene-level statistics used to rank the genes in the first step include correlation of molecular measurements with phenotype (Pavlidis et al. 2004; Al-Shahrour et al. 2005), ANOVA (Al-Shahrour et al. 2005), Q statistic (Goeman et al. 2004), Hotelling's T^2 (Kong et al. 2006), signal-to-noise ratio (Subramanian et al. 2005), *t*-test (Al-Shahrour et al. 2005; Tian et al. 2005), and *Z* score (Chung et al. 2005). The gene-level statistic is then used to compute a pathway-level statistic. The pathway-level statistic used by the current approaches includes the Kolmogorov–Smirnov statistic (Mootha et al. 2003; Subramanian et al. 2005); the sum, mean, or median of gene-level statistic (Jiang and Gentleman 2007); Wilcoxon rank sum (Zeeberg et al. 2005); and the maxmean statistic (Efron 2007). The final step in FCS is assessing the statistical significance of the pathway-level statistic, which can be computed by permuting either class labels (phenotypes) for each sample or gene labels for each pathway, both of which correspond to related but different null hypotheses (Tian et al. 2005; Efron 2007).

Limitations

Although FCS-based approaches are improvements over ORA-based approaches, they also have several limitations. First, similar to ORA, FCS analyzes each pathway independently. However, genes are often important in more than one pathway. Consequently, in an experiment, a given pathway may be truly affected by a certain treatment, but other pathways may appear to be significantly affected as well, owing to the set of overlapping genes (that is, genes common to the different pathways). Such a phenomenon is very common when the pathway under study is defined using the GO terms due to the hierarchical nature of the GO. For example, the cytokine TNF (tumor necrosis factor) is involved in several pathways. If, in our experiment, its expression is significantly up-regulated, that will affect the statistics in all of the pathways in which it is involved, biasing the results of the analysis.

Second, many FCS approaches use a sum of the statistics used to rank genes for calculating a score for a pathway. Because pathways are of different sizes, it is important to normalize the score for the pathway size. Otherwise, larger pathways will have higher scores. However, some FCS approaches do not perform this normalization (Pavlidis et al. 2002, 2004).

Third, although FCS approaches consider correlations among molecular measurements in a given pathway, similar to ORA, they give equal weight to each measurement. The change in gene expression is only used to rank the genes in a given pathway and is discarded from further analysis.

THIRD GENERATION: PATHWAY TOPOLOGY–BASED APPROACHES

Neither ORA nor FCS methods exploit the full knowledge embedded into pathways. Additional information available from the public knowledge bases includes not only which gene products interact with each other in a given pathway, but also how they interact (e.g., activation, inhibition, etc.) and where they interact (e.g., cytoplasm, nucleus, etc.). ORA and FCS methods only consider the number of genes in a pathway or the correlation between them to identify significant pathways.

Removing or adding genes would affect the list of genes with a given term and would be accounted for by ORA/FCS methods. However, if the pathways are completely redrawn with new links among the genes, as long as they contain the same set of genes, ORA and FCS will continue to produce the same results.

Recently a number of pathway analysis approaches have been proposed that consider topology (i.e., shape and structure) when analyzing pathways in addition to just gene lists. However, unlike ORA and FCS approaches, no common themes have yet emerged among the existing PT-based approaches. Rahnenführer et al. (2004) proposed integrating PT as an extension to other previous methods to score a pathway. Instead of giving equal weight to all pairwise correlations, they propose dividing it by the number of reactions needed to connect the two genes in a given pathway (Fig. 1, lower panel). Although this approach is proposed to only analyze metabolic pathways, in theory, it is also applicable to signaling pathways.

A recent analysis examines signaling pathways from a systems biology perspective by incorporating a number of important biological factors in addition to the number of differentially expressed genes in a pathway (Fig. 1, lower panel) (Draghici et al. 2007; Khatri et al. 2007). These factors include change in gene expression, the type of interaction, and the positions of genes in a pathway. The analysis models a signaling pathway as a graph in which nodes represent genes and edges represent interactions among genes. Further, it defines a perturbation factor (PF) of a gene as a sum of its measured change in expression and a linear aggregate of the PFs of all genes in a pathway, thus allowing the incorporation of pathway structure into the analysis.

FCS and ORA methods assume that the underlying network does not change as the experimental conditions change. However, both the genes involved and the topologies of pathways are likely to change in some scenarios. For example, it has been shown that the correlation structure between *ARG2* and other genes in the urea-cycle pathway changes with alterations in expression of *ARG2* (Li 2002), suggesting change in the underlying pathway topology. Shojaie and Michailidis (2009) proposed a method that takes into account the change in correlation as well as that in network structure, as experimental conditions change. Their proposed approach models expression of a gene as a linear function of other genes in the network similar to that above, but it also accounts for the baseline expression of a gene and imposes structural requirements on the pathway representation.

COMPARISON OF EXISTING PATHWAY ANALYSIS TOOLS

Table 1 summarizes the currently available tools for knowledge base–driven pathway analysis of high-throughput data separated by class of method. In the following section, we discuss some of the aspects of analysis that one should consider before using a specific tool.

TABLE 1. Pathway analysis tools for overrepresentation analysis

Name	Scope of analysis	*p* Value	Correction for multiple hypotheses	Availability
Onto-Express	GO	Hypergeometric, binomial, χ^2	FDR, Bonferroni, Sidak, Holm	Web
GenMAPP/ MAPPFinder	GO, KEGG, MAPP	Percentage/Z score	None	Stand alone
(High throughput) GoMiner	GO	Relative enrichment, hypergeometric	None	Stand alone, Web
FatiGO	GO, KEGG	Hypergeometric	None	Web
GOstat	GO	χ^2	FDR	R package
GOTree Machine	GO	Hypergeometric	None	Web
FuncAssociate	GO	Hypergeometric	Bootstrap	Web
GOToolBox	GO	Hypergeometric	Bonferroni, Holm, FDR, Hommel, Hochberg	Web
GeneMerge	GO	Hypergeometric	Bonferroni	Web
GOEAST	GO	Hypergeometric, χ^2	Benjamini–Yekutieli	Web
ClueGO	GO, KEGG, BioCarta, user defined	Hypergeometric	Bonferroni, Bonferroni step-down, Benjamini–Hochberg	Stand alone

Gene-Level Statistic

As shown in Table 2, a large number of gene-level statistics have been proposed for ranking individual genes before computing the pathway-level statistic in FCS methods. However, the choice of a gene-level statistic is found to have negligible effect on identifying significantly enriched pathways (Ackermann and Strimmer 2009). When a study has a small sample size (i.e., a few samples or conditions being tested), standardized versions of gene-level statistics are preferable (Jiang and Gentleman 2007; Ackermann and Strimmer 2009) (see Chapter 6 for more details).

Pathway-Level Statistic

The choice of appropriate pathway-level statistics has been the focus of a number of recent papers (Goeman et al. 2004; Ackermann and Strimmer 2009; Glazko and Emmert-Streib 2009). Tables 1–3 list the options available in existing tools for pathway-level statistics. A comparison of six pathway-level statistics on simulated data found that each had difficulty in identifying pathways in which only a subset of genes was differentially expressed (Ackermann and Strimmer 2009).

Furthermore, it is commonly believed that multivariate tests (accounting for multiple variables) have more power than univariate tests (based on a single variable), because univariate tests assume that the genes in a pathway are independent, whereas multivariate tests account for the interdependence in molecular measurements. Other caveats to consider are the power of the statistical method used,

TABLE 2. Pathway analysis tools for functional class sorting

Name	Scope of analysis	Gene-level statistic	Gene set statistic	*p* Value	Correction for multiple hypotheses	Availability
GSEA	GO, KEGG, BioCarta, MAPP, transcription factors, microRNA, cancer molecules	Signal-to-noise ratio, *t*-test, cosine, Euclidean and Manhattan distance, Pearson correlation, (log2) fold change, log difference	Kolmogorov–Smirnov	Phenotype permutation, gene set permutation	FDR	Stand alone, R package
sigPathway	GO, KEGG, BioCarta, humanpaths	*t* statistic	Wilcoxon rank sum	Phenotype permutation, gene set permutation	FDR (NPMLE)	R package
Category	GO, KEGG	*t* statistic		Phenotype permutation	NA	R package
SAFE	GO, KEGG, PFAM	Student's *t*-test, Welch's *t*-test, SAM *t*-test, *f* statistic, Cox proportional hazards model, linear regression	Wilcoxon rank sum, Fisher's exact test statistic, Pearson's test, *t*-test of average difference	Phenotype permutation	FWER (Bonferroni, Holm's step-up), FDR (Benjamini–Hochberg, Yekutieli–Benjamini)	R package
GlobalTest	GO, KEGG	NA	Simple and multinomial logistic regression, Q-statistics mean	Phenotype permutation, asymptotic distribution, γ distribution	NA	R package
PCOT2	User specified	Hotelling's T^2		Phenotype permutation, gene set permutation	FDR (Benjamini–Hochberg, Yekutieli–Benjamini), FWER (Bonferroni, Holm, Hochberg, Hommel)	R package
SAM-GS	User specified	***d*** statistic	Sum of squared ***d*** statistic	Phenotype permutation	FDR	Excel plug-in

FDR, false discovery rate; NPMLE, nonparametric maximum-likelihood estimate; NA, not applicable; FWER, family-wise error rate.

TABLE 3. Pathway analysis tools for topology analysis

Name	Scope of analysis	Gene-level statistic	*p* Value	Correction for multiple hypotheses	Availability
ScorePAGE	KEGG (metabolic)	(Correlation, covariance, cosine, dot product) + number of reactions	Gene set permutation	FDR (Benjamini–Hochberg)	
Pathway-Express/ SPIA	KEGG (signaling)	Number and type of interactions, fold change	Hypergeometric, binomial	FDR	R package, Web

whether it accounts for the variability in the data, etc. (see Chapter 3 for more details). No one method or statistic is superior, and it is useful to analyze biological data using multiple pathway-level statistics (Glazko and Emmert-Streib 2009).

Assessing Statistical Significance of Pathways

In ORA, the probability of x genes being involved in a given pathway just by chance from a list of differentially expressed genes can be modeled by a hypergeometric distribution (Khatri and Draghici 2005) or, in a case in which the background number of genes is large, a binomial distribution (see Chapter 3 for more details). Other possible models for assessing statistical significance in ORA are the χ^2 test for equality of proportions and Fisher's exact test. Table 1 lists the options available with different tools for assessing significance of pathways in ORA tools.

FCS approaches access significance of pathways by permuting either phenotype labels or gene labels (Table 2). There are benefits and drawbacks to using each of these on their own, as well the combination of the two, but this discussion falls outside the scope of this chapter. GSEA, sigPathway, and PCOT2 support both phenotype and gene label permutations, whereas Category, SAFE, and GlobalTest only support phenotype permutation. In addition, GlobalTest also supports asymptotic and γ distributions for identifying significant pathways.

Correction for Multiple Hypotheses

Multiple hypothesis correction (see Chapters 3 and 6 for more details) must be applied on several levels of analysis when dealing with pathway enrichment. It needs to be applied to correct for the number of genes or proteins in the original experiment (microarray, proteomics, etc.) as well as the number of pathways considered in the pathway enrichment calculation. Despite the generally accepted importance of correction for multiple hypotheses, several ORA tools do not provide correction for multiple hypotheses, which include GoMiner, FatiGO, GenMAPP/MAPPFinder, and GOTM (Khatri and Draghici 2005). Tables 1 and 2 list the choices available with existing tools for multiple hypotheses correction. Among the available choices, Bonferroni, Sidak, and Holm's corrections are very conservative and not suitable when the number of significant pathways is large. Furthermore, they incorrectly assume that each pathway is independent of all other pathways. When it is known a priori that there are some dependencies in the data, false discovery rate (FDR) is considered to be more appropriate. Another approach for multiple hypotheses correction is bootstrapping, which generates a null distribution by resampling many times from the data. However, care must be taken when only a few pathways are involved because the number of resamplings may be insufficient for a reliable conclusion.

CURRENT CHALLENGES IN KNOWLEDGE BASE–DRIVEN PATHWAY ANALYSIS

Low-Resolution Knowledge Bases

Although current pathway analysis approaches account for correlation in molecular measurement profiles and pathway topology, there are still a number of important unaddressed challenges. Many of these challenges are primarily owing to a lack of appropriate information in existing knowledge bases despite a large number of annotation knowledge bases with proven usefulness. For instance, the majority of human genes give rise to many distinct transcripts through alternative splicing, and most of the alternatively spliced transcripts may be active in different pathways. However, instead of specifying which protein-coded *transcript* of a gene is active in a given pathway, all knowledge bases only specify which protein-coded *genes* are parts of a given pathway (Fig. 2). Without pathway knowledge bases that annotate exact transcripts of a gene, pathway analysis will not be able to take advantage of current and future measurement capabilities. Furthermore, although genome-wide association studies (GWASs) have started to identify the single-nucleotide polymorphisms (SNPs) involved in different conditions and diseases, the effects of SNPs on pathways are unknown (Fig. 2).

Incomplete and Inaccurate Annotations

Despite the enormous amount of annotations available in the public domain, a large number of genes are not yet annotated. For instance, KEGG only annotates 3587 genes out of 40,641 genes in the human genome, according to Entrez Gene. Similarly, the November 2009 release of GO contained 18,587 genes annotated with at least one GO term (Fig. 2); note that as of December 31, 2010, the GO website reports 45,576 human gene products annotated. Recently, a case study of the epidermal growth factor receptor (EGFR) signaling pathway showed that although the

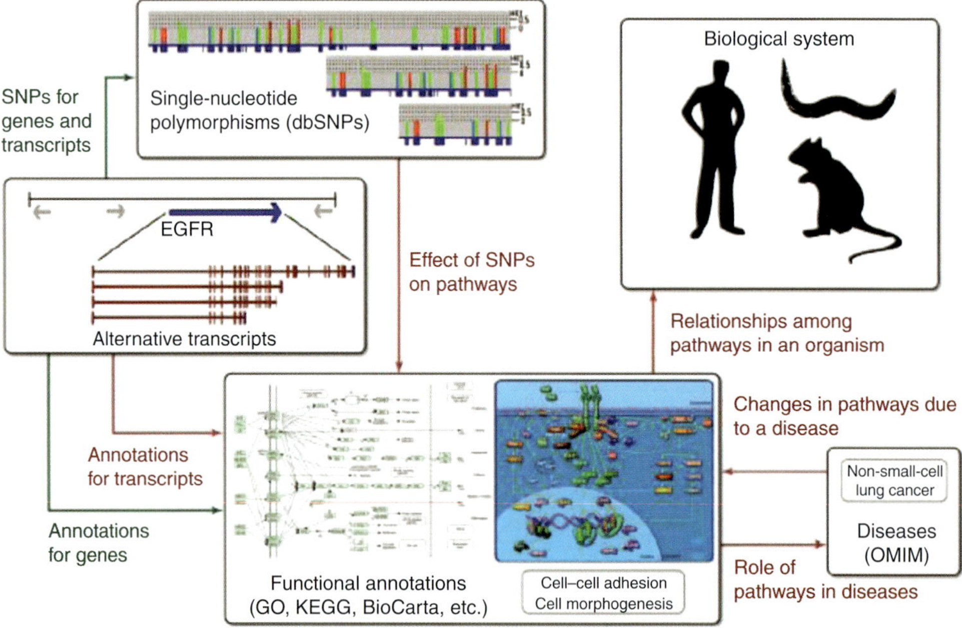

FIGURE 2. Overview of low-resolution, missing, and incomplete information. (Green arrows) Abundantly available information; (red arrows) missing and/or incomplete information. The ultimate goal of pathway analysis is to analyze a biological system as a large, single network. However, the links among smaller individual pathways are not yet well known. Furthermore, the effects of an SNP on a given pathway are also missing from current knowledge bases. Although some pathways are known to be related to a few diseases, it is not clear whether the changes in pathways are the cause for those diseases or the downstream effects of the diseases. (Redrawn from Khatri et al. 2012.)

current pathway databases contain a lot of detailed information, incomplete and inaccurate information may be critical for reactions that are part of regulatory feedback loops, which in turn determine the dynamic behavior of a signaling pathway (Bauer-Mehren et al. 2009). In addition, the available annotations are of low quality and may be inaccurate. For instance, the October 2007 release of GO annotations contained more than 95% of its annotations with the evidence code "inferred from electronic annotations (IEA)," which refers to the only type of annotation in GO that is not curated manually (Rhee et al. 2008). As expected, when filtering out the annotations with the IEA code, the number of genes with good-quality annotations in the November 2009 release of GO reduces from 18,587 to 11,890 (Fig. 3).

Missing Condition- and Cell-Specific Information

Virtually all pathway knowledge bases are built by adding one reaction (or interaction) at a time, such that each interaction is typically confirmed by an individual experiment. These experiments are usually performed in different cells or tissues, at different time points. However, the details of these experimental conditions are

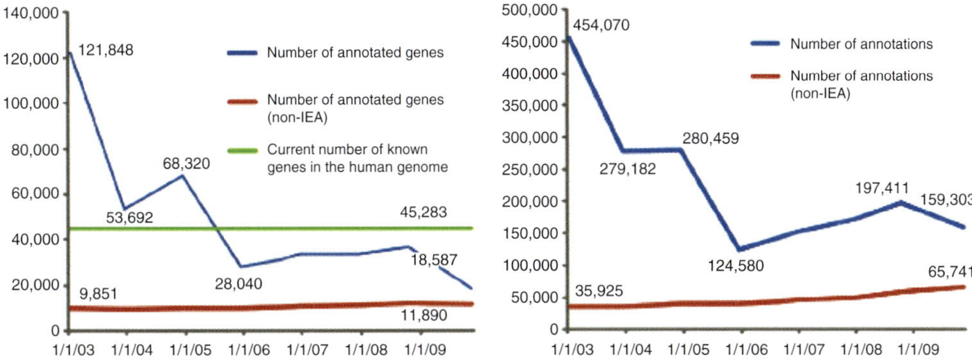

FIGURE 3. Number of GO-annotated genes (*left*) and GO annotations (*right*) for human from January 2003 to November 2009. As the estimated number of known genes in the human genome is adjusted (between January 2003 and December 2003) and annotation practices are modified (between December 2004 and December 2005, and between October 2008 and November 2009), one can argue that, although the number of annotated genes and the annotations are decreasing (mainly owing to the adjusted number of genes in the human genome and changes in the annotation process), the quality of annotations is improving, as shown by the steady increase in non-IEA annotations and the number of genes with non-IEA annotations. However, the increase in the number of genes with non-IEA annotations is very slow. In almost 7 years, between January 2003 and November 2009, only 2039 new genes received non-IEA annotations. At the same time, the number of non-IEA annotations increased from 35,925 to 65,741, indicating a strong research bias for a small number of genes. (Redrawn from Khatri et al. 2012.)

usually ignored, which results in multiple, independent genes being part of the same interaction on a pathway. For instance, in the **regulation of actin cytoskeleton** pathway in KEGG, the symbol *GF* represents 27 different growth factors derived from epidermis, fibroblasts, and platelets (see http://www.genome.jp/kegg/pathway/hsa/hsa04810.html). Another example of this lack of condition-specific information is the ***Wnt*/β-catenin pathway** in the signal transduction knowledge environment (STKE), where the node labeled "*Genes*" represents 19 genes directly targeted by the *Wnt* gene (http://stke.sciencemag.org/cgi/cm/stkecm;CMC_6032). However, these 19 genes are targeted by *Wnt* in different organisms (*Xenopus* [Brannon et al. 1997] and human [Pennica et al. 1998]) and in different cells and tissues, such as colon carcinoma cells (Mann et al. 1999) and epithelial cells (Haertel-Wiesmann et al. 2000). These nonspecific interactions introduce bias for these pathways in all analytical approaches. One can imagine a more biologically realistic solution in which each edge (i.e., interaction between two genes, proteins, or other molecules) is annotated with condition- and cell-specific information, when available, improving the specificity of the analysis.

Furthermore, the existing knowledge bases do not describe the effects of an abnormal condition (e.g., a disease) on a pathway (Fig. 2). For instance, when analyzing the Alzheimer's disease pathway in KEGG (http://www.genome.jp/kegg/pathway/hsa/hsa05010.html), it is not clear how this pathway is different from a normal condition and which set of reactions leads to Alzheimer's disease. It could be possible to assign more weight to these condition-specific interactions, which in turn can allow development of more sensitive analytical approaches.

Weak Interpathway Links

All pathway knowledge bases describe each pathway as an independent component. However, an organism's biology is not a collection of isolated pathways, but a network in which pathways influence one another. Current knowledge bases describe transitions from one pathway to another by inserting a name of the next pathway at the end of the current pathway (e.g., see B-cell survival pathway in PID or MAPK signaling pathway in KEGG). However, they do not provide any information about which specific genes on different pathways interact with each other and the type of interaction that takes place between them (Fig. 2). Consequently, all existing analysis approaches are forced to assume that each pathway is independent of the other pathways. Until these interpathway links are described in detail, it is not possible to develop analysis methods that can analyze gene expression and other high-throughput data from a global perspective.

Inability to Model and Analyze Dynamic Response

Although missing information in the pathway knowledge bases limits analysis from a systems biology perspective, none of the existing approaches are able to collectively model and analyze high-throughput data as a dynamic system. The existing pathway analysis approaches are only designed to analyze a snapshot of a biological system, which is often not representative of what is actually going on.

Inability to Model Effects of an External Stimulus

Current approaches only consider genes and completely ignore the effect of other molecules participating in a pathway (e.g., calcium in a calcium signaling pathway). Although PT-based analysis methods potentially have the ability to consider some of these, none of the existing approaches fully incorporates this information in their models. For instance, rate-limiting steps and other similar characteristics are still missing.

UTILITY AND CONFIDENCE OF PATHWAY ANALYSES

Today, almost every bioinformatics study involves looking for statistically significant pathways, as either biological "interpretation" or "validation" of computationally derived results. Despite the advantages of knowledge base–driven pathway analysis, statistical significance alone is generally deemed insufficient to convince researchers of the biological validity of a result, thus requiring further validation.

One way to predict the future of pathway analysis is to consider another field in which a large number of quantitative measurements are made, differences are calculated, and the significance of these differences is interpreted. One such field would be that of the markets of publicly traded stocks. When comparing pathway analysis with publicly traded stocks, global market indicators are analogous to phenotypes, sectors and mutual funds are analogous to pathways, and individual stocks are analogous to genes. Global changes in major market indicators (i.e., changes in phenotype) are frequently explained by changes within a single sector (i.e., pathway, such as "the economy went into recession as housing foreclosures increased"). However, although both small and large changes in the market are frequently explained by changes in the sectors, one's interest as an investor typically moves toward those stocks (i.e., genes) that are performing differently than their sector, if not better than their sector (e.g., "a biotech company is heartily outperforming its sector"). Highlighting those genes and proteins that are moving contrary to their pathway is an example of a type of analysis not performed today. There are many semi-orthogonal methods to partition the space of stocks, such as industrial sector, class of risk, growth, price/earnings ratio, and value. Many of these characteristics are exploited as stocks and are collected into sets to form mutual funds. The same stock can exist in two separate mutual funds for totally different reasons, just as the same gene can be a part of multiple pathways with different functions. Although we characterize genes and proteins by many parameters (e.g., protein kinase A [PKA], size, charge, tissue expression), few of these are currently represented as pathways.

It is interesting to note that today there are more mutual funds than stocks. However, as shown in Figure 3, we are not yet close to this level of characterization and utility in genomics, proteomics, and other high-throughput technologies. In the past 7 years, the total number of annotations and annotated genes are decreasing (Fig. 3). Furthermore, as of November 2009, ~30% of known genes have good quality (i.e., non-IEA) annotations, and ~46% of the genes has at least one annotation. It can be argued that the reduction in the number of annotations and annotated genes since January 2003 is an indicator of improving quality of annotations as the number of genes in a genome is adjusted and the functional annotation algorithms are modified. Indeed, the number of non-IEA annotations is continuously increasing (Fig. 3). However, the rate of increase for non-IEA annotations is very slow, and manual curation of the entire genome is expected to take a very long time (Baumgartner et al. 2007). In addition, the effect of genes without any annotations on statistical analysis is unknown and requires further systematic study. However, despite these hurdles (and similar to the stock market), as the number and type of functional annotations increase over time, along with analysis methods that provide better guidance for strategic planning for subsequent biological experiments, the utility of knowledge base–driven pathway analysis and confidence in its results are surely expected to improve.

SUMMARY

In this chapter, we discuss the development of knowledge base–driven pathways analysis methods of high-throughput molecular measurements in the last decade, distinctly divided into three generations. Knowledge base–driven pathway analysis has evolved from a classical, and rather simplistic, overrepresentation analysis to more sophisticated topology-based approaches that aim to combine molecular measurements and functional annotations at a global level, by considering the dependencies among genes using correlations in their expressions and the type of interactions among them. However, despite the progress made in pathway analysis of high-throughput molecular measurements in the last decade, there are still significant challenges in terms of high-resolution knowledge bases, pathway models, and relevance to experimental context and disease, that must be addressed by the next generation of methods to improve the specificity, sensitivity, relevance of functional pathway analysis, and consequently, its utility.

REFERENCES

Ackermann M, Strimmer K. 2009. A general modular framework for gene set enrichment analysis. *BMC Bioinformatics* **10:** 47.

Al-Shahrour F, Diaz-Uriarte R, Dopazo J. 2005. Discovering molecular functions significantly related to phenotypes by combining gene expression data and biological information. *Bioinformatics* **21:** 2988–2993.

Bauer-Mehren A, Furlong LI, Sanz F. 2009. Pathway databases and tools for their exploitation: Benefits, current limitations and challenges. *Mol Syst Biol* **5:** 290.

Baumgartner WAJ, Cohen KB, Fox LM, Acquaah-Mensah G, Hunter L. 2007. Manual curation is not sufficient for annotation of genomic databases. *Bioinformatics* **23:** 41–48.

Beissbarth T, Speed T. 2004. GOstat: Find statistically overrepresented Gene Ontologies within a group of genes. *Bioinformatics* **20:** 1464–1465.

Berriz GF, King OD, Bryant B, Sander C, Roth FP. 2003. Characterizing gene sets with FuncAssociate. *Bioinformatics* **19:** 2502–2504.

Boyle EI, Weng S, Gollub J, Jin H, Botstein D, Cherry JM, Sherlock G. 2004. GO::TermFinder—Open source software for accession Gene Ontology information and finding significantly enriched Gene Ontology terms associated with a list of genes. *Bioinformatics* **20:** 3710–3715.

Brannon M, Gomperts M, Sumoy L, Moon RT, Kimelman D. 1997. A β-catenin/XTcf-3 complex binds to the siamois promoter to regulate dorsal axis specification in *Xenopus*. *Genes Dev* **11:** 2359–2370.

Castillo-Davis CI, Hartl DL. 2002. GeneMerge: Post-genomic analysis, data mining, and hypothesis testing. *Bioinformatics* **19:** 891–892.

Chung HJ, Park CH, Han MR, Lee S, Ohn JH, Kim J, Kim JH. 2005. ArrayXPath II: Mapping and visualizing microarray gene-expression data with biomedical ontologies and integrated biological pathway resources using scalable vector graphics. *Nucleic Acids Res* **33:** W621–W626.

Doniger SW, Salomonis N, Dahlquist KD, Vranizan K, Lawlor SC, Conklin BR. 2003. MAPPFinder: Using Gene Ontology and GenMAPP to create a global gene expression profile from microarray data. *Genome Biol* **4:** R7.

Draghici S, Khatri P, Martins RP, Ostermeier GC, Krawetz SA. 2003. Global functional profiling of gene expression. *Genomics* **81:** 98–104.

Draghici S, Khatri P, Tarca AL, Amin K, Done A, Voichita C, Georgescu C, Romero R. 2007. A systems biology approach for pathway level analysis. *Genome Res* **17:** 1537–1545.

Efron B. 2007. Correlation and large-scale simultaneous significance testing. *J Am Stat Assoc* **102:** 93–103.

Goeman JJ, van de Geer SA, de Kort F, van Houwelingen HC. 2004. A global test for groups of genes: Testing association with a clinical outcome. *Bioinformatics* **20:** 93–99.

Glazko GV, Emmert-Streib F. 2009. Unite and conquer: Univariate and multivariate approaches for finding differentially expressed gene sets. *Bioinformatics* **25:** 2348–2354.

Haertel-Wiesmann M, Liang Y, Fantl WJ, Williams LT. 2000. Regulation of cyclooxygenase-2 and periostin by Wnt-3 in mouse mammary epithelial cells. *J Biol Chem* **275:** 32046–32051.

Jiang Z, Gentleman R. 2007. Extensions to gene set enrichment. *Bioinformatics* **23:** 306–313.

Khatri P, Draghici S. 2005. Ontological analysis of gene expression data: Current tools, limitations, and open problems. *Bioinformatics* **21:** 3587–3595.

Khatri P, Draghici S, Ostermeier GC, Krawetz SA. 2002. Profiling gene expression using onto-express. *Genomics* **79:** 266–270.

Khatri P, Draghici S, Tarca AL, Hassan SS, Romero R. 2007. A system biology approach for the steady-state analysis of gene signaling networks. *In CIARP 2007 Proceedings of the 12th Ibero-american Congress on pattern recognition, image analysis and applications* (ed. Rueda L, et al.), pp. 32–41. Springer-Verlag, Heidelberg.

Khatri P, Sirota M, Butte AJ. 2012. Ten years of pathway analysis: Current approaches and outstanding challenges. *PLoS Comput Biol* **8:** e1002375.

Kong SW, Pu WT, Park PJ. 2006. A multivariate approach for integrating genome-wide expression data and biological knowledge. *Bioinformatics* **22:** 2373–2380.

Li K-C. 2002. Genome-wide coexpression dynamics: Theory and application. *Proc Natl Acad Sci* **99:** 16875–16880.

Mann B, Gelos M, Siedow A, Hanski ML, Gratchev A, Ilyas M, Bodmer WF, Moyer MP, Riecken EO, Buhr HJ, et al. 1999. Target genes of β-catenin-T cell-factor/lymphoid-enhancer-factor signaling in human colorectal carcinomas. *Proc Natl Acad Sci* **96:** 1603–1608.

Martin D, Brun C, Remy E, Mouren P, Thieffry D, Jacq B. 2004. GOToolBox: Functional analysis of gene datasets based on Gene Ontology. *Genome Biol* **5:** R101.

Mootha VK, Lindgren CM, Eriksson KF, Subramanian A, Sihag S, Lehar J, Puigserver P, Carlsson E, Ridderstråle M, Laurila E, et al. 2003. PGC-1 α-responsive genes involved in oxidative phosphorylation are coordinately downregulated in human diabetes. *Nat Genet* **34:** 267–273.

Pavlidis P, Lewis DP, Noble WS. 2002. Exploring gene expression data with class scores. *Pac Symp Biocomput* **7:** 474–485.

Pavlidis P, Qin J, Arango V, Mann J, Sibille E. 2004. Using the Gene Ontology for microarray data mining: A comparison of methods and application to age effects in human prefrontal cortex. *Neurochem Res* **29:** 1213–1222.

Pennica D, Swanson TA, Welsh JW, Roy MA, Lawrence DA, Lee J, Brush J, Taneyhill LA, Deuel B, Lew M, et al. 1998. WISP genes are members of the connective tissue growth factor family that are up-regulated in wnt-1-transformed cells and aberrantly expressed in human colon tumors. *Proc Natl Acad Sci* **95:** 14717–14722.

Rahnenführer J, Domingues FS, Maydt J, Lengauer T. 2004. Calculating the statistical significance of changes in pathway activity from gene expression data. *Stat Appl Genet Mol Biol* **3:** 1544–6115.

Rhee SY, Wood V, Dolinski K, Draghici S. 2008. Use and misuse of the gene ontology annotations. *Nat Rev Genet* **9:** 509–515.
Shojaie A, Michailidis G. 2009. Analysis of gene sets based on the underlying regulatory network. *J Comput Biol* **16:** 407–426.
Subramanian A, Tamayo P, Mootha VK, Mukherjee S, Ebert BL, Gillette MA, Paulovich A, Pomeroy SL, Golub TR, Lander ES, et al. 2005. Gene set enrichment analysis: A knowledge-based approach for interpreting genome-wide expression profiles. *Proc Natl Acad Sci* **102:** 15545–15550.
Tian L, Greenberg SA, Kong SW, Altschuler J, Kohane IS, Park PJ. 2005. Discovering statistically significant pathways in expression profiling studies. *Proc Natl Acad Sci* **102:** 13544–13549.
Zeeberg BR, Qin H, Narasimhan S, Sunshine M, Cao H, Kane DW, Reimers M, Stephens RM, Bryant D, Burt SK, et al. 2005. High-throughput gominer, an "industrial-strength" integrative gene ontology tool for interpretation of multiple-microarray experiments, with application to studies of common variable immune deficiency (CVID). *BMC Bioinformatics* **6:** 168.

WWW RESOURCES

http://www.genome.jp/kegg/pathway/hsa/hsa04110.html KEGG (Kyoto Encyclopedia of Genes and Genomes), Cell cycle—Homo sapiens (human), 04110 8/7/13, Kanehisa Laboratories, Japan.
http://www.genome.jp/kegg/pathway/hsa/hsa04810.html KEGG, Regulation of actin cytoskeleton—Homo sapiens (human), 04810 8/8/13, Kanehisa Laboratories, Japan.
http://www.genome.jp/kegg/pathway/hsa/hsa05010.html KEGG, Alzheimer's disease—Homo sapiens (human), 05010 11/16/10, Kanehisa Laboratories, Japan.
http://merriam-webster.com/dictionary/pathway Merriam-Webster Dictionary.
http://stke.sciencemag.org/cgi/cm/stkecm;CMC_6032 Moon RT. Wnt direct target genes. *Sci Signal* (Connections map component in the database of cell signaling).

11

Learning Biomolecular Pathways from Data

Karen Sachs and Gabriela K. Fragiadakis

Stanford University School of Medicine, Stanford, California 94305

High-throughput technologies in biomedical research produce a plethora of data, allowing us to examine many facets of biological systems, including their metabolic, signaling, and genetic regulatory pathways.

Network inference is a class of approaches that uses experimental data to create models of complex biomolecular systems of various types. These models are often probabilistic, giving researchers the ability to account for the noise in complex systems. **Probabilistic** means that the uncertainty in the statistical relationships among the variables is directly incorporated into the model, as explained in this chapter.

Models of molecular pathways must accommodate two main features of the experimental data set. The first corresponds to dimensionality: the number of observations (data points) available to learn the model versus the number of entities (biomolecules) measured. The challenge is that if many entities are considered but the number of data points is small, real statistical relationships will be indistinguishable from spurious ones. The second pertains to the number of biomolecules that are measured using a particular assay versus the number of total entities directly involved in the biological system. Many unobserved entities pose modeling challenges that must be taken into consideration when formulating the modeling approach or during interpretation of modeling results. For example, data reflecting genetic regulatory pathways typically consist of measurements of messenger RNA (mRNA) species under dozens to hundreds of different conditions, as might be generated in a microarray or RNA sequencing (RNA-seq) experiment. In contrast, data for signaling pathways at the protein level generated by a flow cytometry experiment may involve thousands of individual observations, but each observation includes only a small fraction of relevant proteins. Each of these data forms poses distinct modeling challenges requiring variations of the modeling algorithms.

In this chapter, we give an overview of pathway models that can be generated from data. We begin with a short introduction to the types of models used to represent pathways. We then focus on probabilistic models called Bayesian networks. Finally, we present an introduction to algorithms that learn these networks and then explain how algorithms can be used to learn models of signaling pathways and of genetic regulatory pathways.

PATHWAY MODELS

Modeling of biological pathways can proceed at various levels of detail and resolution. In general, there is a trade-off between the amount of detail that can be included in the model and the amount of prior knowledge needed to build the model. More detailed models will also generally require stronger **modeling assumptions**. The modeling approach must attempt to identify the detail needed to address the biological questions while realistically considering the available data and prior knowledge to build the model. Our data often consist of snapshots of dynamic systems captured at specific point(s) in time; most of this chapter therefore describes efforts to learn pathway structures in this context. We also provide an overview of a few methods used to directly model the dynamics of these systems. We present two approaches that are representative of methods used in systems biology: Boolean networks and Bayesian networks. A third model, describing differential equation models, is presented in Box 3 at the end of the chapter.

Boolean Networks

Signaling pathways can be represented visually as a "graph" consisting of "nodes" and "edges" that connect nodes together (Fig. 1) In this setting, nodes represent proteins, genes, or other biomolecules, and an edge from protein *A* to *B* means

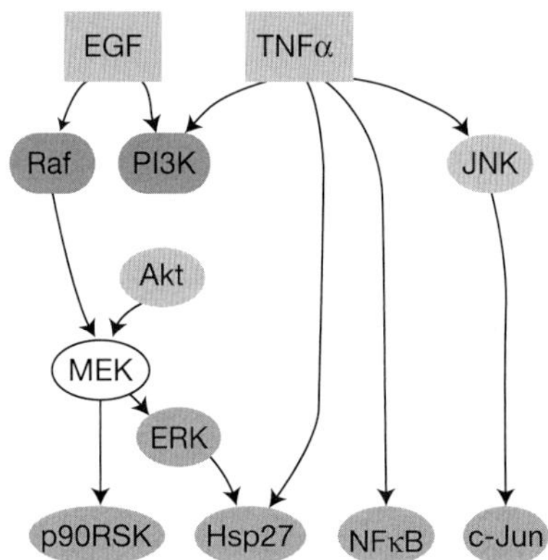

FIGURE 1. A Boolean network. (Adapted from Morris et al. 2013, with permission from Springer Science and Business Media.)

that the amount of active protein A directly or indirectly affects the amount of active protein B. In a "Boolean" (essentially "on/off" or binary) network, there are two types of edges: activating edges (→) and inhibitory edges (--|). $A \rightarrow B$ means that A activates B, and A--|B means that A deactivates B. When there is an arrow from A to B (activating or deactivating), A is called a "parent" of B.

In the Boolean network representation, protein A is either "on" ($x_A = 1$) or "off" ($x_A = 0$), depending on whether the measured amount of active A is higher than a certain threshold (p_A). Parents of a certain protein can be activating or deactivating. In some Boolean networks, one key assumption is that protein A is on if and only if at least one of its activating parents is "on" and all of its deactivating parents are "off."

The advantage of constructing a graph to represent a signaling pathway is that it facilitates a formal, mathematical treatment that subsequently allows us to estimate how the signaling pathway may behave in different conditions or at different time points. For example, let P_+^A denote the activating parents of A, and P_-^A the deactivating parents. In this case of Boolean networks, A is on if one or more of P_+^A are active and all of P_-^A are inactive. To determine if A is active in a given condition, we can use the following equation:

$$x_A = 1\left(\sum_{B \epsilon P_+^A} x_B\right)\left(\prod_{C \epsilon P_-^A} (1 - x_c)\right).$$

In words, this equation takes the product of two sets. The first is the sum of the values of the activating parents. The second is the product of one minus each deactivating parent. Then, if that product is >0, set the value of x_A to 1; otherwise, set x_A to zero.

So the equation for the activity of the proteins PI3K and MEK in Figure 1 is

$$\begin{aligned} x_{\mathrm{PI3K}} &= 1(x_{\mathrm{EGF}} + x_{\mathrm{TNF}\alpha}), \\ x_{\mathrm{MEK}} &= 1(x_{\mathrm{Raf}})(1 - x_{\mathrm{Akt}}). \end{aligned}$$

The following table shows the state of MEK for all possible states of its parents Raf and Akt:

Raf	Akt	MEK
0	0	0
1	0	1
0	1	0
1	1	0

In this representation, MEK is only active when the activator (Raf) is active and the repressor (Akt) is inactive. Remember that all of the values in these equations are Boolean (1 or 0), and the equations can be written for all of the nodes and solved to study the dynamics and the steady state(s) of the system.

Boolean networks have received a lot of attention because of their mathematical simplicity and the abundance of low-resolution (high noise) data that has to be discretized (binned) into two levels. Boolean networks offer a great amount of simplicity because of their rule-based nature. The two main problems with this type of model are that (1) it fails to account for shifts in activity levels because it can only distinguish between "on" and "off" and (2) it is not robust in the presence of noise (i.e., it will generally fail when any of the nodes are inaccurately measured).

In reality, both activating and deactivating parents of a certain protein might be active, with the ultimate activity level of the protein being determined by the combined potencies and abundances of its parents. In our transcriptional example, even in the presence of the repressor (Akt), the activator (Raf) could still bind with some affinity; in a Boolean model, however, we assume this not to be the case.

To account for the nonbinary nature of biological systems, a family of graphs (rather than one graph) can be used, with corresponding probabilities that represent our "confidence" in a certain graph (Shmulevich et al. 2002).

Therefore, one can think of the dynamics of the network in two steps:

1. Randomly choose a graph according to the given probabilities.
2. Calculate the values that the variables take given the chosen graph (just like in a regular Boolean network).

In structure learning, this makes the model much more **robust**, meaning that it is less sensitive to noise and variability in the data set. It also introduces the notions of **influence** and **sensitivity** of a certain variable to another. The sensitivity of variable x to variable y in probabilistic Boolean networks can be thought of as the probability that x changes if y changes (Shmulevich et al. 2002). Conversely, the influence of a variable is the probability that a change in x results in a change in y. Both of these notions can be generalized to collections of variables, so we can think of the sensitivity of a variable to a given set of variables and its influence on another set.

Dynamic Bayesian Networks

Dynamic Bayesian networks (DBNs) are a class of graphical models that target the dynamics of the modeled system based on the statistical dependencies of the entities in question. DBNs can be thought of as an extension of Bayesian networks (discussed in the next section) that can handle dynamical data and temporal models. In a DBN, each node represents the abundance of a biomolecule *at a particular time point*. Bayesian networks and DBNs are **directed acyclic graphs** (DAGs), meaning that the edges are directed such that each edge specifies parent and child, and they are acyclic such that no path can be drawn from a node to itself. Therefore, inference and structure learning are relatively easy. A thorough discussion and tutorial of Bayesian networks follows in the next section.

Boolean networks are a special case of DBNs, where each variable can take only two values and the conditional probabilities are deterministic Boolean functions, functions that return a Boolean output, with deterministic meaning that for a given input, we always get the same output.

DBNs are very descriptive, relatively simple, and very useful for inference and activity prediction. The main challenge with using DBNs is that they require data from several time points (Luna et al. 2007) with appropriately sized time steps.

BAYESIAN NETWORKS

Bayesian networks come from a family of models called graphical models, flexible and interpretable models in which probabilistic relationships among variables are represented in a graph. In our context, the Bayesian network represents relationships among variables in a signaling pathway, where variables can represent signaling molecules, small molecules, lipids, or any biologically relevant molecule.

Bayesian networks can uncover statistical relationships among variables from a set of data. Because statistical relationships may imply a physical or functional connection, we can use Bayesian networks to refine existing knowledge or uncover potential relationships in signaling pathways. When **interventional data** (i.e., data in which specific biomolecules have been manipulated or perturbed) are available, we can begin to add a causal interpretation to our model. (For a further discussion of interventional data, see Box 2 below.) This means that we can orient edges, identifying the parent and the child in a pair in which we see a statistical dependence.

The probabilistic approach determines dependencies and conditional independencies among variables. Thus, given sufficient data, a Bayesian network can provide a map of a signaling pathway and serve as an in silico generator of testable hypotheses.

Bayesian Network Tutorial

In this section, we present a Bayesian network tutorial intended to serve as a basis for understanding Bayesian networks in their general context, with a focus on use in the analysis of signaling pathways. Because it is not a comprehensive description, readers intending to learn more about Bayesian networks are referred to more in-depth references (Pearl 1988; Heckerman 1995; Pe'er 2005).

Model Semantics

In a Bayesian network, probabilistic relationships are represented by a qualitative description—a graph (G)—and a quantitative description—an underlying **joint probability distribution**: the probability of each variable described in the system taking on a specific value. In the graph, the nodes represent variables. In the examples discussed, these are biomolecules that comprise protein signaling pathways. The edges represent dependencies; more precisely, the lack of edges indicates a **conditional independency**, meaning that when all other variable values in the system are known, having the value of one protein gives you no more information about the probable value of a protein that is not connected to it by an edge. As a reminder, these graphs are directed and acyclic, although the acyclicity constraint can be problematic in the context of biological pathways.

For each biomolecule, we want a quantitative description of how its activation state depends on the other biomolecules. This description is called the **conditional probability distribution** (CPD) and must be consistent with the conditional independencies implied by the graph of the signaling network. In general, variables in a Bayesian network may be continuous or discrete. For example, we can think of biomolecules as taking on different states such as activated or deactivated (discrete) or as taking many level of activation across a continuum (continuous). Joint probability distributions may take on any form that specifies a valid probability distribution. However, in this discussion, we handle only discrete variables and multinomial distributions. When discrete variables are used, each variable may take on one of a finite set of states. For example, a protein variable may be in state low, medium, or high, corresponding to protein abundance.

Notation

In general, a Bayesian network represents the joint probability distribution (the probability of each variable described in the system taking on a specific value) for a finite set $X = \{X_1, \ldots, X_n\}$ of random variables, in which each variable X_i may take on a value x_i from the domain Val X_i.[1] Val X is the set of possible values that X could take on. We denote the parents of X_i in G as Par X_i. Par X is the set of possible values that the parents of X could take on.

Let us consider a biological example. The Ras kinase cascade is central to cellular function and is often mutated in cancer. Here is a portion of this pathway in which a Bayesian network represents the joint probability distribution between the activation of the proteins Raf, MEK, and ERK (Fig. 2). In this graph, MEK is the child of Raf, ERK is the child of MEK, and the node Raf has no parents (such nodes are called **root nodes**). Assume that each variable can take on the value 0 (inactive) or 1 (active).

In our "toy" example, ERK is dependent on MEK. ERK is dependent on Raf as well; however, when the value of MEK attack is known, ERK and Raf become

[1] In our notation, we use capital letters such as X, Y, and Z for variable names and lowercase letters such as x, y, and z to denote specific values taken by those variables. Sets of variables are denoted by boldface capital letters such as $\mathbf{X}$, $\mathbf{Y}$, and $\mathbf{Z}$, and assignments of values to the variables in these sets are denoted by boldface lowercase letters such as $\mathbf{x}$, $\mathbf{y}$, and $\mathbf{z}$.

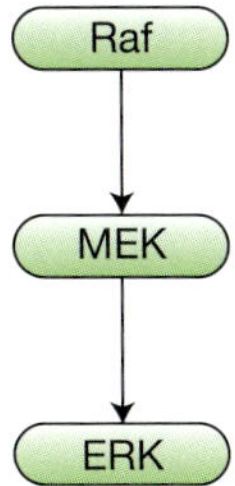

FIGURE 2. A Bayesian network structure for the variables Raf, MEK, and ERK.

independent. If we already know that MEK is active (mek = 1) or inactive (mek = 0), knowing something about the activation state of Raf will not help us to determine the value of ERK. Therefore, ERK is **conditionally independent** of Raf given MEK.

Formally, we say that X is conditionally independent of Y given Z if the probability of X given Y and Z is equal to the probability of X given Z only:

$$P(X \mid Y, Z) = P(X \mid Z)$$

and we denote this statement by $(X \perp Y \mid Z)$.

This graph of the pathway encodes the conditional statement Raf $\perp$ ERK | MEK.

Generally, the graph G encodes the **Markov assumptions**: Each variable X_i is independent of its nondescendants, given its parents in G. For our simple example, this means that when we know the parent of ERK (MEK), then ERK is independent of Raf, but the Markov assumptions can extend to more complicated networks. For example, if additional proteins directly phosphorylated MEK and/or Raf, ERK would be independent of those proteins (its nondescendants) given its parent MEK.

In mathematical notation,

$$\forall X_i (X_i \perp nondescendants_{Xi} \mid PaX_i),$$

where $\forall$ means "for all," as in "the statement that follows applies for all variables X_i." For all values of X, the value of X is conditionally independent of its nondescendants given the value of its parents. As a consequence of the Markov assumption, the joint probability distribution over the variables represented by the Bayesian network can be factored into a product over variables, in which each term is the local CPD of that variable, conditioned only on its direct parent variables, as opposed to a statement that depends on every biomolecule that comes earlier in the pathway[2]:

$$P(X_1, \ldots, X_n) = \prod_{i=1}^{n} P(X_i \mid \mathbf{Pa}_{X_i}).$$

[2]This is called the **chain rule** for Bayesian networks, and it follows directly from the chain rule of probabilities, which states that the joint probability of independent entities is the product of their individual probabilities.

A key advantage of the Bayesian network is its compact representation of the joint probability distribution. With no independence assumptions, the joint probability distribution over the variables Raf, MEK, and ERK is

$$P(\text{Raf, MEK, ERK}) = P(\text{Raf})P(\text{MEK} \mid \text{Raf})P(\text{ERK} \mid \text{MEK, Raf}).$$

Using conditional independencies, the joint probability distribution becomes

$$P(\text{Raf, MEK, ERK}) = P(\text{Raf})P(\text{MEK} \mid \text{Raf})P(\text{ERK} \mid \text{MEK}).$$

The sparser a graph structure, the fewer edges it contains and the more conditional independencies it encodes, yielding a greater savings in parameters (for a further discussion of parameters, see Box 1 below).

What about the joint probability distribution representation? In the case of the multinomial distributions, each CPD can be presented in a **conditional probability table** (CPT), in which each row in the table corresponds to a specific joint assignment of the parents pa_{X^i} to Pa_{X^i} and specifies the probability vector for X_i conditioned on pa_{X^i}.

Returning to our biological example, we assume that the parameters of the CPTs are known:

$P(\text{Raf} = 0)$	$P(\text{Raf} = 1)$
0.7	0.3

Raf	$P(\text{MEK} = 0)$	$P(\text{MEK} = 1)$
Raf = 0	0.99	0.01
Raf = 1	0.4	0.6

MEK	$P(\text{ERK} = 0)$	$P(\text{ERK} = 1)$
MEK = 0	0.95	0.05
MEK = 1	0.2	0.8

These indicate that, for instance, the probability of Raf being active is 0.3. The probability of MEK being active when Raf is active is 0.6.

$$P(\text{MEK} = 1 \mid \text{Raf} = 1) = 0.6,$$
$$P(\text{MEK} = 1 \mid \text{Raf} = 0) = 0.01.$$

Whereas the graph reveals the conditional independencies, the CPTs show the strength of dependencies.

Notice that, as expected, each row in the CPT sums to 1, so the second column of probabilities is redundant. In addition, note that we do not include a table for ERK conditioned on both Raf and MEK, because ERK and Raf are conditionally independent given MEK.

Structure Learning

There are many applications that use Bayesian networks. Bayesian networks can be used to determine the probability of unknown events (e.g., for an insurance company) or of a particular diagnosis in a medical context. In these applications, the values of a subset of the nodes are known, and this information is used to find probability distributions for the unknown nodes.

In our case, however, the graph structure itself, G, is not known, and in fact our goal is to elucidate this structure from the experimental data. In our example, this means that we have measured Raf, MEK, and ERK, and we want to learn the structure of the signaling pathway. We find G by searching for a structure that is consistent with the statistical relationships present in a data set. Finding the graph structure that is most likely to have generated the observed data is called **structure learning**.

Our strategy for structure learning is as follows: First, we define a Bayesian score that indicates, for a given graph structure, how well the structure reflects the dependencies and conditional independencies present in the data (i.e., how well the model matches our measurements of the biomolecules). This score allows us to assess individual structures. Armed with this ability, we can now search over possible model structures until we find the best one (i.e., the one with the highest score). More accurately, we find a set of high-scoring model structures. Finally, we take our set of high-scoring models and average them, to avoid **overfitting** that is, when a model describes noise as opposed to real relationships, and remain consistently Bayesian in our approach (for further details, see Box 1 below).

1. ***Scoring.*** To formulate the Bayesian score, we start with a probability metric:

$$\text{score}_{\mathcal{B}}(\mathcal{G}:\mathcal{D})\ \alpha\ P(\mathcal{G}\mid\mathcal{D}) = \frac{P(\mathcal{D}\mid\mathcal{G})P(\mathcal{G})}{P(\mathcal{D})}.$$

This score expresses the probability of the graph given the data, using the formulation of **Bayes' Rule** (i.e., the likelihood that our measurements imply a given signaling network structure; see Chapter 3 for more details on Bayes' Rule). The motivation behind this score is very intuitive: It expresses the probability of the graph (G) under consideration with respect to the data set at hand (D). Let us examine the components of the score on the right-hand side of the equation.

$P(D)$, called the marginal likelihood, would not be straightforward to assess. However, as the data set remains constant—we are searching for the best graph given that a *specific data set*—$P(D)$—will be the same for all possible structures that will be evaluated. Therefore, we are able to assume that $P(D)$ is constant and have no need to estimate it for the sake of defining the relative score of one structure to another.

$P(G)$ is the structural prior, allowing us to express the a priori probability of each graph structure. Typically, to avoid biasing toward one structure or another, we use a uniform prior. However, when we have constraints or prior biological knowledge, we can incorporate these. For example, if a previous experiment has shown definitively that A influences B, we can encode this constraint into the prior, effectively rendering the probability of any graph missing this edge equal to zero. For example, looking at our cancer signaling pathway, if we have data showing that MEK directly binds ERK, we would assign any structure that does not contain an edge between MEK and ERK a value of zero.

Finally, let us consider $P(D \mid G)$, meaning the probability that we would get the data that we generated (D) given that the true graph of the data is G. Because our data set is a noisy and finite sample from the true distribution, we are uncertain about the value of our parameters (see Box 1). Therefore, we average over all possible parameter assignments when we calculate $P(D \mid G)$.

2. ***Searching the space of possible graph structures.*** For our next step, we must consider possible graph structures and assess them using the Bayesian score, until we find one (or ones) that score well. We cannot exhaustively examine every graph, because the number of possible graph structures is superexponential in the number of variables. Instead, we use a **heuristic search** (a technique designed to find an optimal solution more quickly than an exhaustive search) to find high-scoring models, such as a **greedy random search**: We begin with a possible graph selected at random. Then, at each iteration, we select at random an edge to add, delete, or reverse, while keeping within necessary constraints of the graph, such as acyclicity and known edges. We score the new graph structure

BOX 1. What Are the Parameters?

Parameters express the quantitative aspect of the dependence of each variable on its parent variables in the graph. The prior is critical for two important reasons. First, it allows us to smooth the data found in the data set. For instance, it allows us to assume that even if a particular data point is not found, its probability may be (small but) nonzero. Why is this important? In the limit of small data sets, even if we never, for example, observe protein A = high with protein B = low, we are wary of assuming that such a scenario can never exist. The prior is a natural and convenient way to express that we believe scenarios may exist (i.e., they have a nonzero probability) even if we do not observe them in our limited data set. Aside from this smoothing effect, the prior is useful for incorporating a **complexity penalty**, which ensures that the Bayesian score will prefer simpler models, unless a more complex model is supported by a sufficiently large data set. Because of this complexity penalty, the Bayesian approach avoids overfitting, a process by which we fit our model to noise in the data, rather than a true signal.

and keep the change if we find that the score improves. Otherwise, we revert to the previous structure. We then go to the next iteration and again make a random change, repeating the process. In this way, we can only improve the score from one iteration to the next or stay the same, a trick that helps to limit sampling and testing within the large set of theoretically possible, but nonsensical, graphs that are poorly suited to the data.

A drawback to this technique is that, because we change only one edge at a time, we may encounter a situation in which any single-edge change yields a worse score, even if a two-edge change may lead to a score that is better than the original. We can never get to this two-edge difference, because the first edge change would be rejected. Thus, we are stuck in a score that is locally good (better than any structure that is one edge different) but not very good overall. This is commonly referred to as a **local maximum**. We avoid this problem in two ways. First, we allow a "bad" edge change (one that leads to a lower score) with some probability. As a second measure, we also repeat the search multiple times, each time starting from a different point in the search space (a different random graph). This is called **random restart**.

3. ***Model averaging.*** At the end of our search procedure, we have a collection of high-scoring (relative to their neighbors) models, one for each random restart. We could simply select the highest-scoring model from this collection. However, we are once again concerned about overfitting. For this reason, rather than selecting the single highest-scoring model, we consider the collection of all high-scoring models and choose those features that are common to many of them. Each edge is assigned a confidence score based on a weighted average: the sum of the model scores of those models containing that edge, divided by the sum of all the model scores.

Model Properties

So far, we have considered Bayesian networks in their general context and discussed how to learn the Bayesian network structure from data. In this section, we delve a bit more into the model structure to examine how different structures can be extracted from data, even when they represent similar dependencies; what kinds of dependencies are represented; and when a Bayesian network structure can be interpreted as a causal structure.

Dependencies and Independencies in the Graph Structure

We have briefly touched on this topic before in our discussion of Markov assumptions; here, we revisit it for a more thorough treatment.

Let us look at this situation with Raf, MEK, and ERK, which would all be correlated in our data set. Our data would suggest that Raf and MEK are statistically

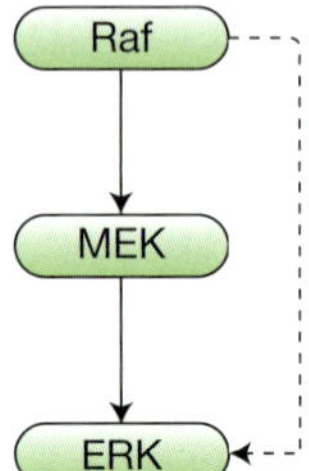

FIGURE 3. Possible dependence between Raf and ERK.

dependent, as are MEK and ERK. We would also see a dependence between Raf and ERK and may be tempted to put an edge between them (Fig. 3).

However, when we know MEK, we gain no more information about ERK from knowing Raf, explaining away the edge between Raf and ERK. Therefore, the edge between ERK and Raf is eliminated.

How will the conditional independence implied by the ability of MEK to explain away ERK's dependence on Raf manifest itself in the Bayesian score? Consider a CPT for this structure: In the correct structure, for the variable ERK, the CPT will specify $P(\text{ERK}|\text{MEK})$. If we also connect Raf to ERK, the CPT must now specify $P(\text{ERK}|\text{Raf, MEK})$, so it will contain twice as many parameters. Recall that the Bayesian score penalizes complexity. For the score to allow the extra parameters, the parameters must fit more precisely (as they would if Raf did affect ERK in some additional way—if it too, for instance, phosphorylated ERK). If Raf affects ERK only via its effect on MEK, the complexity penalty will result in the more connected model scoring more poorly; thus, it will tend to select a structure that is (in this case, appropriately) sparse.

What other structures encode conditional independencies? Consider the structure $A \leftarrow B \rightarrow C$, sometimes called a **fork**, in which A and C depend on B. We have not seen such a structure before, so let us devise an example; say, puddles, rain, and umbrellas. In the model puddles←rain→umbrellas, we see that puddles and umbrellas depend on rain. Because they share a parent, we expect puddles and umbrellas to also be dependent: Seeing people with umbrellas might lead us to suspect that there are puddles on the ground, and seeing puddles on the ground may increase our belief that people will be carrying umbrellas (the puddles may have an alternate cause, such as a sprinkler nearby, that does not induce people to carry umbrellas, so although our belief in umbrellas is increased, it is not necessarily certain). What happens when we know that it is raining? As in the chain structure, knowing the value of B (rain) renders A (puddles) and C (umbrellas) independent, because if we know it is raining, the presence of umbrellas no longer informs us as to the possibility of puddles. In other words, the dependence of each variable on rain *explains* away their dependence on one another, rendering them conditionally independent. As before, the Bayesian score will prefer this structure to one in which an additional edge connects puddles to umbrellas directly.

Let us examine one final graph structure, an important one called a **v-structure**, that has this configuration: $A \rightarrow B \leftarrow C$ (with no edge between A and C). We can construct an example for this situation: cold→sneeze←allergy. The v-structure is quite unique because in a v-structure, in contrast to the other structures we have seen, two otherwise independent variables may become dependent. Consider cold and allergy. Both affect sneeze, but this does not confer a dependence between them. In fact, they are completely independent: The presence of a cold virus does not affect the possibility of an allergy attack and vice versa. What happens if we know that the person is sneezing? Suddenly, allergy and cold become dependent; because we know one must be the cause of the sneezing, knowing the value of one helps us to determine the value of the other. If we knew that the person had a cold virus, we would attribute less likelihood to the sneeze being caused by an allergy attack. V-structures become incredibly useful in adding causal interpretation to the network (more discussion on causal interpretation can be found in the Causality section proceeding this section). In identifying two variables that are independent but then become dependent on knowing a third variable, we can orient edges in a v-structure using only observational data.

For a further discussion of conditional independencies and equivalence classes, see Box 2.

BOX 2. Equivalence Classes and the Utility of Interventional Data

The task of finding the correct model structure relies on assessment of conditional independencies in the domain. Consider our biological example, in which three kinases affect one another: Raf phosphorylates MEK, which then phosphorylates ERK. From this description, the accurate underlying Bayesian network structure is Raf→MEK→ERK. As discussed, the data shows high correlation between Raf and MEK, MEK and ERK, and Raf and ERK as well as a statistical independency between Raf and ERK. However, how can we decide between the two valid structures Raf→MEK→ERK and ERK→MEK→Raf?

Despite their usefulness, conditional independencies often provide insufficient information for selection of a unique model, because the same set of conditional independencies can often be mapped to multiple models. In the kinase example above, the model ERK→MEK→Raf depicts the conditional independencies in the domain equally as well as the true model (Raf→MEK→ERK). This is an example of an **equivalence class**. Models with the same set of conditional independencies form an equivalence class. Such models will always receive the same Bayesian score and, therefore, are indistinguishable for a Bayesian network using only observational data. The simplest such example is the class consisting of MEK→ERK and ERK→MEK. Because of equivalence classes, when we perform structure learning using observational data, we search for the best equivalence class rather than the best model. Typically, this means that we can find a graph with only some of the edges directed (called a PDAG or partially directed acyclic graph).

Equivalence classes are a big problem in practice, but there is a conceptually easy way to get around them: using **interventional data**. Interventional data refers to data obtained by experimentally perturbing the system, such as knocking out a protein (thus setting it equal

(Continued.)

to zero) using RNAi or pharmacological activators and inhibitors. This is in contrast to observational data, in which the system may be generally stimulated, but no specific variable is forced on or off. The power of interventional data is easy to understand. Assume that we have our two signaling molecules, MEK and ERK. MEK and ERK are highly correlated, so if we attempt structure learning, we will end up with the equivalence class consisting of MEK→ERK and ERK→MEK. Now assume we have interventional data: We add a MEK inhibitor, setting its activity to zero. In the observational data, generally MEK and ERK were both 0 or both 1. Now MEK is forced to be 0, and we see that ERK is also zero. This makes us suspect that MEK might be affecting ERK, that is, MEK→ERK is the correct model. If we also have interventional data in which we add an ERK inhibitor, and we see that when ERK is set to zero, MEK is sometimes 0 and sometimes 1, we can select the model MEK→ERK with high confidence, because the data show that ERK is not affecting MEK.

Causality

Edges in the Bayesian network graph represent statistical dependencies, although we are often interested in a **causal interpretation**, that is, orienting the edges to define cause and effect. In general, a statistical dependence may be due to a causal relationship among the variables (the parent influences the child, either directly or via other variables that are not modeled, called **hidden variables**) or due to a shared coparent, a hidden variable that influences both, inducing a statistical relationship between them.[3] Causal interpretations for Bayesian networks have been proposed, along with methods for learning causal networks from data (Cooper and Yoo 1999; Heckerman et al. 1999; Pearl 2000). These both rely on edges that can be directed based only on observational data in which no particular variable has been specifically perturbed, although they are far more effective when interventional data are used as described in the previous section. A more in-depth discussion of causal networks can be found elsewhere (Pearl 2000; Pe'er 2005). In brief, edges in a causal network can be interpreted as causal when they are directed in the corresponding Bayesian network as a result of conditional independencies from observational data (in a v-structure, as discussed), from the use of interventional data, or due to constraints in the domain. However, it must be noted that the causal interpretation relies on assumptions that do not hold in many biological systems, for instance, the absence of cycles in the underlying structure as well as the absence of hidden variables. For this reason, we must treat causal interpretations of the Bayesian network structure with caution. The absence of cycles in the underlying structure means that in no case in the underlying biology is there a set of biomolecules whose influence forms

[3]Such a variable does not need to be a physical entity such as a protein. It can instead be a parameter such as time, cell size, temperature, or anything that might affect the measured level of variables. Note that the coparent must be a hidden variable to induce an edge between the variables, because if the coparent is *not* hidden, the model will include an edge from the parent to each child, rather than an edge between the two child nodes.

a cycle (i.e., RAF→MEK→ERK→RAF). Many such cycles exist in biological pathways and networks.

BAYESIAN NETWORKS IN ACTION: SUCCESSES, CHALLENGES, AND WHERE WE ARE HEADED

Among the first applications of Bayesian networks to molecular biology, the (then fairly new) technology of microarrays was essentially mined to find putative regulatory relationships in genetic regulatory pathways in yeast (Friedman et al. 2000). Bayesian network structure learning was performed multiple times, and features common to multiple graphs were extracted using model-averaging approaches. Why were features extracted, in lieu of learning a full network? The application of Bayesian networks to microarray data suffered from two critical problems that prevented researchers from learning a full network with high confidence.

First, the amount of data was very limited. Because the number of data points numbered in the hundreds, whereas the number of variables in the network numbered in the thousands (i.e., hundreds of microarrays being used to learn a network over thousands of genes), *spurious statistical relationships were likely to appear equally supported by data* as true relationships. In other words, the number of false-positive relationships would be very high.

The second problem is one of causality discussed above. Although interventional data exists, in knockout experiments, for example, it is generally insufficient to orient most of the edges in the network. Because the primary interest is in learning causal networks, orienting the edges is of critical importance.

Subsequent work in learning genetic regulatory networks using Bayesian networks (and similar models) addressed these two issues in a number of ways. One approach took advantage of existing data on transcriptional binding sites and used this information to determine those genes that may be expected to influence other genes, aiding in the determination of both edges and the directionality of these edges (Hartemink et al. 2002). This approach is very useful when informative binding data is available, although it does have the shortcoming that it requires binding data for all relevant transcription regulators, and it *fails to identify any regulators that function indirectly*, as might happen, for example, when a transcription factor regulates other transcription factors that then regulate their own targets.

Another approach that elegantly addresses the two problems presented was developed initially by Segal et al. (2003), although many modifications and elaborations have been published subsequently (see, e.g., Lee et al. 2006). This approach, called **module networks**, solves the data shortage problem by grouping sets of genes together into clusters or "modules," thus allowing multiple genes to indicate the expression level of the module under different conditions. This approach lends statistical power to the inferences of regulator to module edges. It solves the problem of directing edges in the graph by prespecifying a set of putative regulators, has received wide recognition, and has been used in various studies subsequent to the initial paper (see, e.g., Gentles et al. 2009).

LEARNING SIGNALING PATHWAY STRUCTURE FROM FLOW CYTOMETRY DATA

To investigate the statistical interactions of proteins, it is necessary to measure them in their active state, sometimes meaning in their phosphorylated state. Those measurements may be performed by western blot, protein array, or mass spectrometry, but these approaches have the same shortcomings as the microarray data, in that they lack sufficient numbers of observations to provide the statistical power necessary to differentiate spurious statistical relationships from true relationships, as described by Sachs et al. (2002). One approach that strives to circumvent these issues, developed by Sachs et al. (2005), uses high-throughput single-cell data from multidimensional flow cytometry.

Why is single-cell data helpful? The assumption used is that each cell is a unit of biological computation and, as such, each cell can be seen as a single observation of the biological system. This means each cell can serve as a single data point, rendering flow cytometry, or any high-throughput single-cell approach, *capable of providing thousands of data points* for a probabilistic algorithm.

Data type	Microarray	Flow cytometry
Data points	Arrays (hundreds)	Cell events (thousands)
Variables	Genes (thousands)	Signaling proteins (10–30)

Thus, in this approach, the problem of insufficient data is solved directly by collecting thousands of data points. The problem of orienting the edges can also be addressed by, for example, using small-molecule inhibitors to modulate the activity of proteins in the network or techniques such as short-interfering RNA (siRNA) knockdown to reduce a protein's abundance. The effects of these perturbations on the network can be incorporated very effectively into the structure learning effort partially because the number of cells observed under each perturbation is large, lending statistical power to each perturbation. This approach was first applied by Sachs et al. (2005) to primary human T cells.

CONCLUDING THOUGHTS

The study of signaling pathways with probabilistic models is a field still in its infancy, due largely to the shortage of appropriate data sources. In this chapter, we discuss two candidate data sources, microarray (equally applicable to other sources of mRNA abundance data, such as RNA-seq) and flow cytometry data, with many advantages as well as some limitations. One method for improving on these approaches would be to incorporate information from heterogeneous data sources, such as protein–protein interaction predictions. The integration of multiple data sources can increase the validity of the inferred structure dramatically, particularly when no single data source is sufficient.

Protein–protein interaction studies, as well as signaling interaction data sources, attempt to find all potential interactions (for the latter studies, all potential signaling interactions), rather than specific interactions occurring in a particular cell or cell type in response to specific stimuli or conditions. In this way, they differ markedly from the approach using flow cytometry, which extracts interactions in a particular data set and can be catered to a specific biological condition, cell type, disease, or other cell state. The set of influence connections in a particular data set and the set of potential signaling interactions can be extremely complementary data sets. One can take advantage of predicted signaling interactions to aid in the selection of an optimal Bayesian network by using priors on potential edges, as described above, and to select an initial set of molecules to model based on which molecules are thought to interact. Note that the dynamics of interactions among proteins can be described by a set of **ordinary differential equations** (ODEs), described in Box 3.

BOX 3. Ordinary and Partial Differential Equations Models

A more detailed description of the interaction among proteins (or between proteins and DNA) can be in the form of a dynamical system. In this description, the amounts of active proteins represent the "states" of the system, and they are related by a set of ODEs. These are equations that describe the dynamics of a biomolecule by expressing the change of its abundance as a function of time. The choice of equations depends on the levels of accuracy and simplicity required from the model, the availability of data to fix the parameters of the model, and assumptions or prior knowledge of how the proteins interact.

A simple first-order rate equation for the amount of protein A would look like the following:

$$dx_{\alpha}(t)/dt = \alpha_A f\left(P_A^+(t), P_A^-(t)\right) - \gamma_A x_A(t).$$

This equation simply means that the rate of change in the active form of protein A is a function of the parents of A and that there is degradation that is proportional to the amount of protein A that was already active. Note that there is also a natural dependence on the amount of inactive A, but it is hidden here for convenience.

A class of ODEs that has been used extensively is the **mass kinetic differential equation model**, a model based on the physiochemical properties of the chemical reactions that activate the proteins. Therefore, the parents of a certain variable A will be collected in groups, where the parents in each group react with each other to produce or activate A. On the

(Continued.)

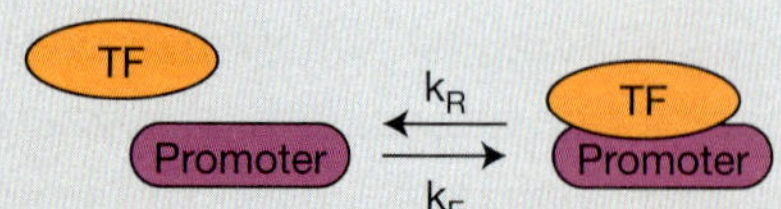

BOX 3, FIGURE 1. Binding of a transcription factor (TF) to a promoter.

other hand, the reverse chemical reactions are also taken into account, and so the variables with which *A* would react (and be deactivated by) are also collected in small groups. The effect of each group (parents or reactants) on the rate of change in *A* is a product of the concentrations of all of the elements in the group. Thus, the differential equations have the following form:

$$dm_\alpha(t)/dt = \sum_i k_i^A \prod_{B_j \in \pi_i(A)} m_{Bj} - \sum_i l_i^A \prod_{C_j \in \sigma_i(A)} m_{Cj}.$$

Here, k_i^A and l_i^A correspond to the rates of the reactions of the particular group, and the *m*s represent the concentrations of the variables. This complicated expression becomes clearer with an example. Suppose that we examine a transcription example in which we consider a transcription factor that can bind a promoter (Box 3, Fig. 1).

We would write the reaction and ordinary differential equation (ODE) model for the bound promoter in the following way:

$$[\textit{transcription factor}] + [\textit{promoter}_{unbound}] \rightleftharpoons [\textit{promoter}_{bound}],$$
$$d[\textit{promoter}_{bound}]/dt = k_f[\textit{promoter}_{unbound}][\textit{transcription factor}] + k_r[\textit{promoter}_{bound}].$$

The top equation reflects the fact that the concentrations of the transcription factor and the unbound promoter must sum to the concentration of the bound promoter. The bottom equation states that the rate of change in concentration of the bound promoter is equal to the product of the concentrations of the unbound promoter and transcription factor (by some proportionality k_f) plus the concentration of the bound promoter (multiplied by proportionality k_r).

Differential equations offer a dynamic system representation of the transcription factor–promoter binding, and thus they can be very detailed and expressive. The downside to this modeling method is that it requires a lot of information about the variables, their connectivity, their reactions, and the corresponding rates, which are often hard to measure. Thus, it can usually only be applied to systems that have been studied extensively and usually uses a previously known **connectivity graph** and variables that affect others, which must be determined either by expert knowledge or via another computational method such as Bayesian networks.

Important generalizations of ODE models are partial differential equation models that study the effect of the location along with the amount of time on the signaling pathway. Because of the existence of gradients in concentrations of different molecules in the cell, the location of a certain protein affects its rate of change (Amonlirdviman et al. 2005). This kind of model is very detailed and descriptive, but similar to the situation with ODEs, it is difficult to identify all of the important proteins/reactants in a certain system and harder to collect enough information and data to get a very detailed model.

One can also use potential signaling interaction data to elucidate the signaling events that underlie newly discovered connections in the Bayesian network. This is because many connections in the Bayesian network structure are indirect—in other words, they do not include an enzyme and its direct substrate, but rather an upstream regulator and a molecule that is eventually affected via several other intermediaries. Such a relationship will occur whenever the intermediate molecules that mediate the effect of the upstream regulator on the downstream target are not measured as part of the data set, a common problem in this data-limited domain.

One way to overcome this problem is to include more molecules in the data set when experimentally feasible. However, in cases with insufficient prior biological knowledge, we may not have a good guess regarding which molecules may act as intermediates in an interaction or indeed if the interaction is indirect or direct. In such cases, predicted signaling interactions that connect the upstream molecule to the downstream molecule can be used to complete the hypothesis of pathway structure. It can also direct experiments for future Bayesian network analyses with data that includes candidate intermediate molecules or for direct wet laboratory verification. This too is an exciting direction for future extensions of these approaches.

Ultimately, appropriate computational and experimental tools that extend learning biological pathways from data may lead to improvements in the understanding of cellular biology and disease states, in accurate and specific diagnostics, and in more specifically tailored and perhaps personalized therapies for human disease.

REFERENCES

Amonlirdviman K, Khare NA, Tree DR, Chen WS, Axelrod JD, Tomlin CJ. 2005. Mathematical modeling of planar cell polarity to understand domineering nonautonomy. *Science* **307:** 423–426.

Cooper G, Yoo C. 1999. Causal discovery from a mixture of experimental and observational data. In *Proceedings of the Fifteenth Annual Conference on Uncertainty in Artificial Intelligence*, July 30, 1999, Stockholm, pp. 116–125. Morgan Kaufmann Publishers, San Francisco.

Friedman N, Linial M, Nachman I, Pe'er D. 2000. Using Bayesian networks to analyze expression data. *J Comput Biol* **7:** 601–620.

Gentles AJ, Alizadeh AA, Lee SI, Myklebust JH, Shachaf CM, Shahbaba B, Levy R, Koller D, Plevritis SK. 2009. A pluripotency signature predicts histologic transformation and influences survival in follicular lymphoma patients. *Blood* **114:** 3158–3166.

Hartemink AJ, Gifford DK, Jaakkola TS, Young RA. 2002. Combining location and expression data for principled discovery of genetic regulatory network models. *Pac Symp Biocomput* **2002:** 437–449.

Heckerman D. 1995. A tutorial on learning with Bayesian networks. Technical Report MSR-TR-95-06, March 1995 (revised November 1996). Microsoft Corportaion, Redmond, WA.

Heckerman D, Meek C, Cooper G. 1999. A Bayesian approach to causal discovery. In *Computation, causation, and discovery* (ed. Glymour C, Cooper GF), pp. 141–166. MIT Press, Cambridge, MA.

Lee SI, Pe'er D, Dudley AM, Church GM, Koller D. 2006. Identifying regulatory mechanisms using individual variation reveals key role for chromatin modification. *Proc Natl Acad Sci* **103:** 14062–14067.

Luna IT, Huang Y, Yin Y, Padillo DP, Perez MC. 2007. Uncovering gene regulatory networks from time-series microarray data with variational Bayesian structural expectation maximization. *EURASIP J Bioinform Syst Biol* **2007:** 71312.

Morris M, Melas I, Saez-Rodriguez J. 2013. Construction of cell-type specific logic models of signaling networks using CellNOpt. *Methods Mol Biol* **930:** 179–214.

Pearl J, 1988. *Probabilistic reasoning in intelligent systems: Networks of plausible inference.* Morgan Kaufmann Publishers, San Mateo, CA.

Pearl J. 2000. *Causality: Models, reasoning and inference.* Cambridge University Press, Cambridge, UK.

Pe'er D. 2005. Bayesian network analysis of signaling networks: A primer. *Sci STKE* **2005:** l4.

Sachs K, Gifford D, Jaakkola T, Sorger P, Lauffenburger DA. 2002. Bayesian network approach to cell signaling pathway modeling. *Sci STKE* **2002:** pe38.

Sachs K, Perez O, Pe'er D, Lauffenburger DA, Nolan GP. 2005. Causal protein-signaling networks derived from multiparameter single-cell data. *Science* **308:** 523–529.

Segal E, Shapira M, Regev A, Pe'er D, Botstein D, Koller D, Friedman N. 2003. Module networks: Identifying regulatory modules and their condition-specific regulators from gene expression data. *Nat Genet* **34:** 166–176.

Shmulevich I, Dougherty ER, Kim S, Zhang W. 2002. Probabilistic Boolean Networks: A rule-based uncertainty model for gene regulatory networks. *Bioinformatics* **18:** 261–274.

12

Meta-Analysis and Data Integration of Gene Expression Experiments

Chirag J. Patel and Andrew H. Beck

Stanford University School of Medicine, Biomedical Informatics Training Program, Stanford, California 94305

The completion of several major genome sequencing projects in the last decade of the 20th century led to the development of relatively inexpensive, high-throughput genomic technologies, which have been widely used to study a broad range of biomedical specimens and conditions. Much of this data is stored in publicly available databases. The Gene Expression Omnibus (GEO) (Barrett et al. 2007) and ArrayExpress (Parkinson et al. 2009) are prime examples, having the results of more than 450,000 and 393,000 individual hybridizations, respectively. These numbers correspond to more than 18,000 (GEO) and 13,000 (ArrayExpress) individual experiments.

A major limitation to progress in biomedical research is that, historically, single investigators have not possessed the tools to access or generate enough data from single experiments to gain the "power" to permit the establishment of valid reproducible conclusions. In contrast, the vast majority of findings by single investigators require confirmation by others before acceptance (Ntzani and Ioannidis 2003; Ein-Dor et al. 2006). This section considers how one may integrate publicly available gene expression data for the discovery of robust biomedical knowledge.

We introduce the "meta-"representation of gene expression experiments and data, and we use this representation to facilitate integration. To illustrate, we consider the meta-analysis of gene expression data and the integration of gene expression data with other genome-wide measurement modalities, such as genetic variants (i.e., single-nucleotide polymorphisms) and protein networks. The concepts illustrated here will reinforce concepts covered in Chapters 10, 12, and 13 and will assume knowledge of differential gene expression, analysis of paired disease–control data sets (presented in Chapter 6), and statistical analysis techniques (presented in Chapter 3), including the R statistical computing environment (http://www.r-project.org/).

Furthermore, knowledge of certain computer science concepts, such as array data structures (described in Chapter 2) will also be helpful. Finally, knowledge regarding the Gene Expression Omnibus (http://www.ncbi.nlm.nih.gov/geo/) and how data sets are stored will be essential.

FORMULATING A QUESTION: INTRODUCTION TO INTEGRATION

Research questions involving integration are often borne out of a "metaview" (or higher-level) view of the biological processes under investigation. We summarize the investigative process in its entirety in Figure 1. Our first step involves formulation of a question that might be answered using multiple data sets. We might be interested in how well the trends observed in single experiments replicate. For example, how might a set of differentially expressed genes from a cancer-based microarray experiment replicate in the hands of other investigators? Alternatively, we might be interested in how variation in different genetic states reflects changes in expression. An example question could be how might genetic variants in a gene change its

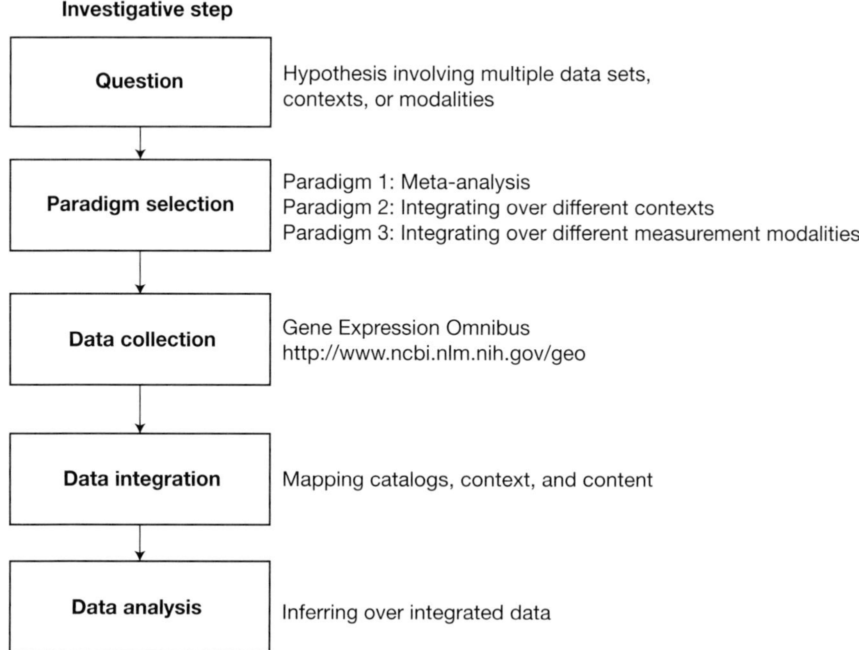

FIGURE 1. Typical investigative steps of an integrative bioinformatics analysis. As in any investigation, we start with a question that may involve multiple data sets (multiple microarray experiments), contexts (multiple diseases), or measurement modalities (genetic variants and gene expression). Next, we select an integration paradigm that fits such a question and collect data (e.g., for expression data, we may query the Gene Expression Omnibus). Data integration involves mapping between catalogs, context, and content. Finally, we infer over our newly constructed integrated data to draw conclusions.

expression? Further still, we may wish to investigate the joint variation of these states in the context of specific phenotypes, such as disease. Here, an example question may be how might differential gene expression reflect changes in functional protein interaction networks in breast cancer? Answering these types of higher-level questions requires the "integration" of data sets from multiple experiments and modalities. In this work, the process of "integration" is not a passive exercise but requires the researcher to develop appropriate tools for the collection, representation, and analysis of diverse genomic data.

After formulating a question, we choose a "paradigm" in which such a question falls (Fig. 1). Selecting a paradigm gives us a framework by which we may integrate multiple data sets. In Paradigm 1, we wish to understand how multiple experiments of the same type are similar (or different). In Paradigm 2, we wish to understand how experiments conducted in different contexts are related. In Paradigm 3, we wish to learn how experiments under the same context, but measuring different things (such as gene expression or genetic variation), may be related to one another. In this chapter, we discuss these three paradigms in detail.

First, however, we must deal with a notion called *data representation*. When proposing a research question requiring integration, we need to learn how to best "represent" data sets such that they are compatible. By standardizing how data sets are represented and making them compatible, we then are able to draw conclusions from them jointly. In the remainder of this section, we present techniques for addressing this task.

REPRESENTATION OF DIFFERENTIAL GENE EXPRESSION DATA

Many gene expression experiments in the public domain are "differential" in design, meaning that they compare the genome-wide expression between a "test" sample (i.e., a disease or exposed state) and a "control" sample (a healthy state). Therefore, the result of a gene expression experiment can often be represented compactly as a "difference vector," in which each element represents a "probe" or "gene" on the microarray of "differences" between cases and controls (for a review of gene expression data analysis, see Chapter 6). The values in the difference vector may represent the true raw expression differences between expression in a disease sample and a control sample. Alternatively, if we use replicates per sample type, these differences will be an aggregated value, such as a median difference, mean difference, rank difference, or a test statistic, such as a t statistic, p value, or q value (see Chapter 3).

For integration of these data, we must consider a higher level of vector representation (Fig. 2). We introduce concepts for describing the representation and describe a simple example to give motivation for such a description. First, each differential gene expression vector resides in a "context." The context includes experimental concepts

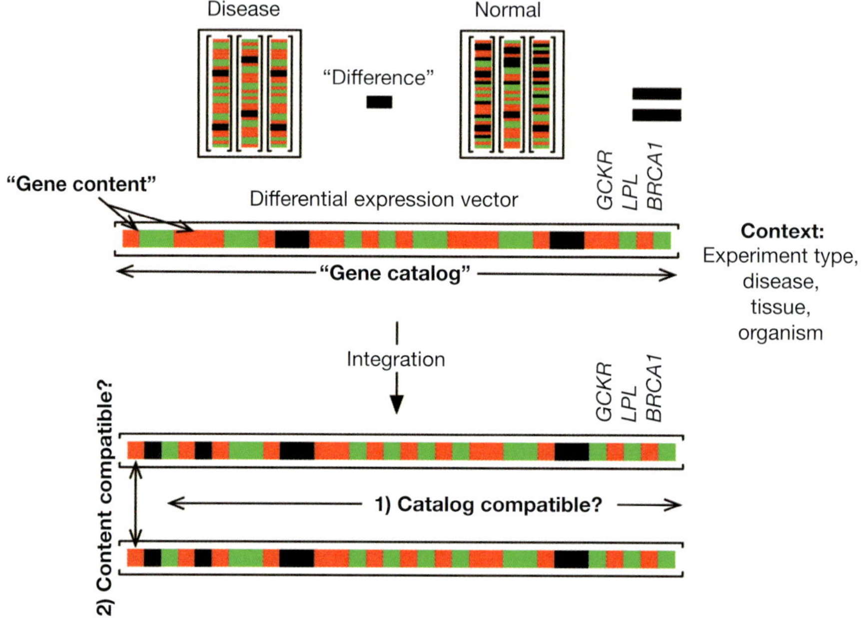

FIGURE 2. An example representation of a gene expression experiment involving disease and control samples. Transformation of a paired ("disease" vs. "control") gene expression experiment to a differential gene expression vector. As described in Chapter 6, multiple samples often make up a disease versus normal experimental context, and "difference" must be executed between these disease and normal samples. Difference functions include a *p* value, *t* statistic, mean of raw subtracted values, or "fold change" and reflect eventual "content"; these different values are represented by different colors. We are left with a single vector for an experimental context, in which each element of the vector represents a gene or probe of a "catalog" (e.g., different genes, such as *GCKR*, *LPL*, or *BRCA1*), and the difference value corresponds to the "content" (different colors). When integrating these expression vectors with other vectors, we need to account for the "context" and ensure that the "catalog" and "content" are compatible and the same.

such as sample or tissue types and experiment scenario (disease vs. control). The elements of the vector can be described as an "index" or "catalog," which is the collection of what the measurement represents, such as a gene- or probe-level expression difference. The quantities of each element are known as the "content," which are the types of specific measurements mentioned above, such as aggregated value or a rank.

Why do we represent this data structure at a higher level? Suppose we are to "integrate" gene expression experiments to find genes associated with sugar metabolism for those with diabetes versus those without. We collect gene expression data from GEO of biopsies of (1) pancreatic β cells from diabetic and nondiabetic mouse models from an Illumina Mouse Array after a high-sugar diet using average difference of probe intensity; and (2) human skeletal muscle tissue from diabetics versus nondiabetics from an Affymetrix Human Array using average fold change as in Sreekumar et al. (2002).

First, notice that our selection of samples is driven by our research question; that is, we are interested in selecting samples that have similar "contexts" (diabetes).

However, during analysis, we must keep in mind where the contexts differ, for example, tissue (skeletal muscle tissue vs. β cells) and organism type (mouse vs. human).

Next, the "catalog" and "content" differs: Mouse Affymetrix probes might not be represented as Affymetrix human probes in the vector. The "content" differs in measurement type of average probe intensity and fold change; we will need to recompute the difference of gene expression to a comparable measurement. Because we are interested in genes and not individual microarray probes, we need to find a correspondence between the mouse probes to mouse genes and human probes to human genes. Often, many probes represent a gene and require additional aggregation such as taking their mean. Finally, we need to establish correspondence between mouse genes from the mouse vector to homologous human genes. These mappings to a shared catalog and content ensure compatibility for comparison and avoid analytic confounding by differences in measurement scale.

In Chapter 2, we described concepts of databases and correspondence between keys in a database table. These concepts form the basis of "mapping" between catalogs, such as finding the correspondence between microarray probes and genes, or between genes in different species. Off-the-shelf tools exist to compute these mappings, such as the annotation packages in Bioconductor in the R statistical environment, or the online databases AILUN (Chen et al. 2007) or DAVID (da Huang et al. 2007). One may also go "directly to the source" and use the source catalogs provided by the National Center for Biotechnology Information (NCBI) or European Molecular Biology Library (EMBL). An example of such a catalog mapping includes HomoloGene, which is used to map homologous genes between species (National Center for Biotechnology Information; http://www.ncbi.nlm.nih.gov/homologene).

Ensuring common content is less straightforward than attaining a mapping between catalogs. For microarrays from similar platforms and experimental designs, we should aim to compute the differential gene expression vector using a method that preserves a maximal amount of information summarizing relative differences in expression and that is consistent across the vectors. Specifically, each entry of each vector should represent the same units of measurement (such as fold change or a p value summarizing the statistical significance of differential expression). If this cannot be achieved, we often must represent our content at a "higher level of abstraction" such as categorical ("expressed" vs. "not expressed") or ordinal data, such as a rank.

Finally, once we have made our data conform to the representation above, the ways in which we analyze our data differ depending on the question that we wish to answer and the type and number of contexts that we wish to integrate.

In the following section, we highlight paradigms and examples from the literature that can be used as starting points for extension or inspiration for novel methods. We end with a hands-on programming and analysis exercise to show how data integration can lead to novel scientific hypotheses.

PARADIGMS AND ILLUSTRATIONS OF DATA INTEGRATION

Having described a representation of expression data compatible for integration, we now describe (1) the integration of gene expression data with other gene expression data measured in similar contexts ("meta-analyses"); (2) the conducting of multiple "meta-analyses" (integrating gene expression data over different contexts); and (3) the integration of genome-wide expression data with other genome-wide measurement modalities, such as protein–protein networks, genetic variants, and clinical measurements. Figure 3A provides an overview of each paradigm by describing prototypical questions with specific examples for each of the questions. Figure 3B shows the general investigative steps described in Figure 1; parts C–E use the specific example of lung cancer to address each step of the integration process, from asking a question, choosing a paradigm, to collecting, integrating, and analyzing the data. Also shown here are prototypical tools for data integration and analysis (depending on the paradigm), such as the R Bioconductor query tool, GEOquery. We describe the steps of the integration process and the use of these tools in the following section, in which we consider three examples: the general pattern of gene expression in lung cancer, an epidemiological study of prostate cancer, and the pattern of expression response in disease due to environmental exposures.

Paradigm 1: Meta-Analysis of Gene Expression Data

In meta-analysis, we seek to combine gene expression data from related contexts. We consider here three examples: the general pattern of gene expression in lung cancer, an epidemiological study of prostate cancer, and the pattern of expression response in disease due to environmental exposures.

FIGURE 3. (*See facing page*) (*A*) Matrix comparing Paradigms 1, 2, and 3. The second column shows a prototypical question that can be asked under the corresponding paradigm. The columns right of the red line display specific examples of these prototypical questions. (*B*) An example investigative process (from Fig. 1), from hypothesis derivation to data analysis. (*C*) In the first step, we may ask whether there are genes that are common in multiple gene expression data sets of lung cancer. From such a hypothesis, we select Paradigm 1 to address it. Then, we collect appropriate data sets to address the question from the Gene Expression Omnibus (e.g., *GSE1037*, *GSE4824*, and *GSE1969*). Next, we must integrate the data, making sure that the identifiers are similar in all of the expression arrays, using a tool such as the GEOquery Bioconductor library for R. We must also ensure that their contents are compatible, such as logged expression values. Finally, in the last step, we conduct a meta-analysis using R statistical software. (*D*) In the second example, we ask how the drug response is connected with lung cancer. To answer this, we select Paradigm 2 and select Connectivity Map (CMap) (Lamb et al. 2006) and lung cancer disease data sets (*GSE1037*). Finally, a gene set enrichment analysis is performed to correlate the two contexts. (*E*) In Paradigm 3, we ask if variants found in a genome-wide association study of lung cancer are in genes that are differentially expressed in a lung cancer data set. Our first modality includes variants found in association to individuals with lung cancer, such as in Amos et al. (2008). Our second modality includes the gene expression data set *GSE1037*. We merge findings from these two modalities by mapping the variants to genes through NCBI's dbSNP database and mapping the gene expression probe to gene identifiers using the GEOquery library.

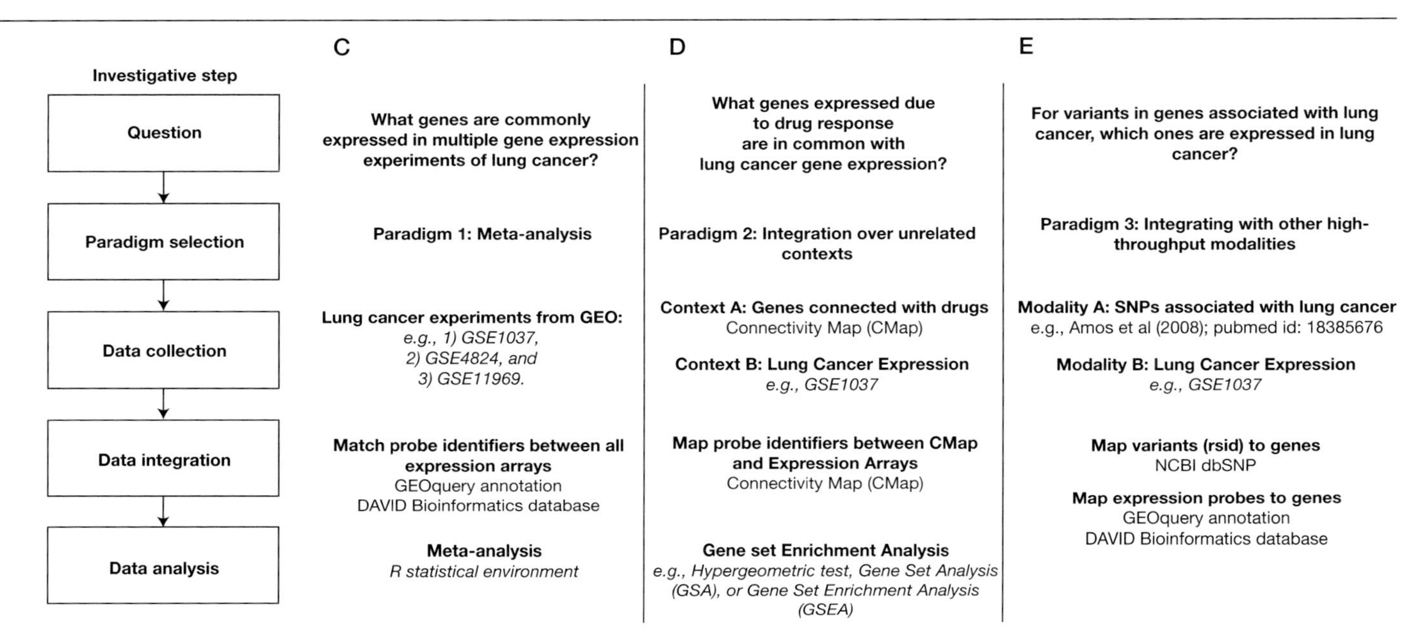

A
Integration paradigm
Prototypical questions
Paradigm 1:
Meta-analysis
What genes are expressed in a similar context?
Paradigm 2:
Integration over unrelated contexts
How is an experimental context "A" related to another distinct context "B"?
Paradigm 3:
Integrating with other high-throughput modalities
How is a context under modality "A" related to a similar context in modality "B"?
B
Investigative step
Question
Paradigm selection
Data collection
Data integration
Data analysis
C
What genes are commonly expressed in multiple gene expression experiments of lung cancer?
Paradigm 1: Meta-analysis
Lung cancer experiments from GEO:
e.g., 1) GSE1037, 2) GSE4824, and 3) GSE11969.
Match probe identifiers between all expression arrays
GEOquery annotation
DAVID Bioinformatics database
Meta-analysis
R statistical environment
D
What genes expressed due to drug response are in common with lung cancer gene expression?
Paradigm 2: Integration over unrelated contexts
Context A: Genes connected with drugs
Connectivity Map (CMap)
Context B: Lung Cancer Expression
e.g., GSE1037
Map probe identifiers between CMap and Expression Arrays
Connectivity Map (CMap)
Gene set Enrichment Analysis
e.g., Hypergeometric test, Gene Set Analysis (GSA), or Gene Set Enrichment Analysis (GSEA)
E
For variants in genes associated with lung cancer, which ones are expressed in lung cancer?
Paradigm 3: Integrating with other high-throughput modalities
Modality A: SNPs associated with lung cancer
e.g., Amos et al (2008); pubmed id: 18385676
Modality B: Lung Cancer Expression
e.g., GSE1037
Map variants (rsid) to genes
NCBI dbSNP
Map expression probes to genes
GEOquery annotation
DAVID Bioinformatics database

FIGURE 3. *(See facing page for caption.)*

In the first example (described in detail in Fig. 3), if we are interested in a general pattern of gene expression in lung cancer over multiple experiments, we may collect gene expression data of lung cancers for meta-analysis (see Fig. 3D). However, meta-analysis has also long been used in epidemiology, and similar methods are applied when analyzing gene expression data. We consider here a second example, using a landmark meta-analytic study (Rhodes et al. 2002) to illustrate these methods using our representation defined above.

In this study, the context of interest included reliably finding genes related to prostate cancer (Fig. 4). The investigators collected four raw data sets from localized prostate cancer versus benign samples, all of which were measured on human Affymetrix-based arrays. To address the "catalog" and "content," the investigators assigned uniform ENTREZ gene identifiers (NCBI) to each probe and performed a similar transformation to all of the raw data from each study. Finally, to represent each differential expression vector, the investigators used the same summary test statistic, *p* values from permutation tests (see Chapter 3). Following application of these techniques, each study was reduced to a vector of *p* values, in which each entry corresponded to a gene.

Rhodes et al. (2002) used a well-established meta-analytic method known as "Fisher's method" to determine their final gene signature related to prostate cancer. Specifically, their "null hypothesis" was that a gene was not expressed in prostate cancer and, therefore, in the null state, the corresponding *p* values would be relatively "high" across all study vectors. To reject the null hypothesis, *p* values across study vectors were required to be uniformly low. Furthermore, because this hypothesis test occurred for every gene in the vector, the researchers controlled for multiple hypotheses by computing the False Discovery Rate (FDR) (see Chapter 3) and considered genes associated with prostate cancer gene dysregulation as those with low FDRs.

Combining *p* values is but one way of conducting meta-analyses. One may also use ranks, degree of effect (such as fold change), or even number of vectors in which a

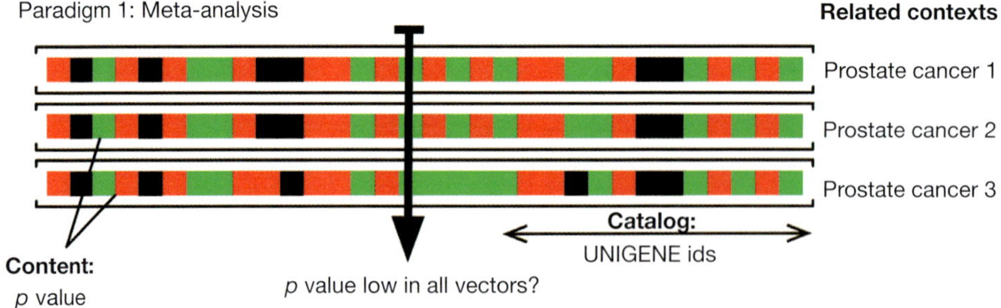

FIGURE 4. "Meta-analysis" of prostate cancer as presented in Rhodes et al. (2002) representing integration Paradigm 1. Each differential expression vector represents a distinct experiment on prostate cancer versus normal tissue. Each element of the vector has been mapped to a UNIGENE gene identifier and the content to a *p* value of differential expression. Fisher's method is used to test whether the *p* values for each gene across all experimental contexts are smaller than expected. (The figure represents a visual example of the methods described in Rhodes et al. 2002.)

gene is differentially expressed (vote counting). For further reading, see Ramaswamy et al. (2008) for a detailed review and stepwise process on conducting gene expression meta-analyses.

Paradigm 2: Integrating Expression Data over "Unrelated" Contexts

We may move beyond consideration of related contexts as in Paradigm 1 and assimilate expression data over many contexts, such as many diseases, effectively conducting multiple meta-analyses. For example, Rhodes et al. (2002) investigated the dysregulation of genes to malignant prostate cancer samples. Using Paradigm 2, we can extend this idea to examine differential gene expression over multiple cancers or diseases in an attempt to learn about the global similarities and differences in disease.

As an example, Dudley et al. (2009) developed an analysis pipeline to make statements regarding the general association between and within disease and tissue expression states. Specifically, the researchers investigated whether differential gene expression in response to disease (regardless of tissue of origin, such as blood or brain) is more concordant than different in expression owing to tissue state. Here, the investigators iterated over all of the possible contexts, identifying *all* disease-associated data sets in GEO, amounting to 429 different contexts composed of 238 diseases and 122 tissues. As in Rhodes et al. (2002), the investigators considered the catalog through extensive annotation of the expression probes and mapped these probes to universal gene identifiers. Furthermore, they executed methods to compute the same content measurement for each element in the differential gene expression vector, such as raw expression or rank normalized difference. Finally, once they generated these 429 vectors, they executed concordance analyses on them, computing the pairwise correlations within vectors associated with the same and different diseases, within vectors associated with the same and different tissues, and between vectors of disease and tissue. Assimilating these pairwise correlations, they were able to conclude that the gene expression within disease states was more concordant than random.

One may also link specific contexts, such as a particular disease and drug responses, using sets or a compendium of differential gene expression vectors (see, e.g., Fig. 3D). An established off-the-shelf method is the "Connectivity Map" (Lamb et al. 2006) that can be thought of as a suite of annotated differential gene expression data and an analytical integration method known as "Gene Set Enrichment Analysis" (GSEA) (Subramanian et al. 2005). The creators of the Connectivity Map sought to integrate gene expression response due to drugs and that of disease, hypothesizing that a reverse association of response phenotypes might point to disease therapy (i.e., "reversing" the phenotype). The investigators compiled a large set of differential gene expression vectors (human cell line samples on Affymetrix arrays) in response to drugs, resulting in an expression vector for each drug treatment. Given a disease phenotype, the investigators query their compendium of drug expression vectors, testing *reverse* association between the expression due to drug treatment and the disease expression using GSEA. Specifically, GSEA compares the rank order of expression of all of the genes between the drug expression vectors and the disease expression vector and computes the degree to which expression rankings are "not alike" after control for multiple hypotheses.

As an exercise, what are the *contexts* of the Connectivity Map? The *catalog*? Can you think of other gene expression data contexts that can be integrated using an analytical tool like GSEA?

In our third related example, Patel and Butte (2010) investigated the correlation between expression response due to environmental exposures and disease (Fig. 5). The investigators created a compendium of binary gene expression vectors associated with environmental exposures from the Comparative Toxicogenomics Database (CTD) (Davis et al. 2009). The researchers represented each context, or chemical exposure, as a vector in which the content was the set of categorical values representing "up-regulated" and "down-regulated" associations with a catalog of human ENTREZ gene identifiers. The investigators then built expression vectors of differentially expressed genes due to disease state via Significance Analysis of Microarrays (SAM, Chapter 6) (Tusher et al. 2001): Each entry of the vector corresponded to whether a gene (represented by a mapped ENTREZ gene ID) had been significantly

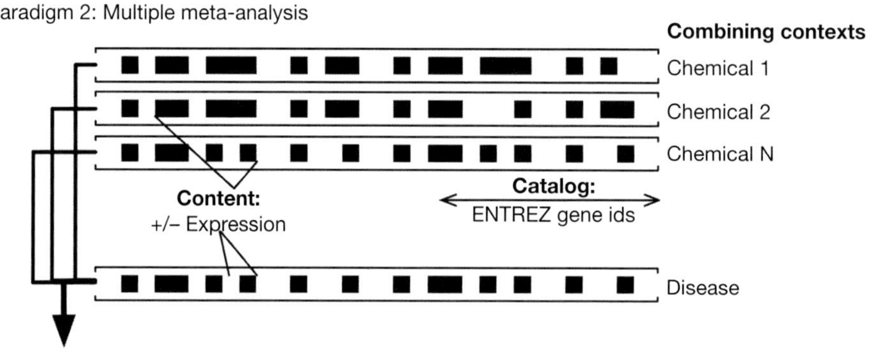

FIGURE 5. Combining multiple contexts for many "meta-analyses" as in Patel and Butte (2010). Differential expression vectors corresponding to "*N*" number of chemical exposures were assembled from publicly available data. Each entry in the vector corresponded to a gene and a binary value ([black] present; [white] not present), "up-regulated" (+ symbols) or "down-regulated" (–). A disease expression vector was computed by executing Significance Analysis of Microarrays and ascertaining genes that were "up-regulated" (+) or "down-regulated" (–). The investigators correlated each of the chemical vectors against a disease vector, using the hypergeometric test of "enrichment." The chemicals that had the highest frequency of genes in the disease vector were predicted to be associated with disease. (The figure represents a visual example of the methods described in Patel and Butte 2010.)

down- or up-regulated, providing content and a catalog compatible with chemical vectors. The investigators then checked for high correlation between each chemical expression vector in the compendium and the disease vector using a test of association similar to GSEA, called the hypergeometric test (Chapter 3). Briefly, the null hypothesis is that the disease expression vector contains an "average" number of over- or underexpressed genes represented in the chemical expression vector. In other words, there is no correlation between the disease and chemical gene sets. Specifically, the null distribution for these tests describes how many elements of the chemical vector are present in the disease vector in a random sampling. The rejection of this hypothesis indicates that the disease gene vector contains more genes from the chemical vector than random and is evidence of association between gene sets corresponding to the environmental factor and gene sets corresponding to disease. Recently, another method has been developed to test associations between sets of genes, called gene set association (GSA), which is an improvement in sensitivity over both GSEA and hypergeometric tests (Efron and Tibshirani 2007). Conceptually, this method delivers the same result: associating sets of genes from different contexts. We apply this method below.

We now take an applied detour and show a step-by-step example of such an integrative exercise, similar to that described by Patel et al. above. Here, we are interested in how environmental factors may be connected with ulcerative colitis. This is especially important because the causes of ulcerative colitis are largely unknown and we hypothesize that environmental factors may play a role in this disease by modulating gene expression. Specifically, we ask whether genes that change because of environmental factors are similar to those genes that are differentially expressed in ulcerative colitis (vs. healthy control subjects)? We shall conduct such an exercise in the R statistical programming language.

Our contexts here include (a) gene expression associated with colitis and (b) gene expression changes associated with environmental factors. For a, we search for gene expression data sets in GEO linked with ulcerative colitis and we find "GDS3268," a study investigating changes in colon gene expression between 67 ulcerative colitis patients versus 31 control subjects (Noble et al. 2008).

We download this data set by installing and using the GEOquery set of tools in R:

```
1. source("http://www.bioconductor.org/biocLite.R")
2. biocLite("GEOquery")
3. library(GEOquery)
4. gds <- getGEO('GDS3268', destdir='.')
```

The GEOquery set of tools allows one to retrieve data sets from GEO directly into R and map the expression identifiers to gene identifiers (the catalog). In lines 1 and 2, we install the GEOquery set of tools in R, which we only need to do once. Then, in lines 3 and 4, we download the ulcerative colitis data set.

```
 5. annot <- getGEO (Meta(gds)$platform)
 6. eset <- GDS2eSet(gds)
 7. pDat <- pData(eset)
 8. y <- as.numeric(pDat[ ,'disease.state'] )
 9. x <- exprs(eset)
10. annotTable <- Table(annot)
11. genenames <- as.character(annotTable$GENE_SYMBOL)
12. genenamesNew <- genenames[genenames != ""]
13. x <- x[genenames != ","]
```

In the lines above, we download the annotation corresponding to the gene expression arrays used in this investigation. In lines 7–9, we extract the relevant data expression measures ("`pData`," short for "phenotype data"), the labels that indicate whether a sample is from an ulcerative colitis patient or control (line 8), the expression matrix (line 9), and finally, the table corresponding to the gene annotation for each probe on the microarray. In line 11, we specifically extract the ENTREZ gene symbols corresponding to each probe. However, not all of the probes on the arrays correspond to a gene; some are used for quality and positive controls and thus have an empty string as a gene symbol (" "). Because they are not biologically relevant, we remove them from consideration (lines 12 and 13).

Now, we are ready for integration. We have mapped each probe on the microarray to a gene, and now we need to figure out what gene sets are common with those that correspond to environmental factors. We have decided to use the GSA method described by Efron and Tibshirani for this analysis; thus, we need to install and load that library into R (lines 14 and 15):

```
14. install.packages('GSA')
15. library(GSA)
16. install.packages('impute')
17. library(impute)
18. xx <- impute.knn(x, k = 10)
```

In lines 16–18, we install and use a package called "`impute.`" Imputation, or filling in estimates to missing data, is required before GSA analysis and ensures that the content is consistent between microarray samples. In short, the algorithm imputes values via a nearest-neighbors clustering method (see Chapter 3).

Now we need to attain gene sets associated with environmental factors. We have prepared such a data set, which is downloaded into R as such:

```
19. geneset.obj    <- GSA.read.gmt('http://www.stanford.edu/~cjp/cshl/
    ctd_expression_interactions_021912.gmt')
```

This loads 1220 environmental factor gene signatures into R from a file located at http://www.stanford.edu/~cjp/cshl/ctd_expression_interactions_021912.gmt, specifically for use by the GSA algorithm. Each entry in the file looks like the following:

```
"Tetrachloroethylene_down", "Tetrachloroethylene_down" BCL2 CASP10
CRYAA FOS GFAP IFNG INSIG1 ISG15 JUN MMP3 NEFL
```

a tab-delimited list, whereby the first two entries correspond to the environmental factor and the direction of expression regulation (i.e., "up" or "down") and represent the "vector" of genes corresponding to each chemical (Fig. 5, top). The remainder of the lines are ENTREZ gene symbols that correspond to genes that are modulated by the environmental factor. For example, this is a list of genes (e.g., BCL2, CASP10, etc.) *down*-regulated by the environmental factor *Tetrachloroethylene*. Such a list is derived from the CTD as described in Patel and Butte (2010).

We are almost finished. Now let us run the final step, the GSA algorithm:

```
20. GSA.obj<-GSA(xx$data,y,     genenames=genenamesNew,     genesets=
    geneset.obj$genesets,resp.type="Two class unpaired," nperms= 100)
21. sets <- GSA.listsets(GSA.obj, geneset.names=geneset.obj$geneset.
    names,FDRcut=0.05)
```

In lines 20–21, we run GSA in R, which is the correlation step shown in Figure 5, specifying the disease expression data set ("`xx`," line 18), the labels for each (ulcerative colitis or control, "`y`"), the gene names for each of the rows ("`genenamesNew`," line 12), and the environmental factor gene sets ("`genesets`," line 20). In line 21, we extract the top environmental factors that are associated with the disease expression data set at a low false discovery rate (FDR = 0.05; see Chapter 3). We show the results here and discuss them briefly:

```
> sets
$FDRcut
[ 1] 0.05

$negative
     Gene_set Gene_set_name Score p-value FDR
[ 1,] "17" "\"2,6-diaminotoluene_" "-0.7209" "0" "0"
[ 2,] "925" "\"gefitinib_down\"" "-0.4655" "0" "0"

$positive
     Gene_set Gene_set_name Score p-value FDR
[ 1,] "192" "\"Coal Ash_up\"" "1.4827" "0" "0"
[2,] "253" "\"Dinitrochlorobenzen" "0.4093" "0" "0"
[ 3,] "318" "\"gallium arsenide_up" "0.9239" "0" "0"
[ 4,] "340" "\"Hexachlorobenzene_u" "0.5227" "0" "0"
[5,] "384" "\"lipopolysaccharide," "1.4326" "0" "0"
[ 6,] "453" "\"nickel sulfate_up\"" "0.2978" "0" "0"
[ 7,] "483" "\"Oxazolone_up\"" "0.5747" "0" "0"
[ 8,] "491" "\"Particulate Matter_" "0.5314" "0" "0"
[ 9,] "499" "\"Peptidoglycan_up\"" "1.1412" "0" "0"
[ 10,] "545" "\"Puromycin Aminonucl" "0.4026" "0" "0"
[ 11,] "579" "\"Asbestos, Serpentin" "0.5189" "0" "0"
[ 12,] "629" "\"Thioacetamide_up\"" "0.2378" "0" "0"
[ 13,] "638" "\"Toluene 2,4-Diisocy" "0.6007" "0" "0"
[ 14,] "647" "\"trimellitic anhydri" "0.9978" "0" "0"
[15,] "649" "\"Trinitrobenzenesulf" "0.6347" "0" "0"
[ 16,] "682" "\"Zymosan_up\"" "0.6008" "0" "0"
[ 17,] "865" "\"Dichlorodiphenyl Di" "0.7689" "0" "0"
[ 18,] "926" "\"gemcitabine_down\"" "0.3387" "0" "0"
[ 19,] "946" "\"Hydrocortisone_down" "1.0696" "0" "0"
[ 20,] "1035" "\"nimesulide_down\"" "0.4169" "0" "0"
[ 21,] "1055" "\"Ozone_down\"" "0.2617" "0" "0"
[ 22,] "1104" "\"Dinoprostone_down\"" "0.8463" "0" "0"
[ 23,] "1161" "\"telmisartan_down\"" "0.5818" "0" "0"

$nsets.neg
[ 1] 2

$nsets.pos
[ 1] 23
```

The "`sets`" variable contains the FDR threshold and the sets that are negatively and positively associated with the disease expression data. A "negative" association denotes that the set of genes corresponding to that environmental factor was, on average, found to be higher in the control samples, and a positive association indicates that the genes corresponding to the environmental factor were found to be higher in the disease samples. The top negative and positive sets are shown; for example, one of the top sets belonged to lipopolysaccharide, or line 5 (in bold) of the positive set (p value and FDR approximated at 0). Another factor includes TBNS, or trinitrobenzenesulfonic acid (bold, line 15) and dinitrochlorobenzene (bold, line 2). Interestingly, factors such as these are used to induce colitis in model organisms (e.g., Scheiffele and Fuss 2002), thus providing some biological plausibility to such an exercise. Specifically, we hypothesize that some of the environmental factors may play a role in ulcerative colitis and are therefore worth further in-depth study. A copy of the R output and scripts to conduct these analyses can be downloaded from http://www.stanford.edu/~cjp/cshl.

Exercise: What drugs might be tested to use as a therapy against ulcerative colitis? Explore this question using the Connectivity Map set of tools (http://www.broadinstitute.org/cmap/).

Exercise: Devise and test an experiment to compute the "similarity" of lung and colon cancer disease expression vectors using tests of correlation and enrichment. What might a high (or low) correlation between these diseases mean biologically?

Paradigm 3: Integrating Expression Data with Other Genome-Wide Modalities

We may also integrate differential gene expression data with other genome-wide modalities, such as protein–protein networks and genetic polymorphisms (e.g., SNPs) (see Fig. 3E). Above, in Paradigms 1 and 2, we were concerned with the compatibility of our differential gene expression vectors with each other. The same consideration applies when integrating with other genome-wide data sets.

An example of integrating expression vectors with protein networks comes in Suthram et al. (2010) and is shown in Figure 6. Here, the investigators asked how changes in expression in multiple disease states could be associated with functions reflected by protein–protein interaction networks. They reasoned that the integration could describe biological mechanisms or processes underlying general disease gene expression states, such as those involved in cancer or infection, and provide a way to assess functional similarity between diseases as described by the protein–protein network. Analogous to the processes described above, Suthram et al. ensured compatibility of the catalog between their disease expression vectors by restricting their analysis to expression to data from a single type of array (human Affymetrix), and they ensured compatibility of content by representing the difference value as a t statistic for all of the expression vectors.

However, Suthram et al. wished to integrate changes in disease expression states with functional protein interaction subnetworks, or "modules." The catalog and content between the expression vectors and the protein–protein modules also

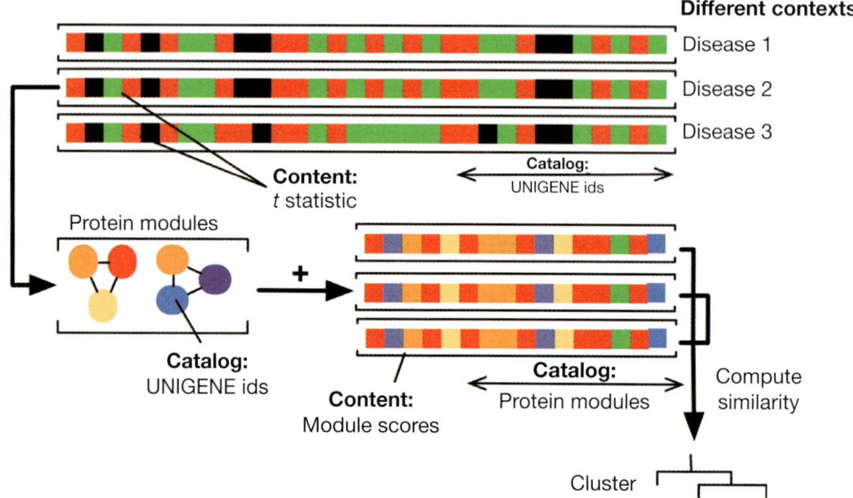

FIGURE 6. Combining multiple differential gene expression vector contexts with other measurement modalities as in Suthram et al. (2010). Functional subnetworks of protein networks are represented as modules, which themselves represent a set of interacting proteins (mapped to UNIGENE identifiers). Each disease expression vector is transformed to a disease protein module vector by "adding" (+ symbol) the set expression content values corresponding to the genes in a module. Finally, diseases are compared through pairwise correlation and hierarchical clustering. (The figure represents a visual example of the methods described in Suthram et al. 2010.)

required mapping (Fig. 6). To this end, the researchers collected human protein–protein interaction data from the Human Protein Reference Database, which contains ENTREZ gene identifiers for pairs of interacting proteins or a table mapping pairs of interacting proteins by their gene identifiers (Keshava Prasad et al. 2009). Furthermore, they identified individual functional modules within these interactions using an off-the-shelf tool called PathBLAST (Kelley et al. 2004). Outputs from these steps were sets of ENTREZ gene IDs that corresponded to distinct functional and interacting protein modules.

To map between the disease expression vectors and the protein modules, the investigators devised a "Module Response Score," which was computed by summing the differential expression *t* statistics over the gene identifiers present in the module, thereby transforming the differential expression vector to a vector of Module Response Scores. They used this transformation to compute correlations between diseases, cluster them accordingly, and identify common functional modules between diseases.

Along with expression arrays, another popular measurement modality is the "SNP chip," which assays the genome-wide variation of individuals. These modalities have enabled the experiment known as the "genome-wide association study" (GWAS, Chapter 7), which associates variation in population allele frequency to disease. Indeed, these data may be integrated with expression vectors. For example,

Chen et al. (2008) investigated the likelihood that genes differentially expressed in disease contain disease-associated SNPs.

The investigators collected multiple disease expression contexts as described above using GEO. Similar to the examples above, each element of the vector represented content as computed by the SAM procedure and a part of the ENTREZ gene identifier catalog. In parallel, they collected variants associated with multiple diseases from the Genetic Association Database (GAD), which contains a mapping between genetic variants (represented by an "RS ID") and ENTREZ gene identifier (Becker et al. 2004). Finally, to test their hypothesis, they examined the trend between the number of contexts in which a gene was differentially expressed and the frequency with which it harbored a disease-associated genetic variant.

Exercise: Perhaps variants in genes induce changes in disease-associated gene expression. Suppose that you wished to investigate this hypothesis in the context of lung cancer. Specifically, you wonder whether genetic variants implicated in a "genome-wide association study" on lung cancer lie within differentially expressed lung cancer genes. Design an integrative investigation to query whether any lung cancer variants lie in differentially expressed genes. Hint: See Figure 3E.

We need not stop with integration of gene expression data with one other modality as in the two examples above. For example, English and Butte (2007) applied the concepts described above using many data sets from many organisms (in addition to differential gene expression data) to discover obesity-associated genes. As context, they selected data annotated with obesity and fat storage phenotypes from worm, mouse, rat, and human samples. The investigators integrated diverse types of molecular data, such as RNA interference, genetic linkage, and proteomics data sets.

Exercise: Create a vector representation (catalog and content) showing how an investigator might be able to integrate gene expression, RNA interference, genetic linkage, and proteomics data sets to identify genes associated with obesity.

Representation, organization, and transformation of data represent three key components of integrative experiments in bioinformatics. In the three paradigms described above, we attempted to show how expression vectors (and other genome-wide measurement modalities) could be represented to consider context, catalog, and content to enable meta-analysis and integration.

Along with devising a research question, finding the "perfect" data set from public, online repositories is crucial to integration. To explore and find more repositories, one is encouraged to review the database issue of *Nucleic Acids Research*, which describes the variety of databases available to "integrative" researchers.

We now provide an example of data integration through a hands-on exercise, examining possible connections between obesity and breast cancer.

Programming Exercise

1. *Formulating a Question*

Obesity is a risk factor for postmenopausal breast cancer; however, the molecular factors linking obesity with breast cancer are not well understood. Therefore, we asked: What genes are significantly highly expressed in both obesity and breast cancer compared with controls?

Exercise 1: Under what "paradigm" does this question fall?

Solution: This falls under Paradigm 2 above because we wish to conduct analysis over two different contexts, obesity and breast cancer.

2. *Finding the Data*

a. Log onto http://insilico.ulb.ac.be/ and search for data sets on obesity and breast cancer.
b. Download the R object associated with data set GSE9624 ("Differential gene expression in omental adipose tissue from obese children").
c. #Download the R object for GSE15852 ("Expression data from human breast tumors and their paired normal tissues").

Exercise 2: What are the "catalog," "content," "and "context" of these data?

Solution: The catalog here is common probes on the "GPL570" Affymetrix platform. The content includes raw, nondifferential expression values taken from diseased cases and controls. The contexts include different tissues (adipose tissue from children vs. breast tumors) and different diseases being studied (breast cancer and obesity).

3. *Representation of Differential Gene Expression Data*

We perform SAM on each data set to identify a list of genes significantly associated with each disease context (obesity and breast cancer). We select a median FDR of

0.10 to generate our list of genes significantly associated with obesity or breast cancer. The FDR, as covered in Chapter 3, is a measure analogous to the *p* value that enables us to estimate the strength of association of a gene to the phenotype, given that we have conducted multiple hypothesis tests. Here, an FDR of 0.10 equates to ~10% of genes that have an FDR <0.1 *might* be a false-positive result, and it is a common threshold used in gene expression studies. We use the delta table to select the delta threshold that results in gene lists with a median FDR of 0.10, and we store these gene lists for both obesity and breast cancer. (Please refer to Chapter 6, Gene Expression Analysis, for additional explanation of Significance Analysis of Microarrays and the FDR.)

4. *Integrating Findings*

We take the intersection, or the genes found in both experiments, of the two lists of significant genes and identify 33 genes positively associated with both obesity and breast cancer:

```
"GATA3" "PARP12" "RBM16" "PHF16" "C8orf4" "KDM3A" "RLF" "ANKRD27"
"KIAA0649" "CBFA2T3" "SIAH1" "N4BP2L2" "ATP6V0E2" "C9orf78" "AGRN"
"NIPBL" "WSCD1" "SLC12A2" "MMP1" "BTBD1" "STK3" "UBE3B" "CDKN1B"
"B3GALT4" "CNOT2" "EVX1" "CHMP4A" "TAX1BP1" "REEP5" "USP20" "XAF1"
"SCUBE2" "RGS12"
```

Exercise 3: What about these data makes their intersection possible?

Solution: The intersection is possible because the experiments use the same catalogs (the same gene expression probes are measured in both experiments).

5. *Interpreting Findings*

Several of these markers known to be expressed in both the normal breast and breast cancer (e.g., GATA3, ANKRD27) are known to be expressed in both the normal breast and in estrogen receptor–positive breast cancer. These data show that these markers are also associated with obesity, suggesting a possible link between obesity and breast cancer mediated through these markers. The role of the markers identified in the analysis may be evaluated in future bioinformatic and mechanistic analyses.

R Programming Solution

```
#1 Think up an interesting question. For example, what genes are signif-
   icantly highly expressed in both obesity and breast cancer compared
   with controls?
#2 Log onto http://insilico.ulb.ac.be/ and search for data sets on
   obesity and breast cancer.
#Download the R object associated with data set GSE9624 ('Differential
gene expression in omental adipose tissue from obese children')
#Download the R object for GSE15852 ('breast cancer paired normal').
#use setwd(CURRENTDIRECTORY) command to set working directory
#3 Load obesity data into R
load('GSE9624GPL570FRMAGENE_4377.RData')
#4 Access obesity phenotype data
pData(GSE9624GPL570FRMAGENE4377)
y.1<-as.numeric(pData(GSE9624GPL570FRMAGENE4377)[ ,'Disease'])
x.1<-exprs(GSE9624GPL570FRMAGENE4377)
#5 Load breast cancer data into R
load('GSE15852GPL96FRMAGENE_4896.RData')
x.2<-exprs(GSE15852GPL96FRMAGENE4896)
#6 Access breast cancer phenotype data
pData(GSE15852GPL96FRMAGENE4896)[ ,'Disease']
#7 Because we are doing a paired analysis, we must
#create a y vector of the form (-1,1,-2,2....) for each control,disease
pair of samples
y.2<-vector(length=ncol(x.2))
pos=c(1,2)
for(i in 1:(length(y.2)/2)){
y.2[ pos]<-c(-1,1)*i
pos<-pos+2}
###Now we are ready to perform SAM to identify differential genes
###in breast cancer and obesity
    library(samr)
#8 Perform SAM on obesity data
d1=list(x=x.1,y=y.1,geneid=row.names(x.1),logged2=F)
samr.obj1<-samr(d1,resp.type="Two class unpaired,"nperms=200)
delta.table1 <- samr.compute.delta.table(samr.obj1)
delta.table1
#Look over the delta table and select a cut-off that gives a median
#FDR of ~10%
delta1=.7
samr.plot(samr.obj1,delta1)
siggenes.table1<-samr.compute.siggenes.table(samr.obj1,delta1, d1,
    delta.table1)
obGenes<-siggenes.table1$genes.up[,"Gene Name"]
#9 Perform SAM on breast cancer data
```

(Continued.)

```
d2=list(x=x.2,y=y.2,geneid=row.names(x.2),logged2=F)
samr.obj2<-samr(d2,resp.type="Two class paired,"nperms=200)
delta.table2 <- samr.compute.delta.table(samr.obj2)
delta.table2

#Look over the delta table and select a cut-off that gives a median
#FDR of ~10%
delta2=0.7
samr.plot(samr.obj2,delta2)
siggenes.table2<-samr.compute.siggenes.table(samr.obj2,
    delta2, d2, delta.table2)
breastCanGenes<-siggenes.table2$genes.up[,"Gene Name"]

#10#Now we look for genes on both lists!
intGenes<-intersect(breastCanGenes,obGenes)
intGenes

#This analysis reveals 33 genes on both lists:
#[1] "GATA3" "PARP12" "RBM16" "PHF16" "C8orf4" "KDM3A" "RLF" "ANKRD27"
"KIAA0649" "CBFA2T3" "SIAH1" "N4BP2L2"
#[13] "ATP6V0E2" "C9orf78" "AGRN" "NIPBL" "WSCD1" "SLC12A2" "MMP1"
"BTBD1" "STK3" "UBE3B" "CDKN1B" "B3GALT4"
#[25] "CNOT2" "EVX1" "CHMP4A" "TAX1BP1" "REEP5" "USP20" "XAF1" "SCUBE2"
"RGS12"

##These genes may be involved in obesity-associated breast
    cancer!
```

ACKNOWLEDGMENTS

We thank Professor Atul J. Butte. Some of the core concepts of the material in this chapter have been extended from Professor Butte's "Translational Bioinformatics" course at Stanford University (http://bmi217.stanford.edu/doku.php) (Butte 2010).

REFERENCES

Amos CI, Wu X, Broderick P, Gorlov IP, Gu J, Eisen T, Dong Q, Zhang Q, Gu X, Vijayakrishnan J, et al. 2008. Genome-wide association scan of tag SNPs identifies a susceptibility locus for lung cancer at 15q25.1. *Nat Genet* **40:** 616–622.

Barrett T, Troup DB, Wilhite SE, Ledoux P, Rudnev D, Evangelista C, Kim IF, Soboleva A, Tomashevsky M, Edgar R. 2007. NCBI GEO: Mining tens of millions of expression profiles—Database and tools update. *Nucleic Acids Res* **35:** D760–D765.

Becker KG, Barnes KC, Bright TJ, Wang SA. 2004. The genetic association database. *Nat Genet* **36:** 431–432.

Butte A. 2010. *Biomedical Informatics 217: Introduction to translational bioinformatics.* Stanford, CA.

Chen R, Li L, Butte AJ. 2007. AILUN: Reannotating gene expression data automatically. *Nat Meth* **4:** 879.

Chen R, Morgan AA, Dudley J, Deshpande T, Li L, Kodama K, Chiang AP, Butte AJ. 2008. FitSNPs: Highly differentially expressed genes are more likely to have variants associated with disease. *Genome Biol* **9:** R170.

da Huang W, Sherman BT, Tan Q, Kir J, Liu D, Bryant D, Guo Y, Stephens R, Baseler MW, Lane HC, et al. 2007. DAVID Bioinformatics Resources: Expanded annotation database and novel algorithms to better extract biology from large gene lists. *Nucleic Acids Res* **35:** W169–W175.

Davis AP, Murphy CG, Saraceni-Richards CA, Rosenstein MC, Wiegers TC, Mattingly CJ. 2009. Comparative Toxicogenomics Database: A knowledge base and discovery tool for chemical–gene–disease networks. *Nucleic Acids Res* **37:** D786–D792.

Dudley JT, Tibshirani R, Deshpande T, Butte AJ. 2009. Disease signatures are robust across tissues and experiments. *Mol Syst Biol* **5:** 307.

Efron B, Tibshirani R. 2007. On testing the significance of sets of genes. *Ann Appl Stat* **1:** 107–129.

Ein-Dor L, Zuk O, Domany E. 2006. Thousands of samples are needed to generate a robust gene list for predicting outcome in cancer. *Proc Natl Acad Sci* **103:** 5923–5928.

English SB, Butte AJ. 2007. Evaluation and integration of 49 genome-wide experiments and the prediction of previously unknown obesity-related genes. *Bioinformatics* **23:** 2910–2917.

Kelley BP, Yuan B, Lewitter F, Sharan R, Stockwell BR, Ideker T. 2004. PathBLAST: A tool for alignment of protein interaction networks. *Nucleic Acids Res* **32:** W83–W88.

Keshava Prasad TS, Goel R, Kandasamy K, Keerthikumar S, Kumar S, Mathivanan S, Telikicherla D, Raju R, Shafreen B, Venugopal A, et al. 2009. Human Protein Reference Database—2009 update. *Nucleic Acids Res* **37:** D767–D772.

Lamb J, Crawford ED, Peck D, Modell JW, Blat IC, Wrobel MJ, Lerner J, Brunet JP, Subramanian A, Ross KN, et al. 2006. The Connectivity Map: Using gene-expression signatures to connect small molecules, genes, and disease. *Science* **313:** 1929–1935.

Noble CL, Abbas AR, Cornelius J, Lees CW, Ho GT, Toy K, Modrusan Z, Pal N, Zhong F, Chalasani S, et al. 2008. Regional variation in gene expression in the healthy colon is dysregulated in ulcerative colitis. *Gut* **57:** 1398–1405.

Ntzani EE, Ioannidis JP. 2003. Predictive ability of DNA microarrays for cancer outcomes and correlates: An empirical assessment. *Lancet* **362:** 1439–1444.

Parkinson H, Kapushesky M, Kolesnikov N, Rustici G, Shojatalab M, Abeygunawardena N, Berube H, Dylag M, Emam I, Farne A, et al. 2009. ArrayExpress update—From an archive of functional genomics experiments to the atlas of gene expression. *Nucleic Acids Res* **37:** D868–D872.

Patel C, Butte A. 2010. Predicting environmental chemical factors associated with disease-related gene expression data. *BMC Med Genomics* **3:** 17.

Ramasamy A, Mondry A, Holmes CC, Altman DG. 2008. Key issues in conducting a meta-analysis of gene expression microarray datasets. *PLoS Med* **5:** e184.

Rhodes DR, Barrette TR, Rubin MA, Ghosh D, Chinnaiyan AM. 2002. Meta-analysis of microarrays. *Cancer Res* **62:** 4427–4433.

Scheiffele F, Fuss IJ. 2002. Induction of TNBS colitis in mice. *Curr Protoc Immunol* **15:** 1–15.

Sreekumar R, Halvatsiotis P, Schimke JC, Nair KS. 2002. Gene expression profile in skeletal muscle of type 2 diabetes and the effect of insulin treatment. *Diabetes* **51:** 1913–1920.

Subramanian A, Tamayo P, Mootha VK, Mukherjee S, Ebert BL, Gillette MA, Paulovich A, Pomeroy SL, Golub TR, Lander ES, et al. 2005. Gene set enrichment analysis: A knowledge-based approach for interpreting genome-wide expression profiles. *Proc Natl Acad Sci* **102:** 15545–15550.

Suthram S, Dudley JT, Chiang AP, Chen R, Hastie TJ, Butte AJ. 2010. Network-based elucidation of human disease similarities reveals common functional modules enriched for pluripotent drug targets. *PLoS Comput Biol* **6:** e1000662.

Tusher VG, Tibshirani R, Chu G. 2001. Significance analysis of microarrays applied to the ionizing radiation response. *Proc Natl Acad Sci* **98:** 5116–5121.

WWW RESOURCES

http://bmi217.stanford.edu/doku.php Translational Bioinformatics Course (Atul Butte, Stanford University)

http://www.bioconductor.org/biocLite.R GEOquery tool set

http://www.broadinstitute.org/cmap Connectivity Map set of tools

http://insilico.ulb.ac.be Obesity and breast cancer data sets

http://www.ncbi.nlm.nih.gov/gene Entrez Gene

http://www.ncbi.nlm.nih.gov/geo Gene Expression Omnibus

http://www.ncbi.nlm.nih.gov/homologene HomoloGene National Center for Biotechnology Information (NCBI), National Library of Medicine (NLM)

http://www.ncbi.nlm.nih.gov/unigene UniGene

http://www.r-project.org R statistical computing environment

http://www.stanford.edu/~cjp/cshl R output and scripts

http://www.stanford.edu/~cjp/cshl/ctd_expression_interactions_021912.gmt Gene sets associated with environmental factors (for use with GSA algorithm)

13

Natural Language Processing: Informatics Techniques and Resources

Bethany Percha and Wei-Nchih Lee

Stanford University School of Medicine, Biomedical Informatics Training Program, Stanford, California 94305

The amount of biomedical knowledge is increasing at an unbelievable pace. An easy way to gauge this increase is the total number of citations in Medline, the largest and most widely used database of biomedical publications; the citation count grows at a compounded rate of ~4.2% per year (Fig. 1) (Hunter and Bretonnel Cohen 2006). As of this writing, Medline contains more than 22 million records from more than 5000 publications covering the years 1948 to the present (Medline 2013). More than 1 million new publications entered the database in the year 2012 alone. This phenomenal growth presents a challenge to the field of biomedical informatics, which operates at the interface of disciplines spanning clinical medicine, computer science, statistics, and the biosciences. A typical bioinformatics researcher must be able to combine research findings and integrate knowledge from many, if not all, of these diverse areas. Because the volume of data within its domain is so large, biomedical informatics relies on computational techniques to integrate and summarize the available data, make novel inferences, and help to generate new hypotheses.

Most biomedical knowledge is contained in the text of scientific publications and clinical documents; thus, text analysis techniques from linguistics, machine learning, and natural language processing have proven to be helpful in addressing a variety of problems in bioinformatics. The central goal of biomedical text mining is to take an unstructured document—anything from a scientific article to a clinical note—and turn it into a set of structured facts that a computer can interpret. A broad set of techniques is used to extract information from the biomedical literature; this chapter presents an overview of some of the most common ones. To augment the ideas

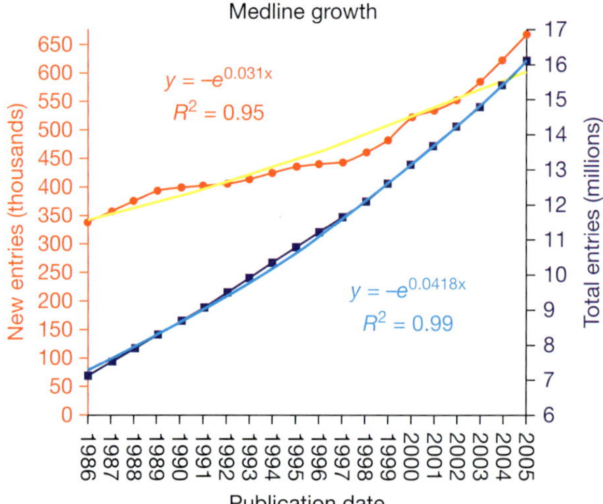

FIGURE 1. Growth of the biomedical literature, 1986–2005. (Orange circles) The number of new articles indexed in MEDLINE with a publication date in each year. (Blue squares) The total number of articles indexed at the end of each year. (Light lines) Exponential curves fitted to each series, and the corresponding formulas show growth rates and goodness-of-fit measures for both curves. (Reprinted from Hunter and Bretonnel Cohen 2006, with permission from Elsevier.)

and methods presented here, we include many links and references to direct interested readers to more detailed information and technical discussion.

The first part of this chapter focuses on methods from natural language processing that can be used to parse text and convert it into machine-interpretable formats. We also discuss statistical techniques for recognizing biomedical entity names in text, and we introduce machine learning methods that are commonly used in text mining. The second part of the chapter focuses on two important resources for processing biomedical text: standard terminologies and ontologies. In a sense, the first half of the chapter describes ways to extract meaning from text without human intervention, and the latter half shows how human knowledge can be incorporated into these analyses to increase our power to process and reason over biomedical text.

CURRENT AND FUTURE APPLICATIONS

We begin this chapter with a list of potential applications for biomedical text mining to provide a sense of the breadth and depth of the field. Integrating textual analysis with research in other areas of biomedical informatics has proven to be extremely fruitful in the past and shows great future promise in areas such as the following.

- ***Document classification.*** A system that could classify documents automatically would be of great use in, for instance, database curation or in the development of a “smart” electronic medical record system that could assign documents automatically to a particular patient or differentiate a laboratory result from a clinic note.

- ***Data integration.*** Integrating data from multiple sources, such as the scientific literature and clinical text records, could help to produce better decision support tools for physicians and enable new interdisciplinary forms of scientific research (Altman et al. 2008).
- ***Biosurveillance.*** Monitoring text-based inputs can alert observers to important epidemiological trends. For example, automatic text mining of hospital patient records could provide a mechanism for more consistent and reliable reporting of adverse events (Bates et al. 2003; Wang et al. 2009). Similarly, analysis of text messages, online postings, or web searches can lead to interesting predictions of future epidemiological events. Google Flu Trends, which predicted influenza incidence throughout the United States with remarkable accuracy, using only Internet search terms, is an example of this approach (Ginsberg et al. 2009).
- ***Consumer health informatics.*** Patients increasingly want access to the latest research and scientific data to inform their health decisions. Biomedical text mining could be used to summarize confusing medical data for patients (e.g., laboratory results, physician notes, etc.), help them search for drug information online, or develop interactive tools that help patients monitor and improve their own health.
- ***Health services delivery.*** The U.S. government has recently devised an incentive scheme, implemented via 2009’s HITECH Act, that rewards physicians and hospitals for their use of electronic medical records (EMRs) (Blumenthal and Tavenner 2010). The involvement of government in EMRs and its push to encourage EMR installation in hospitals and physician offices across the country will likely lead to a wealth of new text-based clinical information over the next few years. Automatically mining these data could lead to the development of better EMRs and facilitate scientific research and clinical trials (Dean et al. 2009).
- ***Drug discovery.*** New drug targets are not usually “discovered” by pharmaceutical companies, but are described and characterized in the scientific literature—often for years—before they enter the drug development pipeline (Agarwal and Searls 2008). Because of this, pharmaceutical companies are extremely interested in techniques for automatically mining the biomedical literature to uncover relevant facts and make connections between diverse scientific fields that can spark the development of new therapeutic targets and agents.

These are just a few of the many applications that can benefit from the bioinformatics resources and techniques discussed in this chapter.

PREPROCESSING RAW TEXT

Biomedical text comes in two main formats: scientific text, which mainly consists of journal articles housed in Medline (PubMed), a large database maintained by the National Library of Medicine; and clinical text, which consists of any text-based documents generated in the process of patient care. First, we must convert raw text into a format that computer software can interpret. In computer science this is called **preprocessing**. In biomedical text mining, preprocessing usually includes some combination of the following steps, as illustrated in Figure 2.

Tokenization

Tokenization is the process of segmenting input text (the sentence) into words or phrases called tokens. Although it sounds simple, tokenization must overcome common complications such as abbreviations, apostrophes, hyphenation, multiple number formats, and sentence boundary ambiguities (Ananiadou and McNaught 2006, Chapter 2). Tokenization is usually accomplished using regular expressions to determine word and sentence boundaries (Jurafsky and Martin 2009, Chapter 2). Jiang and Zhai provide an empirical study of tokenization strategies for biomedical text retrieval (Jiang 2007).

Part of Speech Tagging

Many words take on several different roles in text. Assigning parts of speech to words automatically can help disambiguate, for example, the *noun* "screen" (the monitor

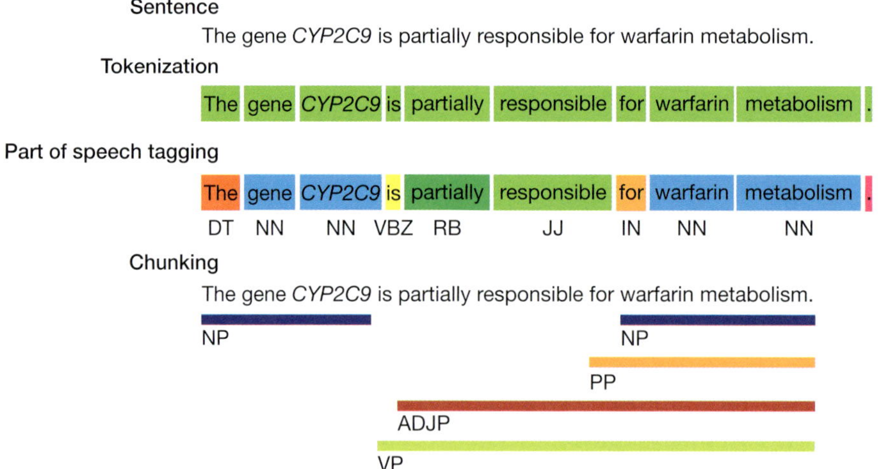

FIGURE 2. Illustration of some important preprocessing steps: tokenization, part-of-speech tagging, and chunking. (NN) noun; (DT) determiner; (VBZ) verb, 3rd person singular present; (RB) adverb; (JJ) adjective; (IN) preposition; (NP) noun phrase; (PP) prepositional phrase; (ADJP) adjective phrase; (VP) verb phrase.

on the desk in front of you) from the *verb* "screen" (to select, reject, or consider, as with subjects for a clinical trial). Preannotating words with part-of-speech (POS) tags increases accuracy in many text-mining applications. There are three main classes of techniques for POS tagging: rule-based methods (Karlsson 1995), statistical methods (Ratnaparkhi 1996; Giménez and Màrquez 2003; Jurafsky and Martin 2009, Chapter 6), and transformation-based approaches (a combination of the previous two) (Brill 1995).

Statistical and transformation-based POS taggers must be trained on a corpus of human-annotated text before they can be used. Note that most statistical POS taggers are trained on general text corpora like the *Wall Street Journal*. Thus, they often misclassify biomedical terms, like the names of genes and proteins. One can retrain a POS tagger on a biomedical corpus, before using it, to increase its accuracy.

Chunking and Parsing

Chunking and parsing processes address the fact that higher-level units of text, such as phrases and sentence clauses, carry meaning. **Chunking**, the process of identifying noun, verb, adjective, or prepositional phrases, can be accomplished using many of the same methods used for part-of-speech assignment. For example, Ramshaw and Marcus (1995) used transformation-based learning to identify noun phrases. A variety of approaches to chunking are described by Tjong et al. (2000); the most successful used a combination of rule-based and statistical approaches.

A deeper version of chunking is **parsing**, in which the entire grammatical structure of a sentence is decoded. There are two varieties of parsing: constituency parsing, in which the words in a sentence are organized into nested groups; and dependency parsing, in which the parser identifies binary grammatical relationships among pairs of words. An example of the different trees that result from these two types of parsing is shown in Figure 3. These parse trees were generated using the

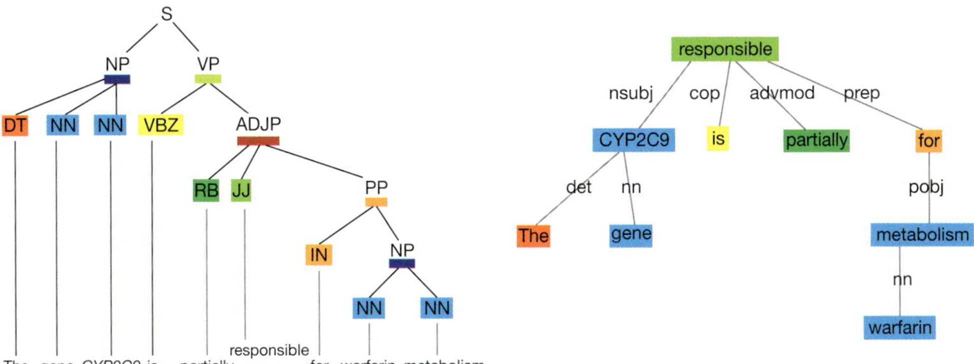

FIGURE 3. Constituency and dependency parse trees. The sentence analyzed in Figure 2, "The gene *CYP2C9* is partially responsible for warfarin metabolism," is represented in a constituency parse tree on the left and in a dependency parse tree on the right.

Stanford Parser (Klein and Manning 2003; de Marneffe et al. 2006). Parse trees can be created using a variety of different statistical algorithms, including the Cocke-Kasami-Younger (CKY) algorithm, the Earley algorithm, and chart parsing (Manning and Schuetze 1999, Chapter 12; Jurafsky and Martin 2009, Chapters 13 and 14). Cer and colleagues describe the results of a head-to-head comparison of the speed and accuracy of many modern statistical parsers (Cer et al. 2010). Although full-sentence parsing was not widely used in biomedical text mining until recently, it has met with much success, in particular, for extracting events from biomedical papers (Yakushiji et al. 2001) and revealing gene–drug, gene–phenotype, and gene–pathway relationships (McDonald et al. 2004; Coulet et al. 2010).

Stemming

Stemming means reducing words to their base forms by removing derivational (e.g., associa-*tion*) and inflectional (e.g., associat-*es*, associat-*ed*) endings. The Porter algorithm (Porter 1980), which uses a predefined set of rules for stripping words of their endings, is the most common approach to stemming. However, the Lovins stemmer (Lovins 1968) and the S stemmer (Harman 1991) are two other useful alternatives. In comparison to the Porter stemmer, the Lovins stemmer tends to be more aggressive in removing endings from words, whereas the S stemmer is less aggressive (Jiang 2007).

Stop Word Removal

Some words, such as *of*, *the*, and *and*, are considered uninformative for many text-mining applications. These words are often removed before further analysis is performed. PubMed provides a list of common stop words for biomedical text-mining purposes. It is also possible to generate a stop word list simply by finding the most common words in the corpus of interest, under the assumption that they carry the least information.

NAMED ENTITY RECOGNITION

One of the most basic uses of text mining is to identify entities of interest in text, a process called named entity recognition (NER). For biomedical text, these entities are often genes, drugs, diseases, or clinical terms. One easy approach is to scan a document to find exact matches with a lexicon of terms. This strategy, however, fails in the biomedical domain for several reasons (Cohen and Hersh 2005). First, no complete lexicon for biomedical entities exists, in part because the rapid pace of biomedical research results in the creation of names for newly discovered genes, proteins, diseases, and so on. In addition, biomedical names often have different meanings depending on context (e.g., creatinine and ferritin are biological substances that are also the names of laboratory tests), and multiword names (e.g., high-density lipoprotein, multiple sclerosis) are common (Cohen and Hersh 2005). Recognizing biomedical entity names in text presents a unique challenge and is currently a very active area of research.

Simply by looking at the words *oxycontin* and *CYP3A4*, we can guess that the former is a drug and the latter is a gene. Many biomedical NER systems seek to identify the features of words and phrases that facilitate those kinds of distinctions. These systems include a set of rules regarding the surface features, morphology, and context of a word or phrase that indicate that it belongs to a particular biomedical type. For example, prefixes ("oxy"), suffixes ("tin," "odone"), and morphological features (placement of capital letters, numbers, punctuation, etc.) can all have a role in identifying entities, as can elements of the surrounding context (words immediately preceeding and following the target word, for example.) Table 1 contains descriptions of some recent and/or highly cited biomedical NER systems, most of which are available for download or for use on the Web.

TABLE 1. Some recent and/or highly cited biomedical named entity recognition systems

Primarily for genes, proteins, cells, molecules		
ABNER	Settles 2004	Uses conditional random fields and a variety of hand-selected features to find protein, DNA, RNA, cell line, and cell type names in text.
BANNER	Leaman and Gonzalez 2008	A machine-learning classifier based on conditional random fields (see Table 4); it is available as an open-source software package for easy deployment.
BioTagger-GM	Torii et al. 2009	Uses a combination of dictionary lookup, machine learning, and postprocessing, along with a voting scheme among several publicly available NER systems, to identify genes and proteins.
LingPipe	Alias-I 2008	A general system for computational linguistics (not just biomedicine), this has been adapted on several occasions to the biomedical domain with much success.
Primarily for diseases, drugs, clinical/medical entities		
MetaMap	Aronson 2001	Provided by the National Library of Medicine, MetaMap allows for the easy annotation of biomedical text with concepts from the Unified Medical Language System (UMLS) Metathesaurus.
MedLEE	Friedman et al. 1995	A rule-based system designed to extract, structure, and encode clinical information in textual patient reports, MedLEE was originally developed by Friedman and colleagues at Columbia University.
cTAKES	Savova et al. 2010	Open-source natural language processing system developed by a team at the Mayo Clinic and built using the UIMA (Unstructured Information Management Architecture) framework, cTAKES processes clinical notes, identifying drugs, diseases/disorders, signs/symptoms, anatomical sites, and procedures.
DrugNer	Segura-Bedmar et al. 2008	Uses word morphology and a set of nomenclature rules to identify drug names in text.

ANNOTATION AND STANDARD TERMINOLOGIES

Let us assume that we have converted our raw text document into a machine-computable format—tokenized and parsed if necessary—and perhaps processed it to label entities of interest, such as genes, drugs, and disease names. At this point, we may want to incorporate an additional step, **annotation**, before proceeding with analysis. An annotation is any descriptive note applied to raw text. For example, the gene names *ABCB1* and *MDR1* are two different names for the same entity, P-glycoprotein, a transporter that helps shuttle substances across cell membranes. We might annotate both "ABCB1" and "MDR1" with the term "P-glycoprotein," to indicate that both gene names refer to that entity.

Annotation extends far beyond synonyms; it can be used to label terms with any descriptor or set of descriptors needed, depending on the task at hand. For example, part-of-speech tagging is a form of annotation. In addition, terms can be annotated with their biomedical category (gene, disease, symptom, etc.), their location within the body (if they are organs or tissues), their location within cells (if they are proteins or subcellular structures), or any other information that can enhance understanding.

Biomedical informatics depends on a set of well-defined terminologies, called **standard terminologies**, to aid in the process of annotation. The naming of objects or processes in biomedicine is as important as the measurement of dimensions or phenomena in the physical sciences. To this end, biomedical researchers have tried to establish conventional nomenclature to suit their needs; however, this approach was never standardized across the many disciplines that comprise biomedicine and has failed as the volume of information has drastically increased. In response, biomedical informatics researchers have developed standard terminologies (also referred to as controlled vocabularies or consensus terminologies) for the purposes of annotation. The goal of these efforts is to provide tools for annotating documents with relevant terms so that researchers from a multitude of perspectives can search them effectively.

Currently, most document annotation is still performed manually by domain experts. Indeed, this is considered the gold standard against which other methods are evaluated. Nevertheless, because of the considerable time and effort required for manual annotation of biomedical documents, there has been a significant amount of research aimed at investigating possibilities for automating the process (Jonquet et al. 2009).

The perspective and domain that govern the annotation process often determine a researcher's choice of standard terminology. For example, a researcher conducting a study of gene expression products in mice may refer to the Gene Ontology (GO), which classifies gene and gene products as *biological processes*, *cellular components*, or *molecular functions*. Alternatively, a meta-analysis of a specific breast cancer therapy may benefit from the use of the National Cancer Institute's (NCI) Enterprise Vocabulary Services, which include the NCI Thesaurus and

Metathesaurus. Table 2 lists some commonly used standard terminologies in the biomedical domain.

Some terminologies such as the GO or the NCI thesaurus are designed for specific biological or clinical domains. Conversely, the unified medical language system (UMLS) provides a broader clinical scope and includes terminologies from multiple sources under a single coding system. The terminologies that fall under the umbrella of the UMLS metathesaurus include the current procedural terminology (CPT), international classification of diseases (ICD-10CM), medical subject headings (MeSH), and the systematized nomenclature of medical terminologies–clinical terms (SNOMED-CT). The UMLS also contains terminologies for pharmaceutical agents (RxNORM) and for laboratory tests that are performed in the clinical setting (LOINC). For example, suppose a researcher wanted to annotate the following hospital discharge summary:

> The patient is a 42-year-old male who was admitted to evaluate chest pain. He was noted to have an elevated troponin level in the Emergency Room, with concurrent EKG changes consistent with STEMI. He immediately underwent a PTCA, which relieved his symptoms of chest pain. After a brief observation in hospital, he is discharged on aspirin, Lopressor, and Plavix.

In this brief example, *chest pain* and *STEMI*, although similar, do not refer to the same entity. The first refers to symptoms experienced by a patient, whereas the latter refers to a specific physiological event, a myocardial infarction characterized by ST elevations on the electrocardiogram. Also present in this summary are laboratory terms, such as *troponin* and *EKG*, as well as procedural terms such as *PTCA*, which refers to a percutaneous trans-coronary angiography. Finally, there are terms that refer to medications such as *aspirin*, *Lopressor*, and *Plavix*.

Another physician may use equivalent, but different terms. For example, *anterior wall myocardial infarction* may be used instead of *STEMI* to refer to a myocardial infarction, or the drug *Plavix* may be called by its generic name, *clopidogrel*. Similarly, different hospitals may have different naming conventions for the same entity. The terminologies mentioned above are developed independently—the UMLS brings them together into a unified vocabulary, linked by the metathesaurus, so that clinical text, which often contains different *types* of information, can be annotated with standard terms.

In addition to the resources of the metathesaurus, the UMLS includes two other useful features. First is the SPECIALIST Lexicon, which contains syntactic, morphological, and spelling information regarding both common English words and specialized biomedical terms (Browne et al. 2000). Second is the UMLS semantic network, which provides broad classifications of the metathesaurus terms, based on semantic types and relations (Table 3). The semantic network delineates rela-

TABLE 2. Some important biomedical ontologies and terminologies

Resource	Terms	Type	Description	Reference
NCI metathesaurus	1,726,164	Terminology	Produced by the US National Cancer Institute's Enterprise Vocabulary Service, it is based on the UMLS Metathesaurus (National Library of Medicine) supplemented with additional cancer-related terms.	Bodenreider 2004
Unified medical language system (UMLS) semantic network		Ontology	The ontological version of the UMLS metathesaurus, this contains all of the terms in the metathesaurus as well as the semantic relationships among them.	McCray 2003
SNOMED clinical terms (SNOMED-CT)	393,075	Ontology	Developed by the College of American Pathologists, this is a structured collection of medical terms covering diseases, pathogens, chemical substances, procedures, and so on, that is designed to facilitate standard reporting in health care. It is organized into a set of acyclic, taxonomic ("is a") hierarchies.	Cornet and de Keizer 2008
Medical subject headings (MeSH)	223,184	Terminology	Produced by the National Library of Medicine, this vocabulary is often used for indexing and search retrieval of biomedical documents.	Nelson et al. 2001
RxNorm	200,739	Terminology	Provided by the NIH as part of the UMLS, RxNorm provides normalized names for clinical drugs and links its names to many of the drug vocabularies commonly used in pharmacy management and drug interaction software.	Liu et al. 2005
Logical observation identifier names and codes (LOINC)	150,500	Terminology	A standard terminology for laboratory tests and test results, developed in 1994 by the Regenstrief Institute, a not-for-profit medical research organization affiliated with Indiana University.	Forrey et al. 1996
International classification of diseases (ICD-10CM)	91,590	Terminology	Developed by the WHO and an international consortium of medical professionals, the ICD-10 is a terminology for the standardized coding of diseases, symptoms, and environmental and social causes of injury and disease.	WHO 2011
Medical dictionary for regulatory activities (MedDRA)	69,389	Terminology	International medical terminology used by regulatory authorities and the pharmaceutical industry. It is maintained by the International Federation of Pharmaceutical Manufacturers and Associations.	MSSO 2011
Online mendelian inheritance in man (OMIM)	63,154	Terminology	Compendium of terms for human genes and genetic phenotypes. The online version represents a collaboration between NLM and Johns Hopkins University.	NCBI 2011
Gene ontology (GO)	34,400	Terminology	A collection of gene-related terms organized in three separate hierarchies: molecular functions, biological processes, and cellular components. It was created in 1998 as a collaboration of three model organism databases, Flybase, the *Saccharomyces* genome database, and the mouse genome database, as a way to make descreptions of gene products more consistent across databases.	Ashburner et al. 2000

TABLE 3. Some common classifications of UMLS Metathesaurus terms found in the UMLS Semantic Network

Entity	Event
Physical object	*Activity*
Organism	Behavior
Anatomic structure	Occupational
Manufactured object	*Phenomenon or process*
Substance	Human caused
Conceptual entity	Natural
Idea or concept	Injury or poisoning
Finding	
Laboratory or test result	
Symptom or sign	

tionships between terms in a vocabulary so that a computer can make simple inferences regarding an entity based on its semantic type. For example, the aforementioned drugs *aspirin* and *Plavix* would both have the semantic type of *Pharmacologic Substance*, suggesting some form of similarity between the two entities that would not be present in *aspirin* and *EKG*. This simple inference leads nicely into the next discussion on the uses of ontology in biomedicine.

ONTOLOGIES IN BIOMEDICINE

A particular limitation of standard terminologies is that the knowledge contained in each term is *implicit*; that is, a human user is required to interpret the meaning(s) of the terms and to annotate documents properly with concepts from the terminology. For any application in which a computer must make inferences automatically based on the text and any information provided by annotations, the information contained in the terms must be made *explicit*, and that is where ontologies have an important role in biomedicine.

Objects and Processes: Looking at Nouns and Verbs

Strictly speaking, Ontology is the branch of philosophy concerned with the nature of objects and the relations among them. In computer science, *ontology* refers to the explicit representation of *items* in the real world—typically *objects* or *processes*. An object can be a physical object, such as an amino acid or an enzyme, or it can be a state, such as a disease or a condition. Processes are descriptors for actions that objects perform, such as phosphorylation or combustion. Note that the descriptor refers to the process as an object, rather than an action; because the terms *phosphorylates* and *combusts* are not objects, they are therefore are not used. Machines can use ontologies to classify data within the scope of its descriptions for objects and

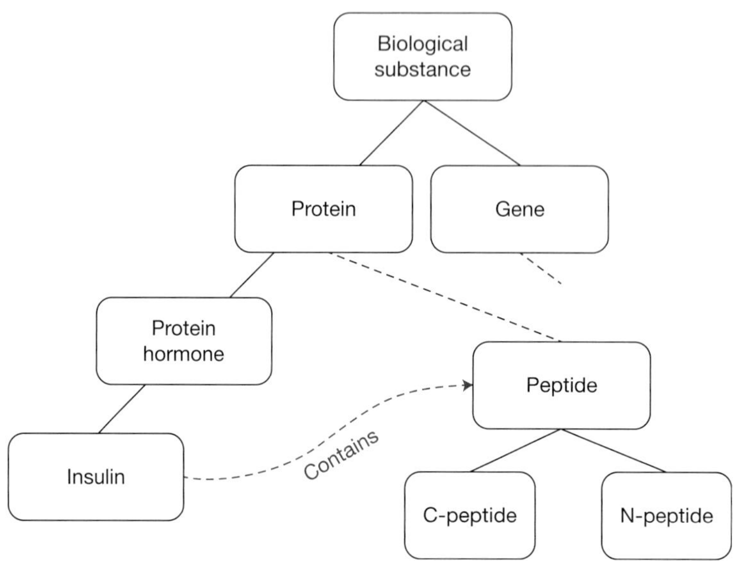

FIGURE 4. Example of an ontology in which the substance insulin is represented. (Solid lines) "is-a" links; (short dotted lines) "is-a" hierarchies that are not shown in the figure; (blue dotted line) a property of the entity insulin.

processes (learning). It can then use those classifications to make conclusions without human direction (inference).

In ontologies, the objects represented are referred to as "instances," whereas the descriptors for the objects and processes are referred to as "concepts." Ontologies explicitly represent domain knowledge through two primary means. The first is through a hierarchical structure. Typically, this hierarchy follows an *is-a* formalism, although other formalisms, such as *part-of* have also been used. For example, as shown in Figure 4, one may represent the "concept" *insulin* as

insulin *is-a* protein_hormone *is-a* protein *is-a* substance.

Note that the concepts become more abstract as we move up the hierarchy. In addition, note that insulin is just one kind of protein hormone. We may choose to represent several, and we can imagine that as subconcepts are added under each concept, the ontology hierarchy will begin to look very much like a hierarchical tree, with numerous branches all leading to a root concept. An advantage of this hierarchical representation is that *is-a* inferences can be performed easily using subsumption—that is, all instances of a child concept are automatically considered part of the parent class. Thus, if we state that bovine insulin and human insulin are both instances of the concept *insulin*, a machine can automatically infer that bovine insulin and human insulin are both *endocrine hormones.*

The second means by which ontologies explicitly represent knowledge is through the *relations* (or *properties*) among the concepts. Properties connect branches of an ontology hierarchy. For example, as shown in Figure 4, the property *contains* may be used with *insulin* to describe two of its components: C-peptide and N-peptide. Properties ascribed to a concept are transmitted via inheritance to any child concepts. Thus, if the concept *hormone* has the property *attaches to receptor* and the object of this property refers to a *protein-receptor* concept, then all child concepts subsumed by the class hormone have the property *attaches to receptor* to the object concept *protein-receptor* (and by extension all the child concepts to *protein receptor*).

Beyond the use of hierarchical structure and properties, several logical formalisms exist to provide greater descriptive expressivity of objects and processes. A full accounting of these formalisms is beyond the scope of this book. Suffice it to say that greater expressivity in the formalism rapidly leads to increases in computational complexity, so it is important to find the proper balance between expressivity that is sufficient to fully capture the domain of interest and computational tractibility. In recent years, there has been a great deal of interest in the biomedical informatics community on the use of the Web Ontology Language (OWL), a description logic formalism that uses decidable fragments of first-order logic. OWL uses set theoretic principles in its expressivity, allowing, for example, references to cardinality (the number of elements within a set) and transitivity (the idea that if set *A* is a subset of set *B*, and *B* is a subset of *C*, *A* is also a subset of *C*.). An excellent reference can be found at the World Wide Web Consortium (http://www.w3.org/TR/owl-guide/), which endorses OWL as a de facto standard for ontology formalisms.

An excellent resource for biomedical terminologies and ontologies is available at NCBO BioPortal (Bioportal 2011), which also includes a feature that searches multiple ontologies simultaneously. A list of some of the largest and most widely used terminologies and ontologies is shown in Table 2. In addition, WordNet, a database developed at Princeton University, is a useful general-language resource that groups words into sets of synonyms ("synsets") and contains information regarding the relationships among these synsets (Felbaum 1998).

MACHINE LEARNING AND BIOMEDICAL TEXT

In recent years, machine learning has emerged as a viable way to approach the process of analyzing biomedical text. In machine learning, a classifier is trained to recognize a set of *features* that tell us something about the problem of interest. For example, if the goal is to classify documents based on topic, the features might consist of all of the words contained in each document (the so-called bag-of-words approach). In addition, careful chunking, parsing, and annotation can extend the list of useful features beyond individual words. Words, phrases, parts of speech, information regarding sentence clauses and dependency relationships inferred from a parse tree, annotations, ontological relationships, ... all tell us something different about our document, and all can serve as features for a machine-learning classifier.

TABLE 4. Machine-learning algorithms commonly used in biomedical text mining

Technique	General information	Past applications in biomedicine
Naive Bayes	Manning et al. 2008, Ch. 13, Text Classification and Naive Bayes; Bird et al. 2009, Section 6.5, Naive Bayes Classifiers	Wilbur 2000; Marcotte et al. 2001; Chiang and Yu 2003
Support vector machines (SVM)	Schoelkopf and Smola 2002; Bishop 2006, Ch. 6, Kernel Methods; and Ch. 7, Sparse Kernel Machines; Manning et al. 2008, Ch. 15, Support Vector Machines and Machine Learning on Documents; Hastie et al. 2009, Ch. 12, Support Vector Machines and Flexible Discriminants	Dumais 1998; Wilbur 2000
k-nearest neighbors	Manning et al. 2008, Section 14.3: *k*-Nearest Neighbor; Hastie et al. 2009, Ch. 13, Prototype Methods and Nearest-Neighbors	De Bruijn et al. 1997
Maximum entropy (multinomial logistic regression) classifiers	Jurafsky and Martin 2009, Ch. 6, Hidden Markov and Maximum Entropy Models	Tsuroka et al. 2005
Conditional random fields	Lafferty et al. 2001; Koller 2009, Sect. 4.6, Partially Directed Models	McDonald and Pereia 2005
Hidden Markov models (HMMs)	Jurafsky and Martin 2009, Ch. 6, Hidden Markov and Maximum Entropy Models	
Decision trees and random forests	Hastie et al. 2009, Ch. 15, Random Forests	Wilcox and Hripcsak 1999

A variety of machine-learning algorithms are discussed in Chapter 3. Some of these algorithms have been much more commonly used for text mining than others. The algorithms most central to biomedical text mining include Naïve Bayes, support vector machines, maximum entropy (multinomial logistic regression) classifiers, conditional random fields, hidden Markov models (HMMs) (discussed in Chapter 4), and random forests. A summary of these algorithms, along with technical references and a list of biomedical text-mining papers implementing each algorithm, is given in Table 4. Some common software packages for text mining include Weka (Hall et al. 2009), R, and the Python NLTK toolkit (Bird et al. 2009).

Example: Clinical Document Classification

Imagine a collection of clinical notes, some of which represent general patient history reports and others that are imaging reports. You wish to distinguish the two classes of documents based on word use alone. The following example of a text-mining pipeline describes how you could do this, using a combination of rule-based annotations and machine learning.

For the purposes of this example, assume that you want to place the following two documents, each of which consists of only a single sentence, in the correct class:

Document 1: "The patient reported that he consumes two pounds of liver daily."
Document 2: "His liver displayed signs of cirrhosis on CT scan."

It would be logical to begin by tokenizing, tagging, and annotating, and conducting named entity recognition on the two documents, as illustrated in Table 5.

At this point, there are two alternative ways to proceed, using either a rule-based or a machine-learning-based approach. A rule-based approach to this type of classification might look at a collection of similar documents and note that only imaging reports tend to include references to scans or other diagnostic procedures. A researcher might then write a rule stating that all notes containing terms tagged with the UMLS category "Diagnostic Procedure" would be classified as imaging

TABLE 5. Tokenization, POS tagging, annotation, and named entity recognition of the example documents

Step	Results (document 1)	Results (document 2)
Tokenization	The patient reported that he consumes two pounds of liver daily.	His liver displayed signs of cirrhosis on CT scan.
Part-of-speech tagging	The/DT patient/NN reported/VBD that/IN he/PRP consumes/VBZ two/CD pounds/NNS of/IN liver/NN daily/RB./.	His/PRP liver/NN displayed/VBD signs/NNS of/IN cirrhosis/NNS on/IN CT/NNP scan/NNP./.
Stop word removal	Patient/NN reported/VBD consumes/VBZ pounds/NNS liver/NN daily/RB	Liver/NN displayed/VBD signs/NNS cirrhosis/NNS CT/NNP scan/NNP
Named entity recognition and annotation with UMLS concept identifiers (MetaMap)	Patient (Patients) [Patient or Disabled Group] Reported (Report (document) [Intellectual Product] Reported (Reporting) [Health Care Activity] Two [Quantitative Concept] Pounds [Quantitative Concept] Liver (Entire liver) [Body Part, Organ, or Organ Component] Liver [Body Part, Organ, or Organ Component] LIVER (Liver (Food)) [Food] LIVER (Liver Flavor) [Food] Daily [Temporal Concept]	Liver (Entire liver) [Body Part, Organ, or Organ Component] Liver [Body Part, Organ, or Organ Component] LIVER (Liver (Food)) [Food] LIVER (Liver Flavor) [Food] displayed (Display - arrangement) [Functional Concept] signs (Aspects of signs) [Functional Concept] signs (Manufactured sign) [Manufactured Object] SIGNS (Physical findings) [Finding] Cirrhosis [Disease or Syndrome] CAT scan (X-Ray Computed Tomography) [Diagnostic Procedure] CAT Scan (CAT Scan Section ID) [Intellectual Product]

TABLE 6. Representation of the example documents as a vector of features

	patient/NN	reported/VBD	consumes/VBZ	pounds/NNS	liver/NN	daily/RB	displayed/VBD	signs/NNS	cirrhosis/NNS	CT/NNP	scan/NNP	[Patient or Disabled Group]	[Body Part, Organ, or Organ Component]	[Food]	[Intellectual Product]	[Disease or Syndrome]	[Health Care Activity]	[Diagnostic Procedure]	[Finding]	[Manufactured Object]	[Quantitative Concept]	[Temporal Concept]	[Functional Concept]
D1	1	1	1	1	1	1	0	0	0	0	0	1	1	1	1	0	1	0	0	0	1	1	0
D2	0	0	0	0	1	0	1	1	1	1	1	0	1	1	1	1	0	1	1	1	0	0	1

reports. Similarly, notes containing terms tagged with the UMLS category "Temporal Concept" would be classified as patient history reports.

The machine-learning-based approach would also require a collection of similar documents and a definition of which features merit consideration. In this case, the features might include all word/part-of-speech pairings from the documents, as well as all UMLS concepts that are referred to in the documents. Each document could then be represented as a vector of features that might look something like Table 6.

By representing a large collection of documents in this way and hand labeling them as "imaging reports" (Class 1) or "patient history" (Class 2), one could then apply a machine-learning classifier to learn which features are most representative of each class. The classifier may report back, for example, that the token "scan/NNP" is highly indicative of imaging reports, whereas "consumes/VBZ" is almost never found in imaging reports and is indicative of patient histories. The classifier could then be applied to our two example documents to determine that D1 is a patient history and D2 is an imaging report.

CONCLUSION: PROCESSING BIOMEDICAL TEXT

Bioinformatics researchers typically begin with a set of documents consisting of raw clinical or scientific text. In this chapter, we have examined many of the options available for converting this text into a machine-computable format. These include tokenization, labeling each word with its part of speech, chunking or parsing to break down sentences, identifying entities of interest, and annotating those entities with a variety of different concepts and properties, depending on which terminology and/

or ontology is most appropriate. In short, using prior knowledge regarding how English text is organized (POS tagging, chunking, parsing) and specific biomedical knowledge (NER, annotation) can bring out hidden information contained within text that enables us to break it down further into machine-computable facts (if we so choose) or use it to infer new facts and generate new knowledge.

There are two ways to approach the process of inference, in which known facts are used to generate new ideas. One is rules based; that is, a set of rules that govern the domain of interest is developed and used to infer new facts. Much of the material presented in this chapter regarding ontologies falls naturally into a rules-based paradigm. The other approach, the statistical or machine-learning approach, uses machine-learning techniques and a training set of relevant documents to learn textual features that can help solve the problem of interest. Both approaches are valid and provide better or worse results in different contexts. Furthermore, most modern biomedical text-mining efforts combine elements of both: We may annotate text using concepts from an ontology, for example, and then use the presence or absence of those concepts as features to train a machine-learning classifier. The choice of the approach that will provide the best research outcome depends on the problem to be solved: one listed at the beginning of this chapter or perhaps one that we have yet to imagine.

REFERENCES

Agarwal P, Searls DB. 2008. Literature mining in support of drug discovery. *Brief Bioinform* **9:** 479–492.

Altman RB, Bergman CM, Blake J, Blaschke C, Cohen A, Gannon F, Grivell L, Hahn U, Hersh W, Hirschman L, et al. 2008. Text mining for biology: The way forward: Opinions from leading scientists. *Genome Biol* **9:** S7.

Ananiadou S, McNaught J. 2006. *Text mining for biology and biomedicine*. Artech House, Boston.

Aronson AR. 2001. Effective mapping of biomedical text to the UMLS Metathesaurus: The MetaMap program. *Proc AMIA Symp* **2001:** 17–21.

Ashburner M, Ball CA, Blake JA, Botstein D, Butler H, Cherry JM, Davis AP, Dolinski K, Dwight SS, Eppig JT, et al. 2000. Gene ontology: Tool for the unification of biology; The Gene Ontology Consortium. *Nat Genet* **25:** 25–29.

Bates DW, Evans RS, Murff H, Stetson PD, Pizziferri L, Hripcsak G. 2003. Detecting adverse events using information technology. *J Am Med Inform Assoc* **10:** 115–228.

Bird S, Klein E, Loper E. 2009. *Natural language processing with Python*. O'Reilly Media, Sebastopol, CA.

Bishop CM. 2006. *Pattern recognition and machine learning*. Springer, London.

Blumenthal D, Tavenner M. 2010. The "meaningful use" regulation for electronic health records. *New Engl J Med* **363:** 501–504.

Bodenreider O. 2004. The Unified Medical Language System (UMLS): Integrating biomedical terminology. *Nucleic Acids Res* **32:** D267–D270.

Brill E. 1995. Transformation-based error-driven learning and natural language processing: A case study in part-of-speech tagging. *Comput Linguist* **21:** 543–565.

Browne AC, McCray AT, Srinivasan S. 2000. *The Specialist lexicon*. Lister Hill National Center for Biomedical Communications, National Library of Medicine, Bethesda, MD.

Cer D, de Marneffe M-C, Jurafsky D, Manning CD. 2010. Parsing to Stanford dependencies: Trade-offs between speed and accuracy. In *Proceedings of the 7th International Conference on Language Resources and Evaluation (LREC 2010)*, May 17–23, Malta. European Language Resources Association, Paris.

Chiang JH, Yu HC. 2003. MeKE: Discovering the functions of gene products from biomedical literature via sentence alignment. *Bioinformatics* **19:** 1417–1422.

Cohen AM, Hersh WR. 2005. A survey of current work in biomedical text mining. *Brief Bioinform* **6:** 57–71.

Cornet R, de Keizer N. 2008. Forty years of SNOMED: A literature review. *BMC Med Inform Decis Mak* **8:** S2.

Coulet A, Shah NH, Garten Y, Musen M, Altman RB. 2010. Using text to build semantic networks for pharmacogenomics. *J Biomed Inform* **43:** 1009–1019.

Dean BB, Lam J, Natoli JL, Butler Q, Aguilar D, Nordyke RJ. 2009. Use of electronic medical records for health outcomes research: A literature review. *Med Care Res Rev* **66:** 611–638.

De Bruijn LM, Hasman A, Arends JW. 1997. Automatic SNOMED classification—A corpus-based method. *Comput Methods Programs Biomed* **54:** 115–122.

de Marneffe M-C, MacCartney B, Manning CD. 2006. Generating typed dependency parses from phrase structure parses. In *Proceedings of the 5th International Language Resources and Evaluation Conference (LREC 2006)*, May 22–28. Genoa, Italy. European Language Resources Association, Paris.

Dumais S. 1998. Using SVMs for text categorization. *IEEE Intelligent Systems* **13:** 21–23.

Felbaum C. 1998. *WordNet: An electronic lexical database, language, speech, and communication*. MIT Press, Cambridge, MA.

Forrey AW, McDonald CJ, DeMoor G, Huff SM, Leavelle D, Leland D, Fiers T, Charles L, Griffin B, Stalling F, et al. 1996. Logical observation identifier names and codes (LOINC) database: A public use set of codes and names for electronic reporting of clinical laboratory test results. *Clin Chem* **42:** 81–90.

Friedman C, Johnson SB, Forman B, Starren J. 1995. Architectural requirements for a multipurpose natural language processor in the clinical environment. *Proc Annu Symp Comput Appl Med Care* **1995:** 347–351.

Giménez J, Màrquez L. 2003. Fast and accurate POS tagging. In *Proceedings of the International Conference on Recent Advances in Natural Language Processing (RANLP 2003) held in Borovets, Bulgaria* (ed. Nicolov NN, et al.), pp. 158–165. In *Current issues in linguistic theory (CILT)*, John Benjamins, Amsterdam/Philadelphia.

Ginsberg J, Mohebbi MH, Patel RS, Brammer L, Smolinski MS, Brilliant L. 2009. Detecting influenza epidemics using search engine query data. *Nature* **457:** 1012–1014.

Hall M, Frank E, Holmes G, Pfahringer B, Reutemann P, Witten IH. 2009. The WEKA data mining software: An update. *ACM SIGKDD Explorations Newslett* (Issue 1) **11:** 10–18.

Harman D. 1991. How effective is suffixing? *J Am Soc Inform Sci* **42:** 7–15.

Hastie T, Tibshirani R, Friedman J. 2009. *The elements of statistical learning: Data mining, inference, and prediction*, 2nd ed. Springer, New York.

Hunter L, Bretonnel Cohen K. 2006. Biomedical language processing: What's beyond PubMed. *Mol Cell* **21:** 589–594.

Jiang J, Zhai CX. 2007. An empirical study of tokenization strategies for biomedical information retrieval. *Info Retriev* **10:** 341–363.

Jonquet C, Shah N, Youn C, Musen MA, Callendar C, Storey M-A. 2009. NCBO Annotator: Semantic annotation of biomedical data. In *8th International Semantic Web Conference*

(ISWC 2009) Posters and Demonstrations, October 25–29, 2009, Chantilly, VA. Springer, New York.

Jurafsky D, Martin JH. 2009. *Speech and language processing: An introduction to natural language processing, computational linguistics, and speech recognition*, 2nd ed. Pearson Prentice-Hall, Upper Saddle River, NJ.

Karlsson F. 1995. *Constraint grammar: A language-independent system for parsing unrestricted text.* De Gruyter, New York/Berlin.

Klein D, Manning CD. 2003. Accurate unlexicalized parsing. In *ACL'03 Proceedings of the 41st Meeting of the Association for Computational Linguistics*, pp. 423–430. ACL, Stroudsburg, PA.

Koller D, Friedman N. 2009. *Probabilistic graphical models: Principles and techniques.* MIT Press, MA.

Lafferty JD, McCallum A, Pereira FCN. 2001. Conditional random fields: Probabilistic models for segmenting and labeling sequence data. In *ICML'01 Proceedings of the 18th International Conference on Machine Learning*, pp. 282–289. Morgan Kaufmann, San Francisco.

Leaman R, Gonzalez G. 2008. BANNER: An executable survey of advances in biomedical named entity recognition. *Pac Symp Biocomput* **13:** 652–663.

Liu S, Ma W, Moore R, Ganesan V, Nelson S. 2005. RxNorm: Prescription for electronic drug information exchange. *IT Professional* **7:** 17–23.

Lovins JB. 1968. Development of a stemming algorithm. In *Mechanical translation and computational linguistics*, Vol. 11, pp. 22–31. University of Chicago Press, Chicago.

Manning CD, Schuetze H. 1999. *Foundations of statistical natural language processing.* MIT Press, Cambridge, MA.

Manning CD, Raghavan P, Schütze H. 2008. *Introduction to information retrieval.* Cambridge University Press, Cambridge, UK.

Marcotte EM, Xenarios I, Eisenberg D. 2001. Mining literature for protein–protein interactions. *Bioinformatics* **17:** 359–363.

McCray AT. 2003. An upper level ontology for the biomedical domain. *Comp Funct Genom* **4:** 80–84.

McDonald R, Pereira F. 2005. Identifying gene and protein mentions in text using conditional random fields. *BMC Bioinformatics* **6:** S6.

McDonald DM, Chen H, Su H, Marshall BB. 2004. Extracting gene-pathway relations using a hybrid grammar: The Arizona parser. *Bioinformatics* **20:** 3370–3378.

Nelson SJ, Johnston D, Humphreys BL. 2001. Relationships in medical subject headings. In *Relationships in the organization of knowledge* (ed. Bean CA, Green R), pp. 171–184. Kluwer, Amsterdam.

Porter MF. 1980. An algorithm for suffix stripping. *Program* **14:** 130–137.

Ramshaw LA, Marcus MP. 1995. *Text chunking using transformation-based learning. ACL Third Workshop on Very Large Corpora*, June 30, 1995, pp. 82–94. Cambridge, MA.

Ratnaparkhi A. 1996. A maximum entropy part-of-speech tagger. In *Proceedings of the 1st Conference on Empirical Methods in NLP Philadelphia, PA*, May 17–18, 1996, pp. 133–142.

Savova GK, Masanz JJ, Ogren PV, Zheng J, Sohn S, Kipper-Schuler KC, Chute CG. 2010. Mayo clinical Text Analysis and Knowledge Extraction System (cTAKES): Architecture, component evaluation and applications. *J Am Med Inform Assoc* **17:** 507–513.

Schoelkopf B, Smola AJ. 2002. *Learning with kernels: Support vector machines, regularization, optimization, and beyond.* MIT Press, Cambridge, MA.

Segura-Bedmar I, Martinez P, Segura-Bedmar M. 2008. Drug name recognition and classification in biomedical texts: A case study outlining approaches underpinning automated systems. *Drug Discov Today* **13:** 816–823.

Settles B. 2004. Biomedical named entity recognition using conditional random fields and rich feature sets. In *Proceedings of the COLING 2004. International Joint Workshop on Natural Language Processing in Biomedicine and Its Applications (NLPBA/BioNLP)*, pp. 104–107. ACL, Stroudsburg, PA.

Tjong EF, Sang K, Buchholz S. 2000. Introduction to the CoNLL-2000 Shared Task: Chunking. In *Proceedings of CoNLL-2000 and LLL-2000*, Lisbon, Portugal, pp. 53–57. ACL, Stroudsburg, PA.

Torii M, Hu Z, Wu CH, Liu H. 2009. BioTagger-GM: A gene–protein name recognition system. *J Am Med Inform Assoc* **16:** 247–255.

Tsuroka Y, Tateishi Y, Kim J-D, Ohta T, McNaught J, Ananiadou S, Tsujii J. 2005. Developing a robust part-of-speech tagger for biomedical text. In *Advances in Informatics—10th Panhellenic Conference on Informatics, Volos, Greece*, Vol. 3746, pp. 382–392. Springer-Verlag, New York. http://www.nactem.ac.uk/aigaion2/index.php?/publications/show/112.

Wang X, Hripcsak G, Markatou M, Friedman C. 2009. Active computerized pharmacovigilance using natural language processing, statistics, and electronic health records: A feasibility study. *J Am Med Inform Assoc* **16:** 328–337.

Wilbur WJ. 2000. Boosting naïve Bayesian learning on a large subset of Medline. *Proc AMIA Symp* **2000:** 918–922.

Wilcox A, Hripcsak G. 1999. Classification algorithms applied to narrative reports. *Proc AMIA Symp* **1999:** 455–459.

Yakushiji A, Tateisi Y, Miyao Y. 2001. Event extraction from biomedical papers using a full parser. *Pac Symp Biocomput* **6:** 408–419.

WWW RESOURCES

http://alias-i.com/lingpipe Alias-i. 2008. LingPipe 4.1.0 (accessed 31 May 2013)

http://bioportal.bioontology.org/ontologies Bioportal. 2013 (accessed 31 May 2013)

http://www.meddramsso.com/index.asp MSSO. 2011 (accessed 31 May 2013)

http://www.ncbi.nlm.nih.gov/omim OMIM (Online Mendelian Inheritance in Man) NCBI. 2011 (accessed 31 May 2013)

http://www.nlm.nih.gov/pubmed Pubmed 2013 (accessed 31 May 2013)

http://www.who.int/classifications/icd/en/ WHO. 2011. "International Classification of Diseases (ICD)." World Health Organization (accessed 31 May 2013)

http://www.w3.org/TR/owl-guide/ OWL

Index

Page references followed by b denote boxes; those followed by f denote figures; those followed by t denote tables.

D

P

Q

R

S

T